Arthur Bolles Lee

The Microtomist's Vade-Mecum

A Handbook of the Methods of Microscopic Anatomy. Second Edition

Arthur Bolles Lee

The Microtomist's Vade-Mecum
A Handbook of the Methods of Microscopic Anatomy. Second Edition

ISBN/EAN: 9783744691987

Printed in Europe, USA, Canada, Australia, Japan

Cover: Foto ©berggeist007 / pixelio.de

More available books at **www.hansebooks.com**

THE
MICROTOMIST'S VADE-MECUM

THE
MICROTOMIST'S VADE-MECUM

A HANDBOOK OF THE METHODS OF MICROSCOPIC ANATOMY

BY

ARTHUR BOLLES LEE

ASSISTANT IN THE RUSSIAN LABORATORY OF ZOOLOGY AT VILLEFRANCHE-SUR-MER (NICE)

SECOND EDITION

LONDON
J. & A. CHURCHILL
11, NEW BURLINGTON STREET
1890

PREFACE

The reader who compares the present work with the first edition will find that it has been not only enlarged, but so fundamentally revised and to so great an extent re-written that it would possibly have been more correct to consider it as a new book, and to give it a new title. It is not based on the original English edition of 1885, but on the French edition of 1887, which I prepared with the valuable aid of Dr. L. F. HENNEGUY, and which was considered when completed to have so far outgrown its English parent that we decided on giving it a dissimilar and independent title (" Traité des Méthodes Techniques de l'Anatomie Microscopique,—Histologie, Embryologie, et Zoologie," par MM. Arthur Bolles Lee et F. Henneguy, Préparateur du cours d'embryogénie au Collège de France. Avec une Préface de M. Ranvier, Professeur au Collège de France. Paris, O. Doin. 1887).

As stated in the *Avant-propos* of that work, the "Traité" is not a mere translation of the "Vade-Mecum;" it differs from it by the compression or omission of many passages dealing with matter of secondary importance; by the suppression of the chapter relating to the demonstration of pathological "Micro-organisms"—a subject which can now no longer be dealt with satisfactorily within the limits of a chapter; and by the far greater development given to certain important subjects, such as the methods of Embryology, of Cytology, of Neurology. I am greatly indebted to my able friend Dr. HENNEGUY for the important aid rendered by him in the preparation of our joint work. His share in it consisted not merely in valuable suggestions as to the choice and treatment of the matter, and careful revision of the whole text, but also in important original contributions. The valuable chapters on the Methods of Cytology and on the Protozoa are from his pen, as is also that on the methods of Embryology, concerning which I am free to say (as I had no share in it myself) that it is probably the most authoritative and masterly exposition of the subject ever written. It is my privilege to be able here to reproduce this admirable little treatise.

I should be ungrateful if I did not take this opportunity of expressing my thanks to Prof. RANVIER for the graceful and kindly Preface in which he was good enough to introduce our work to the French-reading public.

To return to the work now before us—it remains to be said that it stands in much the same position with regard to the "Traité" as the "Traité" stands in with regard to the original "Vade-Mecum." It is not a mere translation of the French work, but has been thoroughly revised, brought up to date, and to a very considerable extent re-written. It goes even a step farther than the "Traité" in the way of throwing less important matter into the background and treating important matter with increased fulness. A still greater number of superannuated or untrustworthy or superfluous processes have been either suppressed or relegated into the background of small type and brief mention. But it also goes a step further than the "Traité" in the way of fuller treatment of important matter. The all-important subject of Fixing and Fixing Agents has been set forth with all the fulness allowed by my knowledge of the matter. A new chapter on Killing has been added. The two chapters on Staining with the Coal-tar colours have been re-written, not translated, and treat the subject very fully. The chapters on Impregnation Methods, on the processes of Paraffin imbedding and Celloidin imbedding, on the methods of Cytology, and on the Central Nervous System have been re-written and brought up to date The methods for Serial Sections are fully given. And new and in many cases very detailed Introductions to each of the other more important branches of the subject have been written.

The result is that, as compared with the first edition, as compared even with the "Traité," the work is much less historical, and much more critical. Much more choice has been exercised—has had to be exercised. The first edition could claim to have brought together a "practically exhaustive" collection of the formulæ relating to the methods of microscopic anatomy. To make such a collection now would be well-nigh impossible and well-nigh useless. It would form not a book but a library, in which the really useful matter would remain smothered in a sea of details of doubtful utility.

VILLEFRANCHE-SUR-MER;
  February, 1890.

PREFACE

TO THE

FIRST EDITION

In its primary intention this work appeals rather to the instructed anatomist than to the beginner. Its aim is to put into his hands a concise but complete account of all the methods of preparation that have been recommended as useful for the purposes of Microscopic Anatomy, and so furnish him with a ready source of information on points of detail as to which his memory or his knowledge may be at fault. This object is attained by the mere collection of Formulæ set out in Part I, and of special methods described in Part II. But the book could obviously be made to subserve a further end —that of a guide to the beginner. To this end I have added a General Introduction and a series of introductory paragraphs prefixed, where needful, to the different chapters. These introductory portions, taken together, go far to make up a formal treatise on the art. And as a further aid to the beginner I have added the collection of examples given in Part II. These examples are of course not intended for servile imitation, but rather as hints suggestive of the most fitting processes.

The collection of Formulæ here brought together is, I believe, practically exhaustive; no process having any claim to scientific status having been rejected, nor any, I trust, unwittingly omitted. It may be useful here to say a word as to the reasons for this—perhaps apparently excessive—catholicity of treatment. Doubtless a large proportion of the formulæ given are quite superseded in modern practice; but that is not a sufficient reason for rejecting them. The inclusion of all of them is justified by the consideration that some one or other of them may perhaps serve, in some way that cannot now be foreseen, to suggest some new method of value. Let me give an example. Who, ten years ago, would have thought that the formula of Blanchard's 'Liqueur saline hydrargyrique' deserved reprinting in a treatise on histologic technic? Yet it is to the disinterment of that forgotten formula by Lang that we owe the establishment of corrosive sublimate as one of the most useful fixing agents in the arsenal of the microtomist. Or who would have deemed Thiersch's lilac borax-carmine (Formula No. 80a), published in 1865, to be of greater importance than any other stain till then made known? Yet that formula it was that directly suggested Woodward's admirable aqueous borax-carmine, and through this, if I am not mistaken, the aqueous and the alcoholic borax-carmines of Grenacher, the latter of which is now to be found on the table of every embryologist.

All my abstracts and translations have been made from the original sources, except where it has been impossible for me to obtain sight of these. References to the sources are given in all cases; but I desire here to make special acknowledgment of the great assistance rendered me by the Journal of the Royal Microscopical Society—in many respects the best-edited periodical known to me.

GENEVA (SWITZERLAND);
February, 1885.

CONTENTS

PART I

CHAPTER I.
INTRODUCTORY.—The General Method . . PARAGRAPHS 1—5

CHAPTER II.
KILLING . . . . 6—21
    Sudden Killing, 6—8; Narcotisation, 9—21.

CHAPTER III.
FIXING . . . . 22—25

CHAPTER IV.
FIXING AGENTS: MINERAL ACIDS AND THEIR SALTS . 26—44

CHAPTER V.
FIXING AGENTS: CHLORIDES, ORGANIC ACIDS, AND OTHERS . . . . 45—66

CHAPTER VI.
HARDENING AGENTS . . . . . . 67—90
    Introduction, 67—69; Mineral Acids, 70—76; Salts, 77—82; Chlorides and others, 83—90.

CHAPTER VII.
STAINING . . . 91—94

CHAPTER VIII.

ANILIN COLOURS GIVING INDIRECT NUCLEAR STAINS (FLEMMING'S METHOD) . . . . . 95—104
PARAGRAPHS
General Directions, 96—99; Victoria, Anilin Green, Gentian, and Dahlia, 100—103; Safranin and others, 103, 104.

CHAPTER IX.

OTHER ANILIN STAINS . . . . . 105—129
Direct Nuclear Stains, 105—107a; Plasmatic Stains, 108—111; other Anilins, 112—129.

CHAPTER X.

CARMINE STAINS . . . . . . 130—166
Aqueous Carmines, Alkaline, 134—139; Neutral and Acid, 140—162; Alcoholic Carmines, 163—166.

CHAPTER XI.

COCHINEAL, HÆMATOXYLIN, AND OTHER ORGANIC STAINS 167—196
Cochineal, 167—171; Hæmatoxylin, 172—187; other Organic Stains, 188—196.

CHAPTER XII.

METALLIC STAINS (IMPREGNATION METHODS) . . 197—223
Silver, 198—203; Gold, 204—213; other Metallic Stains, 214—222.

CHAPTER XIII.

COMBINATION STAINS . . . . . . 224—262
Combinations having Carmine for a Primary Stain, 226—236; Combinations having Hæmatoxylin for a Primary Stain; 237—244; other Combinations, 245—262.

CHAPTER XIV.

IMBEDDING METHODS: INTRODUCTION   263—266

CHAPTER XV.

IMBEDDING METHODS: PARAFFIN AND OTHER FUSION
MASSES . . . . 267—289
    Paraffin, 267—281; Soap, 282—285; Gelatin, 286—289.

CHAPTER XVI.

COLLODION (CELLOIDIN) AND OTHER IMBEDDING METHODS 290—312
    Collodion or Celloidin, 290—299; other Evaporation Masses, 300—312.

CHAPTER XVII.

SERIAL SECTION MOUNTING . . . . 313—327
    Methods for Paraffin Sections, 314—321; Methods for Watery Sections, 322—324; Methods for Celloidin Sections, 325—327.

CHAPTER XVIII.

CLEARING AGENTS . . . 329—341

CHAPTER XIX.

INDIFFERENT LIQUIDS: EXAMINATION AND PRESERVATION MEDIA . . . . . 342—406
    Aqueous Liquids, 342—362; Mercurial Liquids, 363—369; other Fluids, 370—382; Glycerin Media, 383—392; Resinous Media, 396—406.

CHAPTER XX.

CEMENTS AND VARNISHES . . . 407—430

CHAPTER XXI.

INJECTIONS: GELATIN MASSES . . . 431—470
    Carmine, 439—446; Blue, 447—454; other Colours, 455—470.

CHAPTER XXII.

INJECTIONS: OTHER MASSES . . . . . 471—494
    White of Egg, Gum, 471, 472; Glycerin, 473—479; Aqueous, 480—485; Celloidin, 486, 487; other Masses, 488—494.

CHAPTER XXIII.

MACERATION AND DIGESTION . . . . 495—526
    Maceration, 495—520; Digestion, 521—526.

CHAPTER XXIV.

CORROSION, DECALCIFICATION, AND BLEACHING . 527—550
    Corrosion, 527—530; Decalcification and Desilicification, 531—541; Bleaching, 542—550.

PART II

SPECIAL METHODS AND EXAMPLES

CHAPTER XXV.

EMBRYOLOGICAL METHODS . . . . . 551—597

CHAPTER XXVI.

CYTOLOGICAL METHODS . . . . 598—612

CHAPTER XXVII.

TEGUMENTARY ORGANS . . . . . 613—628

CHAPTER XXVIII.

MUSCLE AND TENDON (NERVE-ENDINGS) . . 629—643
    Striated Muscle, 629—634; Tendon, 635—637; Smooth Muscle, 638—643.

CHAPTER XXIX.

RETINA, INNER EAR, NERVES . . . . 644—656
   Retina, 644—647; Inner Ear, 648, 649; Nerves, 650—656.

CHAPTER XXX.

CENTRAL NERVOUS SYSTEM . . . . . 657—681a
   Hardening, 658—671; Imbedding and Cutting, 672; Staining, 673, 674; Mounting, 675, 676; other Methods, 677—681a.

CHAPTER XXXI.

SOME OTHER HISTOLOGICAL METHODS . . . 682—704
   Connective Tissues, 682—691; Blood, 692—696; Glands, 697—704.

CHAPTER XXXII.

SOME ZOOLOGICAL METHODS . . . . . 705—760
   Tunicata, 705; Mollusca, 706—713; Arthropoda, 714—724; Vermes, 725—736; Echinodermata, 737—740; Cœlenterata, 741—748; Porifera, 749; Protozoa, 750—760.

APPENDIX . . . . 761—783

INDEX . . . . . . pp. 391—413

PART I

THE MICROTOMIST'S VADE-MECUM.

CHAPTER I.

INTRODUCTORY.

1. THE methods of modern microscopic anatomy may be roughly classed as General and Special. There is a General or Normal method, known as the method of sections, which consists in carefully *fixing* the structures to be examined, *staining* them with a *nuclear* stain, *dehydrating* with alcohol, and mounting *series of sections* of the structures in balsam. It is by this method that the work is blocked-out and very often finished. Special points are then studied, if necessary, by Special Methods, such as examination of the living tissue elements, *in situ*, or in "indifferent" media; fixation with special fixing agents; staining with special stains; dissociation by teasing or maceration; injection; impregnation; and the like.

2. THE GENERAL METHOD.—The first thing to be done with any structure is to *fix* its histological elements. (This statement applies equally to all classes of objects, whether it be desired to cut them into sections or to treat them in any other special way.) Two things are implied by the word "fixing;" first, the rapid *killing* of the element, so that it may not have time to change the form it had during life, but is fixed in death in the attitude it normally had during life; and second, the *hardening* of it to such a degree as may enable it to resist without further change of form the action of the reagents with which it may subsequently be treated. Too much stress can hardly be laid on this point, which is the most distinctive

feature of modern histological practice; without good fixation it is impossible to get good stains, or good sections, or preparations good in any way.

The structure having been duly fixed by one of the processes described in the chapter on FIXING AGENTS, is *washed* in order to remove from the tissues as far as possible all traces of the fixing re-agent.

The kind of liquid with which washing out is done, is not a matter of indifference. If corrosive sublimate (for instance) or osmic acid, or a solution into which chromic acid or a chromate enters, have been used for fixing, the washing may be done with water. But if picric acid in any form has been used, the washing must be done with alcohol. The reason of this difference is that the first-named reagents (and, indeed, all the compounds of the heavy metals used for fixing) appear to enter into a state of chemical combination with the elements of tissues, rendering them insoluble in water; so that the hardening induced by these agents is not removed by subsequent treatment with water. Picric acid, on the other hand, produces only a very slight hardening of the tissues, and does not appear to enter into any combination whatever with their elements, as it is entirely removable by treating the tissues with water or alcohol. If the removal be effected by means of water, the tissue elements are left in a soft state in which they are obnoxious to all the hurtful effects of water. Alcohol must therefore be taken to remove the picric acid and to effect the necessary hardening at the same time. Instructions for washing out are given, when necessary, in the discussion of the different fixing agents in the following parts of this work.

At the same time that the superfluous fixing agent is being removed from the tissues, or as soon as that is done, the *water of the tissues must be removed*. This is necessary for two reasons; firstly, in the interest of preservation, the presence of water being the condition of all others that most favours post-mortem decomposition; and secondly, because all water must be removed in order to allow the tissues to be impregnated with the imbedding material necessary for section-cutting, or with the balsam with which they are to be finally preserved. (The cases in which aqueous imbedding and preserving media are employed are exceptional, and will be treated of in the proper places.) The *dehydration* is performed as follows: the objects are brought into weak alcohol, and are then passed through successive alcohols of gradually increased strength—for instance, 50 per cent. two hours, 70 per cent. six to twenty-four hours, 80 per cent. several hours, 95 per cent. two or three hours, absolute alcohol, time enough for complete saturation. (Very small objects, so small that section-cutting is not necessary, may be dehydrated much quicker than this. Infusoria may be prepared in a few minutes.)

The water having been thus completely removed, the alcohol is in its turn removed from the tissues, and its place taken by some anhydrous substance, generally an essential oil, which is miscible with the material used for imbedding. This operation is known as *Clearing*. It is very important that the passage from the last alcohol to the clearing agent be made gradual. This is effected by placing the clearing medium *under* the alcohol. A sufficient quantity of alcohol is placed in a tube (a watch-glass will do, but tubes are generally better), and then with a pipette a sufficient quantity of clearing medium is introduced *at the bottom of the alcohol*. Or you may first put the clearing medium into the tube, and then carefully pour the alcohol on to the top of it. The two fluids mingle but slowly. The objects to be cleared being now quietly put into the supernatant alcohol, float at the surface of separation of the two fluids, the exchange of fluids takes place gradually, and the objects slowly sink down into the lower layer. When they have sunk to the bottom, the alcohol may be drawn off with a pipette, and the objects will be found to be completely penetrated by the clearing medium. (It may be noted here that this method of making the passage from one fluid to another applies to all cases in which objects have to be transferred from a lighter to a denser fluid—for instance, from alcohol, or from water, to glycerin. It is a more exact method than that of successive baths of mixture of alcohol and clearing agent.)

The objects are now *imbedded*. They are removed from the clearing medium, and soaked until thoroughly penetrated in the imbedding medium. This is, for small objects, generally paraffin, liquefied by heat, and for large objects generally a solution of collodion or " celloïdin." The imbedding medium containing the object is then made to solidify, as described in the chapter on imbedding processes, and sections are made with a microtome through the imbedding mass and the included objects. The sections are then mounted on a slide by one of the methods described in the chapter on Serial Section Methods, the imbedding material is removed from them (in the case of paraffin), they are stained *in situ* on the slide, dehydrated with alcohol, cleared, and mounted in balsam or damar. Or they may be stained, washed, dehydrated, and cleared in watch-glasses, and afterwards mounted as desired—the imbedding medium being first removed if desirable.

It is not always desirable to remove the imbedding mass; celloidin sections stain well without being freed from it, and are usually even dehydrated, cleared, and mounted without removal of the mass, which becomes quite transparent in balsam. This plan has the advantage, which is a very important one for large sections, of allowing the sections to remain during the whole of the manipulations protected by a supporting mass that holds all their parts together.

The plan of staining sections on the slide is of somewhat recent introduction; before it had been worked out the practice was to stain structures *in toto*, before cutting sections. And in cases in which structures are sufficiently small and permeable to allow of satisfactory staining in this way, and if it be not essential to save time, this plan is sometimes as good as the one described. In this case the object, after having been fixed and washed out, is taken from the water, or while still on its way through the lower alcohols (it should not be allowed to proceed to the higher grades of alcohol before staining, if that can be avoided), and passed through a bath of stain (generally alcoholic borax-carmine or other alcoholic stain) of sufficient duration, then dehydrated with successive alcohols, passed through a clearing medium into paraffin, cut, and treated as above described, the sections in this case being mounted direct from the turpentine, naphtha, or other solvent with which the paraffin is removed. If aqueous staining media be employed (and it is sometimes very desirable for particular purposes to prepare specimens with some aqueous stain) the structures should either be stained *in toto* immediately after fixing and washing out, or sections may be stained on the slide, the objects being passed through successive baths of alcohol of gradually decreasing strength before being put into the aqueous stain (a precaution which will not be necessary for chromic objects (*see* below, § 5)).

It was stated in the first edition of this work that " the great majority of preparations are made by fixing either with sublimate or a picric acid combination, washing out with alcohol, staining with alcoholic borax-carmine, imbedding in chloroform-paraffin, cutting with a sliding microtome, and mounting the sections in series in Canada balsam." That is probably still the case, but the method can no longer claim to be what it then appeared to be, the classical method of microscopic anatomy. I suggest the following, as being quite as easy to carry out, and as giving preparations far richer in detail and more truthfully preserved :—Fix in Flemming's chromo-aceto-osmic mixture; wash out with water; dehydrate; clear with oil of cedar-wood; imbed in paraffin; mount sections on the slide with Mayer's albumen medium; stain with safranin, or double-stain with gentian-violet and eosin; and mount in balsam or damar. That, or something like that, is now the practice of many of the most advanced workers; and I know of no method that seems to have equal claims to be considered a classical method of general morphological investigation.

As regards the method to be used for imbedding, I take it that the paraffin method is the method *par excellence* for small objects (objects up to 5 or 7 — — millimetres diameter); whilst the collodion or "celloïdin" method is the method *par excellence* for large objects.

As regards the rival claims of the method of staining objects *in toto* before section cutting, and that of staining the sections, I confess that I cannot see any reason for preferring the practice of staining *in toto*, which I consider only has a *raison d'être* in the case of objects which are not to be cut into sections.

As regards the rival claims of the practice of staining sections on the slide, and that of staining them in watch-glasses, I think that all small sections may be conveniently stained on the slide, and that all large ones should be stained in watch-glasses; in other words, the watch-glass process is the proper process for celloïdin sections, and the slide process is the most convenient one for paraffin sections.

3. The treatment of objects which can be studied without being cut into sections is identical with that above described, with the omission of those passages that relate to imbedding processes. Its normal course may be described as fixation, washing out with alcohol, staining with alcoholic borax-carmine or some other alcoholic stain, treatment with successive alcohols of gradually increasing strength, final dehydration with absolute alcohol, clearing, and mounting in balsam. This method, which may be termed the dehydration method, is generally preferred, as a general method, to what may be termed the wet methods, by which objects are prepared and preserved in aqueous media. The chief reason for this lies in the great superiority of the dehydration methods as regards the preservation of tissues. The presence of water is the most important factor in the conditions that bring about the decomposition of organic matter, and its complete removal is the chief condition of permanent preservation.

It is of course not intended here to suggest that wet methods of preparation should be altogether discarded. They have great value, they are even indispensable, for special ends; and all that is intended to be suggested is, that they should be regarded not as *general* but as *Special* methods.

4. In the preparation of entire objects or structures that are intact and covered by an integument not easily permeable by liquids, special care must be taken to avoid swelling from endosmosis on the passage of the objects from any of the liquids employed to a liquid of less density, or shrinkage from exosmosis on the passage to a liquid of greater density. This applies most specially to the passage from the last alcohol into the clearing medium. A slit should be made in the integument, if possible, so that the two fluids may mingle without hindrance. And in all cases the passage is made gradual by placing the clearing medium under the alcohol, as above described. Fluids of high diffusibility should be employed as far as possible in all the processes. Fixing agents of great penetrating power (such as picrosulphuric acid or alcoholic sublimate solution) should be employed where the objects present not a easily permeable integument. Washing out is done with successive alcohols, water being used only in the case of fixation by osmic acid, or the chromic mixtures or other fixing solutions that render washing by water imperative. Staining is done by preference with alcoholic staining media. The stains most used are Grenacher's borax-carmine, Mayer's modification of Grenacher's alcoholic carmine, and Kleinenberg's hæmatoxylin (for all these *see* STAINING AGENTS). Anilin stains are rarely applicable to this class of preparations. Aqueous stains are more seldom used, though there are many cases in which they are admissible, and some in which they are preferable.

Minute dissections are best done, if necessary, in a drop of clearing agent. I recommend cedar-wood oil for this purpose, as it gives to the tissues a consistency very favorable for dissection, whilst its viscosity serves to lend support to delicate structures. Clove oil has a tendency to make tissues that have lain in it for some time very brittle. This brittleness is also sometimes very helpful in minute dissections. Another property of clove oil is that it does not easily spread itself over the surface of a slide, but has a tendency to form very convex drops. This property also makes it frequently a very convenient medium for making minute dissections in.

5. Following Paul Mayer, I gave in the first edition the following reasons for employing alcoholic rather than aqueous staining media. Since, in most cases, treatment with alcohol

forms part of the fixing process, alcoholic solutions are logically indicated for staining. For by means of them it is possible to avoid the bad effects that follow on passing delicate tissues from alcohol into water, violent diffusive currents being thereby set up which sometimes carry away whole groups of cells; swellings being caused in the elements of the tissues; and, if the immersion in the aqueous medium be prolonged, as is generally necessary in order to obtain a thorough stain, maceration of the tissues supervening. But alcoholic staining fluids have still other advantages; they are vastly more penetrating; with them alone is it possible to stain through chitinous integuments; and, if it be desired to stain slowly, tissues may be left in them for days without hurt.

Applied to the case now under consideration, the preparation *in toto* of objects protected by not easily permeable investments, this doctrine is evidently a wise one. For such objects must necessarily be fixed by some highly penetrating but not permanently hardening agent such as picric acid, and must necessarily be washed out with alcohol; and it is a good maxim for tissues so fixed that an object that has once been in alcohol should not be allowed to go back into water, if that can possibly be avoided.

But in the case of structures that have been well fixed in a strongly and permanently coagulating medium such as chromic acid, this precaution is much less necessary. Sections of tissues that have been fixed for twenty-four hours in Flemming's solution may be passed with relative impunity from absolute alcohol into an aqueous stain, and from that back again direct into absolute alcohol. It is this property of tissues fixed in chromic solution that determines me to recommend the practice of staining sections, instead of staining objects *in toto*.

For an excellent exposition of the principles underlying the practice above recommended, the reader may consult with advantage the paper of Paul Mayer, in *Mitth. Zool. Stat. Neapel*, ii (1881), p. 1, *et seq.* See also the abstract in *Journ. Roy. Mic. Soc.* (N. S.), ii (1882), pp. 866—881, and that in *Amer. Natural.*, xvi (1882), pp. 697—706, in which two last some improvements are mentioned which have been worked out since the publication of Mayer's paper.

CHAPTER II.

KILLING.

6. In the majority of cases, the first step in the preparation of an organ or organism consists in exposing it as rapidly and as completely as possible to the action of one of the Fixing Agents that are discussed in the next chapter. The organ or organism is taken in the normal living state; the fixing agent serves to bring about at the same time, and with sufficient rapidity, both the death of the organism and that of its histological elements.

But this method is by no means applicable to all cases. There are many animals, especially such as are of a soft consistence, and deprived of any rigid skeleton, but possessing a considerable faculty of contractility—such as many Cœlenterata, Bryozoa, and Serpulida, for instance—which if thus treated contract violently, draw in their tentacles or branchiæ, and die in a state of contraction that renders the preserved object a mere caricature of the living animal. In these cases, special methods of killing must be resorted to.

Sudden Killing.

7. Speaking generally, there are two ways of dealing with these difficult cases. You may kill the animal so suddenly that it has not time to contract; or you may paralyse it by narcotics before killing it.

The application of **heat** is a good means of killing suddenly. It has the great advantage of allowing of good staining subsequently, and of hindering less than any other method the application of chemical tests to the tissues. By it, the tissues are fixed at the same time that somatic death is brought about. And what is more, the fixation thus brought about is extremely faithful, provided the operation be properly performed.

The difficulty consists in hitting off the right temperature, which is of course different for different objects. I think that a temperature of 80° to 90° C. will generally be amply sufficient, and that very frequently it will not be necessary to go beyond 60° C. An exposure to heat of a few seconds will generally suffice.

Small objects (Protozoa, Hydroids, Bryozoa) may be brought into a drop of water in a watch-glass or on a slide and heated over the flame of a spirit lamp. For large objects, the water or other liquid employed as the vehicle of the heat, may be heated beforehand and the animals thrown into it.

As soon as it is supposed that the protoplasm of the tissues is coagulated throughout, the animals should be brought into alcohol (30 to 70 per cent. alcohol) (if water be employed as the heating agent).

An excellent plan for preparing many marine animals is to kill them in hot fresh water. Some of the larger Nemertians are better preserved by this method than by any other with which I am acquainted.

8. Animals that contract but slowly, such as Alcyonium and Veretillum, and some Tunicates, such as Pyrosoma, are very well killed by throwing them into some very quickly acting fixing liquid, used either hot or cold. Glacial or very strong acetic acid is an excellent reagent for this purpose; it may be used, for example, with some Medusæ. After an immersion of a few seconds or a few minutes, according to the size of the animals, they should be brought into alcohol of at least 70 per cent. strength. Corrosive sublimate is another excellent reagent for this purpose. *See* §§ 45, 46.

Narcotisation.

9. The secret of narcotisation consists in adding some anæsthetic substance very gradually, in very small doses, to the water containing the animals, and waiting patiently for it to take effect slowly.

The Tobacco-smoke Method for Actiniæ (Hertwig, *Die Actinien*, 1879) is frequently practised as follows:—A dish containing the animals in water is covered with a bell-glass, under which passes a curved glass or rubber tube, which dips into the water. Tobacco smoke is blown into the water for some time, through the tube, and the animals are then left for some hours in order that narcotisation may become fully established. The animals are irritated from time to time by touching a tentacle with a needle. As soon as it is observed that an animal begins to react slowly, that is to say as soon as it is found that the contraction of the tentacle does not begin until a considerable time after it has been irritated by the needle, the narcotisation may be considered sufficient. A quantity of some fixing liquid sufficient to kill the animals before they have time to contract is then added to the water.

A space of several hours is necessary in order to thoroughly narcotise an Actinia by this method.

10. Nicotin in solution may be used instead of tobacco smoke (ANDRES, *Atti R. Accad. dei Lincei*, v. 1880, p. 9; see *Journ. Roy. Mic. Soc.*, N.S., ii, 1882, p. 881). Andres employs a solution of 1 gramme of nicotin in a litre of sea-water. The animal to be anæsthetised is placed in a jar containing half a litre of sea-water, and the solution of nicotin is gradually conducted into the jar by means of a thread acting as siphon. The thread ought to be of such a thickness as to be capable of carrying over the whole of the solution of nicotin in twenty-four hours.

11. Chloroform may be employed either in the liquid state or in the state of vapour. KOROTNEFF (*Mitth. Zool. Stat. Neapel*, v, Hft. 2, 1884, p. 229; *Zeit. f. wiss. Mik.*, 2, 1885, p. 230) operates in the following manner with Siphonophora. The animals being extended, a watch-glass containing chloroform is floated on the surface of the water in which they are contained, and the whole is covered with a bell-glass. As soon as the animals have become insensible, they are killed by means of hot sublimate or chromic-acid solution plentifully poured on to them.

12. Liquid chloroform is employed by squirting it in small quantities on to the surface of the water containing the animals. A syringe or pipette having a very small orifice, so as to thoroughly pulverise the chloroform, should be employed. Small quantities only should be projected at a time, and the dose should be repeated every five minutes, until the animals are anæsthetised.

I have seen large Medusæ very completely anæsthetised in the state of extension in an hour or two by this method. ANDRES finds that this plan does not succeed with Actiniæ, as with them maceration of the tissues supervenes before anæsthesia is established.

13. Ether and **Alcohol** may be administered in the same way. ANDRES has obtained good results with Actiniæ by the use of a mixture (invented by SALVATORE LO BIANCO) containing 20 parts of glycerin, 40 parts of 70 per cent. alcohol, and 40 parts of sea-water. This mixture should be carefully poured on to the surface of the water containing the animals, and allowed to diffuse quietly through it. Several hours are sometimes necessary for this.

EISIG employs alcohol in the same way.

14. Hydrate of Chloral, which was first recommended, I believe, by Foettinger (*Arch. de Biol.*, vi, 1885, p. 115), gives very good results with some subjects. Foettinger operates by dropping crystals of chloral into the water containing the animals. For Alcyonella he takes 25 to 80 centigrammes of chloral for each hundred grammes of water. It takes about three quarters of an hour to render a colony sufficiently insensible to allow of fixing. Foettinger has obtained satisfactory results with marine and fresh-water Bryozoa, with Annelida, Mollusca, Nemertians, Actiniæ, and with *Asteracanthion*. He did not succeed with Hydroids.

I am bound to state that I have never had the slightest success with Nemertians.

VERWORN (*Zeit. f. wiss. Zool.*, xlvi, 9, 1887, p. 99; see also *Journ. Roy. Mic. Soc.*, 1888, p. 148) operates differently for fresh-water Bryozoa. He puts *Cristatella* for a few minutes into 10 per cent. solution of chloral, in which the animals sooner or later become extended.

KÜKENTHAL (*Zeit. f. wiss. Mik.*, iv, 8, 1887, p. 878; *Journ. Roy. Mic. Soc.*, 1888, p. 509) has obtained good results with some Annelids, by means of a solution of one part of chloral in 1000 parts of sea-water.

15. Cocaïn (RICHARD; *Zool. Anz.*, 196, 1885, p. 332) has been found to give good results. Richard puts a colony of Bryozoa into a watch-glass with 5 cc. of water, and adds gradually 1 per cent. solution of hydrochlorate of cocaïn in water. After five minutes, the animals are somewhat numbed, and half a cubic centimetre of the solution is added, and the tentacles are caused to contract by irritating them with a needle. Ten minutes later the animals should be found to be dead in a state of extension.

This method is stated to succeed with Bryozoa, *Hydra*, and certain worms.

16. Morphia, Curare, Strychnin, Prussic Acid, and other paralysing drugs have also been employed.

17. Carbonic Acid Gas has been recommended (by FOL, *Zool. Anz.*, 128, 1885, p. 698). The water containing the animals should be saturated with the gas. The method is stated to succeed with most Cœlenterata and Echinodermata, but not with Molluscs or Fishes.

18. Asphyxiation may be sometimes successfully practised. Terrestrial Gastropods are best killed for dissection by putting them into a jar quite full of water that has been deprived of its air by boiling, and hermetically closed. After from twelve to twenty-four hours the animals are generally found dead and extended. The effect is obtained somewhat quicker if a little tobacco be added to the water.

19. Poisoning by small doses of some fixing agent is sometimes a good method. SALVATORE LO BIANCO employs the following method for preserving Ascidiæ in an extended state. A 1 per cent. solution of chromic acid acidulated with acetic acid is poured on to the surface of the water containing the animals, and allowed to diffuse slowly through it. The operation takes four or five days (v. GARBINI, *Manuale per la Technica Mod. del Microscopio*, p. 168).

Osmic acid, or Kleinenberg's solution, is sometimes employed in the same way.

I have seen Medusæ killed in a satisfactory manner by means of crystals of corrosive sublimate added to the water containing them.

20. Marine Animals are sometimes successfully killed by simply putting them into fresh water.

21. Warm Water will sometimes serve to immobilise and even kill both marine and fresh-water organisms.

CHAPTER III.

FIXING.

22. The Necessity of Fixing.—The meaning of the term "fixing" has been explained above (§ 2). It only remains here to insist on the absolute necessity of the employment of fixing agents, and to briefly illustrate this necessity by a single example. If a portion of living retina be placed in aqueous humour, serum, or other so-called "indifferent" medium, or in any of the media used for permanent preservation, it will be found that the rods and cones will not preserve the appearance they have during life for more than a very short time; after a few minutes a series of changes begins to take place, by which the outer segments of both rods and cones become split into discs, and finally disintegrate so as to be altogether unrecognisable, even if not totally destroyed. Further, in an equally short time the nerve-fibres become varicose, and appear to be thickly studded with spindle-shaped knots; and other post-mortem changes rapidly occur. If, however, a fresh piece of retina be treated with a strong solution of osmic acid, the whole of the rods and cones will be found perfectly preserved after twenty-four hours' time, and the nerve-fibres will be found not to be varicose. After this preliminary hardening, portions of the retina may be treated with water (which would be ruinous to the structures of a fresh retina), they may even remain in water for days without harm; they may be stained, acidified, hardened, imbedded, cut into sections, and mounted in either aqueous or resinous media without suffering.

23. The Action of Fixing Agents consists in the coagulation of certain of the constituents of tissues, of their albuminoids, their gelatin, their mucin. Some fixing agents seem to have further the property of combining chemically with the tissues, so that they cannot be easily removed from them by washing. This is a consideration of great importance in view of ulterior operations, and most particularly in view of staining. Chromic acid and its salts, osmic acid, the chlorides of palladium, of gold, and of iron, are reagents that seem to combine chemically with the tissues, and render necessary a special after-treatment and special modes of staining, whilst picric acid, nitric acid, and corrosive sublimate do not appear to enter into that kind of combination, and can be entirely removed from the tissues by washing, and leave the tissues in a state in which they are susceptible of any kind of staining.

Practically it amounts to this, that if you fix with a chromic or osmic mixture, you cannot stain with carmine but can only stain with hæmatoxylin, if you wish to stain your objects *in toto;* or you may make sections, and stain them with safranin or some other coal-tar colour. Whilst if you fix with sublimate or a picric acid mixture, you may do as you like in the matter of staining.

The after-treatment appropriate to each fixing-agent is indicated in the special paragraphs.

24. Choice of a Fixing Agent.—Indications concerning the proper fixing agent to employ for the different tissues and organs of the animal kingdom, will be found in Part II. The following remarks are intended as hints for beginners only.

The chief fixing agents for general work are Flemming's mixture (§§ 35 and 36), osmic acid, corrosive sublimate, and picro-sulphuric acid.

I recommend that Flemming's mixture should be used wherever it is possible, as I believe it to be in general by far the best fixing agent yet invented.

But it will not always be found possible to use it. Its low power of penetration, for instance, puts it out of court in the case of very impermeable objects, such as are frequently offered by the Arthropoda. For these, picrosulphuric acid may be recommended.

For very small objects, such as may be mounted whole, osmic acid is nearly as good as Flemming's mixture, and is frequently much more convenient to use.

For objects a little larger, and for much embryological work on objects that it is convenient to have stained *in toto*, corrosive sublimate may be recommended. I think it is in general a much better preservative of the forms of anatomical elements than either Altmann's nitric acid or Kleinenberg's fluid.

25. The Practice of Fixation,—Hints and Cautions.—See that the structures are *perfectly living* at the instant of fixation, otherwise you will only fix pathological states or post-mortem states.

Do all you can to facilitate the *rapid penetration* of the fixing agent. To this end, let the structures be divided into the smallest portions that can conveniently be employed, and if entire organs or organisms are to be fixed whole, let openings, as large as possible, be first made in them.

The penetration of reagents is greatly facilitated by *heat*. You may warm the reagent and put it with the objects to be fixed in the paraffin stove, or you may even employ a fixing agent heated to boiling point (as boiling sublimate solution for certain corals and Hydroids, or boiling absolute alcohol for certain Arthropods with very resistant integuments).

Let the *quantity* of fixing agent employed be at least *many times* the volume of the objects to be fixed. If this precaution be not observed the composition of the fixing liquid may be seriously altered by admixture of the liquids or of the soluble substances of the tissues thrown into it. For a weak and slowly acting fixing agent, such as picric acid, the quantity of liquid employed should be in volume about one hundred times that of the object to be fixed. Reagents that act very energetically, such as Flemming's solution, may be employed in smaller proportions.

Be careful to use the *appropriate liquid for washing out* the fixing agent after fixation. It is frequently by no means a matter of indifference whether water or alcohol be employed for washing out. Sometimes water will undo the whole work of fixation (as with picric acid). Sometimes alcohol causes precipitates that may ruin the preparations. Instructions on this head are given in the paragraphs devoted to the different fixing agents.

Use *liberal quantities* of liquid for washing.

Change the liquid as often as it becomes turbid, if that should happen.

The process of washing out is often greatly facilitated by *heat*. Picric acid, for instance, is nearly twice as soluble in alcohol warmed to 40° C. as in alcohol at the normal temperature (Fol).

In the case of *marine organisms* it may be stated as a general rule that their tissues are more refractory to the action of reagents than are the tissues of corresponding fresh-water or terrestrial forms, and fixing solutions should in consequence be stronger (about two to three times stronger, according to Langerhans).

Marine animals ought to be *freed from the sea-water* adherent to their surfaces before treating them either with alcohol or any fixing reagent that precipitates the salts of sea water. If this be not done, the precipitated salts will form on the surfaces of the organisms a crust that prevents the penetration of reagents to the interior, thus allowing maceration to be set up, and hindering the penetration of staining fluids. Fixing solutions for marine organisms should therefore be such as serve to keep in a state of solution, and finally remove, the salts in question. They should *never be made with sea water* as a menstruum, as some workers have inconsiderately proposed. If alcohol be employed it should be acidified with hydrochloric acid. Picro-sulphuric acid is a reagent that fulfils the conditions here spoken of. (On this subject see Paul Mayer, in *Mitth. Zool. Stat. Neapel*, ii (1881), p. 1, et seq. See also the abstract in *Journ. Roy. Mic. Soc.* (N.S.), ii (1882), pp. 866—881, and that in *Amer. Natural.*, xvi (1882), pp. 697—706.)

CHAPTER IV.

FIXING AGENTS. MINERAL ACIDS AND THEIR SALTS.

26. Osmic Acid.—The tetroxyde of osmium (OsO_4) is the substance commonly known as osmic acid, though it does not possess acid properties. It is a substance that is exceedingly difficult to keep in use for any length of time. It is extremely volatile, and in the form of an aqueous solution becomes partially reduced with great readiness in presence of the slightest contaminating particle of organic matter. (It is generally believed that the aqueous solutions are reduced by light, but this is not the case; they may be exposed to the light with impunity if dust be absolutely denied access to them.) After having carefully tried several of the plans that have been recommended for keeping the working solutions free from dust, I have come to the conclusion that the following is the most practical plan :—The solution of osmic acid in chromic acid solution is not, like the solution in pure water, easily reducible, but may be kept without any special precautions. I therefore keep the bulk of my osmium in the shape of a 2 per cent. solution of osmic acid in 1 per cent. aqueous chromic acid solution. This solution serves for fixation by osmium vapours, and for making up solution of Flemming. A small quantity of osmic acid may also be made up in 1 per cent. solution in distilled water, and kept carefully protected from dust. Those who have to do a great deal of fixing by means of the vapours may also keep a supply of the solid oxide for this purpose.

Great stress is laid by authors on the fact that the vapour of osmium is very irritating to mucous tissues. It is said that the slightest exposure to it is sufficient to give rise to serious catarrh, irritation of the bronchial tubes, laryngeal catarrh, conjunctivitis, &c. I think these statements greatly exaggerated.

27. Fixation by the Vapours.—Osmic acid is frequently employed in the form of vapour, and its employment in this form is indicated in most of the cases in which it is possible to expose the tissues directly to the action of the vapour. The tissues are pinned out on a cork which must fit well into a wide-mouthed bottle in which is contained a little solid osmic acid (or a small quantity of 1 per cent. solution will do). Very small objects, such as isolated cells, are simply placed on a slide, which is inverted over the mouth of the bottle. They remain there until they begin to turn brown (isolated cells will generally be found to be sufficiently fixed in thirty seconds, whilst in order to fix the deeper layers of relatively thick objects, such as retina, an exposure of several hours may be desirable). It is well to wash the objects with water before staining, but a very slight washing will suffice. For staining, methyl-green may be recommended for objects destined for study in an aqueous medium, and, for permanent preparations, alum-carmine, picro-carmine, or hæmatoxylin.

The reasons for preferring the process of fixation by vapour of osmium, where practicable, are that osmium is more highly penetrating when employed in this shape than when employed in solution, and produces a more *equal* fixation, and that the arduous washing out required by the solutions is here done away with. In many cases delicate structures are better preserved, all possibility of deformation through osmosis being here eliminated.

In researches on nuclei, it is possible and may be useful to employ the vapours of a freshly-prepared mixture of osmic and formic or acetic acid (Gilson, *La Cellule*, i, 1885, p. 96).

28. Fixation by Solutions.—When employed in aqueous solutions osmic acid is used in strengths varying from $\frac{1}{20}$ per cent. to 2 per cent. Solutions of $\frac{1}{2}$ per cent. to 1 per cent. have been very largely used, but the tendency of modern practice seems to be towards weaker solutions and longer immersion. For Infusoria $\frac{1}{2}$ per cent. for a few seconds; for Porifera $\frac{1}{20}$ to $\frac{1}{10}$ per cent. for some hours; for Mollusca 1 to 2 per cent. for twenty-four hours; for epithelia $\frac{1}{10}$ to $\frac{1}{2}$ per cent. for an hour or two; for meroblastic ova $\frac{1}{10}$ per cent. for twenty-four hours; for medullated nerve-fibre $\frac{1}{10}$ to 1 per cent. for from twenty minutes to two hours; for tactile corpuscles $\frac{1}{3}$ to 1 per cent. for twenty-four hours; for retina ¼ to 2 per cent. for from ten minutes to twenty-four hours; for nuclei $\frac{1}{10}$ to 2 per cent. for two or three hours. Such figures as these will serve to give a general idea of the practice, whilst more precise instructions will be given when dealing with the tissues in detail. (The durations here quoted appear to me exaggerated, except for very voluminous specimens.)

The osmium must be well washed out before proceeding to any further steps in preparation; water should be used for washing. Notwithstanding the greatest care in soaking, it frequently happens that some of the acid remains in the tissues, and causes them to over-blacken in time. To obviate this it is necessary to wash them out in ammonia-carmine or picro-carmine, or to soak them for twenty-four hours in a solution of bichromate of potash (Müller's solution or Erlicki's will do), or in 0·5 per cent. solution of chromic acid, or in Merkel's solution, or in a weak solution of ferrocyanide of potassium or cyanide of potassium, or to bleach them (see BLEACHING, No. 542). The treatment with bichromate solutions has the great advantage of highly facilitating staining with carmine or hæmatoxylin. Max Schultze recommended washing, and mounting permanently in acetate of potash (see § 359), but I believe the virtues attributed to this method are illusory. Fol has lately recommended treatment with a weak solution of carbonate of ammonia.

The same stains recommended for objects fixed by vapour will be found useful here, with the addition of ammonia carmine, which is really very useful for strongly fixed specimens. For sections, of course in both cases, safranin and the other nuclear anilin stains may be employed with advantage.

Osmic acid stains all fatty structures black; it must therefore be avoided for tissues in which much fat is present; or the fat may afterwards be dissolved out with turpentine (Flemming, see the chapter on Connective Tissues in Part II). All solutions of osmic acid must be kept protected from the light even during the immersion of tissues. If the immersion is to be a long one the tissues must be placed with the solution in well-closed vessels, as osmium is very volatile.

A little acetic or formic acid (0·5 to 1 per cent.) may frequently with advantage be added to the solutions just before using.

KOLOSSOW (*Zeit. f. wis. Mik.*, v, i, 1888, p. 51) recommends a 0·5 per cent. solution of osmium in 2 or 3 per cent. solution of nitrate or acetate of uranium, as having a greatly enhanced penetrating power.

29. Osmic Acid and Alcohol (RANVIER ET VIGNAL. RANVIER, *Leç. d'Anat. Gén.*, "App. term. des muscles de la vie org.,' p. 76; Vignal, *Arch. de Physiol.*,' 1884, p. 181). Equal volumes of 1 per cent. osmic acid and 90 per cent. alcohol (freshly mixed). Allow it to act, for medium-sized objects, such as embryos of a few millimètres diameter, for an hour or two. Wash for some hours in 80 per cent. alcohol. Then wash with water and stain for forty-eight hours in picro-carmine or hæmatoxylin. Viallanes has applied this method to the histology of insects.

30. Chromic Acid.—Chromic anhydride, CrO_3, is found in commerce in the form of red crystals that dissolve readily in water, forming chromic acid, H_2CrO_4. These crystals are very deliquescent, and it is therefore well to keep the acid in stock in the shape of a 1 per cent. solution. Care must be taken not to allow the crystals to be contaminated by organic matter, in the presence of which the anhydride is readily reduced into sesquioxide.

Chromic acid is employed in solution either in water or in alcohol.

The most usual strengths in which it is employed in aqueous solution are from 0·1 to 1·0 per cent. for a period of immersion of a few hours (structure of cells and ova). For nerve-tissues weaker solutions are taken, $\frac{1}{50}$th to $\frac{1}{8}$th per cent. for a few hours. Stronger solutions, such as 5 per cent., should only be allowed to act for a few seconds.

The object should be washed out with water before passing into alcohol or staining fluids. Long washing in water is necessary to prepare them for staining, except an anilin stain be used. It is possible to wash out in alcohol, and this may be useful in special cases, but in general I think the practice is not to be recommended. It is well to wash for many hours in *running* water.

Tissues that have been fixed in chromic acid may be stained in aqueous solutions if desired, as water does not appear to have an injurious effect on them; the acid appears to enter into some chemical combination with the elements of the tissues, forming with them a compound that is not affected either physically or chemically by water. The best stain to follow chromic acid is hæmatoxylin, or, for sections, some anilin stain. But the previous washing out with water must be very thorough if good results are to be insured; it may take days.

Chromic acid is not a very penetrating reagent, and for this reason, as well as for others, is seldom used pure, but plays an important part in the mixtures described below, of which the chief is certainly the mixture of Flemming. A chief objection to the use of chromic acid is that it precipitates certain of the liquid albuminoids of tissues in the form of filaments or networks, which are often of great regularity, and simulate structural elements of the tissues. This objection applies to all mixtures into which chromic acid enters.

Action of light on alcohol containing chromic objects.—When objects that have been treated by chromic acid or a chromate are put into alcohol for hardening or preservation, it is found that after a short time a fine precipitate is thrown down on the surface of the preparations, thus forming a certain obstacle to the further penetration of the alcohol. Previous washing by water does not prevent the formation of this precipitate, and changing the alcohol does not prevent it from forming again and again. It has been found by Hans Virchow (*Arch. f. mik. Anat.*, Bd. xxiv, 1885, p. 117) that the formation of this precipitate may be entirely prevented by simply keeping the preparations in the dark. The alcohol becomes yellow as usual (and should be changed as often as this takes place), but no precipitate is formed. If this precaution be taken, previous washing with water may be omitted, or at all events greatly abridged.

The brownish-green colour of chromic objects may be removed by treating them with peroxide of hydrogen (Unna, in *Arch. f. mik. Anat.*, Bd. xxx, 1887, p. 47; cf. *Journ. Roy. Mic. Soc.*, 1887, p. 1060).

31. Chromo-acetic Acid (Flemming, *Zellsbz. Kern. u. Zellth.*, p. 382).

Chromic acid . 0·2 to 0·25 per cent.
Acetic acid . . 0·1 per cent., in water.

Flemming finds this the best reagent for the study of the *achromatic* elements of karyokinesis. (Flemming wrote this in 1882, and I doubt whether it would now hold good.) Stain with hæmatoxylin (the preparations are *not* favorable for staining with safranin or other coal-tar colours).

32. Chromo-formic Acid (RABL, *Morph. Jahrb.*, x, 1884, pp. 215, 216).—Four or five drops of concentrated formic acid are added to 200 c.c. of 0·33 per cent. chromic acid solution. The mixture must be freshly prepared at the instant of using. Fix for twelve to twenty-four hours, wash out with water, harden in alcohol, stain with hæmatoxylin or safranin. For the study of karyokinesis. This is acknowledged to be one of the very best reagents for the purpose.

33. Chromic Acid and Spirit.—A mixture of 2 parts of $\frac{1}{6}$th per cent. chromic acid solution with 1 part of methylated spirit was much used by Klein in his investigations into the structure of cells and nuclei, and found to give better results than the ordinary reagents (including even osmic acid). Hæmatoxylin was used for staining.

The addition of alcohol to augment the penetrating power of chromic acid seems to be a step in the right direction, and it is matter for surprise that such mixtures are not more used. The alcohol should be added to the acid in aqueous solution, as if strong alcohol be added to crystals of chromic anhydride, a very violent reaction is set up.

34. Chromo-osmic Acid (MAX FLESCH, *Arch. f. mik. Anat.*, xvi, 1878, p. 300).—This mixture (osmium 0·10, chromic acid 0·25, water 100·0), originally introduced for the preparation of the auditory organ of vertebrates, is of general application. It does not require to be kept in the dark. Objects may remain in it for twenty-four or thirty-six hours without risk of the osmic acid over-blackening them. Flemming found it to preserve nuclear figures well, but the preparations are pale, and difficult to stain well. He finds that the action of the mixture is improved (for nuclear figures) by the addition of acetic, formic, or other acid. This addition brings out the figures more sharply, and has the further advantage of allowing of a sharper stain with hæmatoxylin, picro-carmine, or gentian violet. He recommends the following formula, which may be considered to have superseded Max Flesch's.

35. Chromo-aceto-osmic Acid (FLEMMING, FIRST or WEAK formula, *Zellsubstanz, Kern und Zelltheilung*, 1882, p. 381).—

    Chromic acid . . 0·25 per cent. ⎫
    Osmic acid . . 0·1 per cent. ⎬ In water.
    Glacial acetic acid . 0·1 per cent. ⎭

The best results (as regards faithfulness of fixation) are obtained with this mixture when it is allowed to act for only a short time (about half an hour).

But it may, without inconvenience, be allowed to act for many hours or even days. Wash out, very thoroughly, in water. Stain with hæmatoxylin, if you wish to stain *in toto* (staining in this way with other reagents is possible, but very difficult, and not to be recommended). Stain sections with safranin, or other anilin, or with hæmatoxylin or Kernschwarz.

To make up this mixture with the usual stock solutions, you take:

Chromic acid of 1 per cent. . .	25 volumes
Osmic acid of 1 per cent. . .	10 ,,
Acetic acid of 1 per cent. . .	10 ,,
Water	55 ,,

If you keep your osmium in 2 per cent. solution in chromic acid of 1 per cent., as I have recommended, you will have to take only 20 vols. of chromic acid, 5 of your osmium solution, and 65 of water.

It has been already stated more than once that Flemming's solution is probably the very best fixing reagent in general yet discovered. It has, however, been criticised. Faussek (*Zeitschr. f. wiss. Zool.*, Bd. xlv, 1887, pp. 694, *et seq.*) found it totally inapplicable to the histology of the intestine of insects. He states that it caused the intima to disappear, and the cells to run together into a compact mass. Arnold (*Arch. f. Mik. Anat.*, Bd. xxx, 1887, p. 205) states that it does not preserve cell-bodies faithfully. And A. Kotlarewsky (*Mitth. d. naturf. Ges. Bern.*, 1887; cf. *Zeit. f. wiss. Mik.*, iv, 3, 1887, p. 387) found that it preserved the forms of nerve-cells (spinal ganglia) less faithfully than any of the reagents tried. I have not, myself, been struck by any decided defect in the preservation of cystoplasmic structures in my preparations made by this reagent, but think it possible that the observations of these authors may be well founded as regards the present formula, but take it that that is merely a reason for preferring the stronger mixture set forth below.

It is not necessary in all cases to observe the exact proportions of the ingredients in this mixture. FOL (*Lehrb. d. vergl. Mik. Anat.*, 1884, p. 100) recommends the following:

1 per cent. chromic acid	25 vols.
1 per cent. osmic acid	2 ,,
2 per cent. acetic acid	5 ,,
Water	68 ,,

That is to say, a mixture much weaker in osmium than Flemming's. In the *Traité des Méthodes Techniques*, &c., Lee et Henneguy, 1887, I recommended this mixture, as giving better results in general, but am now inclined to think that, at all events as regards fidelity of fixation, it is a step in the wrong direction, and that, on the contrary, the stronger mixture of Flemming (next §) is a step in the right direction. Fol's formula has the advantage of allowing better staining with carmine, that is all.

36. Chromo-aceto-osmic Acid (FLEMMING, SECOND OR STRONG formula, *Zeit. f. wiss. Mik.*, 1, 1884, p. 349).—

1 per cent. chromic acid	15 parts.
2 per cent. osmic acid	4 ,,
Glacial acetic acid	1 ,,

If this mixture be kept in stock in large quantities, it may go bad, probably on account of the large proportion of organic acid contained in it. I therefore recommend that it be made up from time to time from stock solutions, in which the osmium is kept separate from the acetic acid. The proportions being as follows:

CrO_3	0·15
Os.	0·08
Acid. acet.	1·00
Aq.	19·00

You may make up and keep separately—

(A) 1 per cent. chromic acid	11 parts.
Distilled water	4 ,,
Glacial acetic acid	1 ,,

and (B) a 2 per cent. solution of osmium in 1 per cent. chromic acid solution, and when required, mix four parts of A with one of B.

Merk (*Denksch. d. Math. Naturw. Cl. d. K. Acad. d. Wiss. Wien.*, 1887; cf. *Zeit. f. wiss. Mik.*, v, 2, 1888, p. 237) proposes to make up separately (A)

2 per cent. chromic acid	7·5 parts
Water	3·5 ,,
Acetic acid	1 ,,

and (B), some 1 per cent. osmium solution, and to mix for use 12 parts of A with 8 of B. But this plan leaves you in the old difficulty of keeping your osmium in aqueous solution.

It does not appear necessary to observe the exact proportions of the ingredients of these mixtures, a certain latitude is allowable. Thus CARNOY (*La*

Cellule, 1, 2, 1885, p. 211) has employed a mixture one third stronger in osmium and twice as strong in chromic acid, viz.

    Chromic acid of 2 per cent. (or even stronger). 45 parts.
    Osmic acid of 2 per cent. . . . . . 16 „
    Glacial acetic acid . . . . . . 3 „

In the *Traité des Méth. Techniques*, 1887, I treated this formula somewhat coldly, pointing out (what was the case) that Flemming recommended it merely for a very special purpose, the hunting for karyokinetic figures, and that he did not recommend it for general purposes. Further experience has shown that it is applicable to general purposes, and will probably be found for most purposes considerably superior to the weak formula. I should use it by the hogshead if it were not somewhat expensive.

Arnold, in the place quoted in the last paragraph, says that it is to be avoided if you wish to demonstrate the structure of certain nuclei (of wandering cells); and the other objections there quoted as applying to the weak formula are intended to apply more or less to the present formula. It will be well not to attach too much importance to them. Let delicate structures be fixed for twenty-four hours or more, washed in running water for an hour, and in successive alcohols for twenty-four hours, sectioned, and stained with safranin or gentian violet, and there will be little complaint of defective preservation.

The strong mixture does not brown tissues more than the weak mixture, but rather less.

Fat is blackened by these mixtures; but the blackened fat can be entirely dissolved out of the tissues by treating them for a few hours with turpentine that has been exposed to sunlight for an hour or two (*see* Flemming in *Zeit. f. wiss. Mik.*, vi, 1, 1889, p. 39; and vi, 2, 1889, p. 178).

PODWYSSOZKI recommends (for glands especially) the following modification:

    1 per cent. CrO_3 dissolved in 0·5 per cent. solution of corrosive sublimate . . . . 15 cc.
    2 per cent. osmium solution. . . . . . 4 cc.
    Glacial acetic acid . . . . . . 6 to 8 drops.

The sublimate is said to augment the penetration of the osmium, but is unfavorable to staining. The proportion of acetic acid is reduced in order to avoid swelling of the tissue elements (ZIEGLER's *Beiträge z. path. Anat.*, i, 1886; *cf. Zeit. f. wiss. Mik.*, iii, 3, 1886, p. 405).

37. Nitric Acid (ALTMANN, *Arch. Anat. u. Phys.*, 1881, p. 219).—(Of general histological application, but most specially referring to embryology.)

Altmann employs dilute nitric acid, containing from 3 to $3\frac{1}{2}$ per cent. pure acid. Such a solution has a sp. gr. of about 1·02; an aræometer may conveniently be used to determine the concentration of the solution. Stronger solutions have been used, but do not give such good final results.

His (*Ibid.*, 1877, p. 115) recommended a 10 per cent. solution. Altmann tried it, but found he could not demonstrate the nuclear figures. He considers that the strong solutions coagulate the soluble albuminoids of the tissues *too strongly*, which is a hindrance to the optical differentiation of structure. Flemming writes to Altmann that he employs solutions of 40 to 50 per cent. for the ova of invertebrates. This of course has the advantage of a very rapid fixing action.

The embryos, or other objects, are to be put *fresh* into the solution; it is useful, though not necessary, to employ a liquid cooled to zero; the cold stops all molecular processes, and the acid has time to fully complete the fixing process.

The objects must not be left too long in the liquid; for blastoderms and small embryos a quarter to half an hour is enough, for larger ones two to four hours. Only small pieces of tissues other than embryonic should be employed.

I understand the author to say that he then washes out the acid, and completes the hardening in strong alcohol. He points out that the process does not afford a true hardening such as is obtained by the use of chromic acid; but then he considers that by the use of paraffin imbedding such strictly so-called hardening is superfluous. I consider myself that it does not harden enough to faithfully preserve delicate structures.

I have used this reagent extensively, and have at length abandoned it, for general purposes, in favour of "Flemming." It may still be useful for certain ova. It has the valuable property of hardening yolk without making it brittle.

Before imbedding, the objects are stained according to Altmann's instructions, *in toto*, slowly, with dilute hæmatoxylin. By moderately staining (either before or after the hæmatoxylin) with eosin, good double stains of nuclear figures are obtained, the chromatin structures taking up the blue colour. But any other staining process may be used.

38. Silver Nitrate.—Silver nitrate is frequently used in the study of epithelia, not alone for the purpose of demonstrating the outlines of cells by staining the intercellular cement-substance, but also with the totally distinct object of rapidly fixing the cells. Solutions of from $\frac{1}{2}$ to 2 per cent. are employed and allowed to act for merely a few seconds. Solutions of only 3 to 1000 strength may be allowed to act for an hour. Wash out with water. Stain as desired. Weak solutions, rapidly applied, do not hinder subsequent staining; strong solutions do.

For the purpose of fixation, as well as for that of staining, nitrate of silver is at present a reagent too uncertain in its action to be generally recommendable.

39. Chromo-nitric Acid (PERENYI's *formula*) (*Zool. Anzeig.*, v (1882), p. 459).

  4 parts 10 per cent. nitric acid.
  3 parts alcohol.
  3 parts 0·5 per cent. chromic acid.

These are mixed, and after a short time give a fine violet-coloured solution.

The objects (ova) are immersed for four to five hours, and then passed through 70 per cent. alcohol (twenty-four hours) strong alcohol (some days), absolute alcohol (four to five days). They are then fit for cutting. The advantage of the process is stated to be that segmentation spheres and nuclei are perfectly fixed, the ova do not become porous, and cut like cartilage.

Another advantage is that the fixing solution may be combined with a stain. (In this case the albuminous envelopes of the ova must be carefully removed, otherwise the stain will not penetrate.)

Some stains, such as fuchsin or anilin red, may be dissolved directly in the fixing solution. Others, such as eosin, purpurin, anilin violet, must first be " dissolved in three parts of alcohol, and then shaken into the liquid."

Picro-carmine and borax-carmine may be added to the liquid, but they give rise to a precipitate, which must be removed by filtration before using.

Chromo-nitric acid is above all an embryological reagent, and a very important one.

Another formula given by Perenyi (*Zool. Anzeig.*, 274, 1888, p. 139, and 276, p. 196) is as follows:

  3 parts 20 per cent. nitric acid.
  3 parts 1 per cent. chromic acid.
  4 parts absolute alcohol.

For embryos of Lacerta. Fix for twenty minutes. Wash out for an hour with 70 per cent. alcohol, and then with strong alcohol. Stain with Delafield's hæmatoxylin, and treat the stained material for three to five minutes with 1 per cent. chromic acid.

40. Chromo-nitric Acid with Bichromate (KOLLMANN, *Arch. f. Anat. u. Phys.*, 1885, p. 296).

  Bichromate of potash. . . . . 5 per 100
  Chromic acid . . . . . . 2 ",,"
  Concentrated nitric acid . . . . 2 ,,

For ova of Teleostea. Fix for twelve hours, wash with water for twelve hours, then remove the chorion, and put the ova into 70 per cent. alcohol.

41. Picro-chromic Acid (*Fol. Lehrb.*, p. 100).—
    Picric acid, sol. sat. in water. . . . 10 vols.
    1 per cent. chromic acid solution . . . 25 „
    Water . . . . 65 „

At the instant of using, you may add 0·005 of osmic acid, which makes the action more energetic. Wash with water (hot, nearly boiling water is best), and then with alcohol. Fol says, "This reagent hardens tissues admirably, without hindering staining in any way; but it is not very penetrating and fixes slowly."

42. Chromic Acid and Platinum Chloride (*Merkel's solution;* from *Mitth. Zool. Stat. Neapel*, 1881, p. 11).—Equal volumes of 1·400 solution of chromic acid and 1·400 solution of platinum chloride. Objects should remain in it for several hours or even days, as it does not harden very rapidly. After washing out with alcohol of 50 per cent. to 70 per cent., objects stain excellently, notwithstanding the admixture of chromic acid. This is a very delicate and admirable fixative. If objects that have been fixed by osmium be put into it for some hours, blackening is effectually prevented.

Salts.

43. The Chromates are useful as hardening rather than fixing agents. They have a very mild and even action on tissues, but are not at all penetrating and act very slowly. For hardening they are very valuable. They may still be found useful for fixing certain tissues, some of those of Mollusca, for example. It must be borne in mind that (as pointed out by Flemming) they do not preserve the true structure of nuclei, and must absolutely be avoided in all cases in which it is desired to demonstrate that structure.

(1) Potassium Bichromate.—Used in solutions of from 1 per cent. to 2 per cent., or sometimes more, for most classes of objects. It is less of a hindrance to future staining than chromic acid. Wash out with water or alcohol. Altmann strongly recommends the use of a 2 per cent. solution, containing a little free chromic acid, and cooled to zero, followed by washing out in strong alcohol. The cooling of the liquid serves to stop instantly all molecular processes; and the slowly-acting mixture has time to complete the fixing.

(2) Ammonium Chromate.—Appears to be generally used in 5 per cent. solution, for twenty-four hours. Wash out in water and stain in picro-carmine. This salt is a still more delicate reagent than the preceding.

For the mixtures of bichromate with sulphates, see the chapter on Hardening Agents, *Liquid of Müller*, and *Liquid of Erlicki*.

44. Bichromate and Cupric Sulphate Mixture (Kultschitzky, *Zeit. f. wiss. Mik.*, iv, 3, 1887, p. 348).—A saturated solution of bichromate of potash and sulphate of copper in 50 per cent. alcohol, to which is added at the instant of using a little acetic acid, five or six drops per 100 cc.

To make the solution, add the finely powdered salts to the alcohol in excess, and leave them together *in total darkness* for twenty-four hours.

Fix for twelve to twenty-four hours *in the dark*, otherwise the salts will be precipitated. Then treat with strong alcohol for twelve to twenty-four hours, and make sections.

The rationale of this mixture is, that it fixes tissues faithfully, without causing the production of the delusive reticular precipitates of albuminoids which we have mentioned as being produced by chromic acid—that is the part played by the bichromate and sulphate; and that it also fixes faithfully the chromatin of nuclei—that is the part played by the organic acid.

44 a. Cupric Sulphate.—Cupric sulphate was recommended some few years ago in one of the well-known handbooks. It was recommended (in a place which I cannot now find) to be used for marine organisms in saturated solution in sea-water, the organisms to be preserved in the solution itself till wanted. It has been quite recently (*Arch. d. Sci. phys. et. nat.*, Juin, 1889, t. xxi, p. 556) recommended by Bedot for the preparation of Siphonophora and other delicate pelagic animals. Bedot directs that a large quantity of 15 to 20 per cent. solution of the salt be suddenly added to the sea-water containing the animals. As soon as the animals are fixed (which happens in a few minutes), a few drops of nitric acid are to be added and mixed in (this is in order to prevent the formation of precipitates), and the whole is left for four to five hours. The animals are then to be hardened *before* bringing them into alcohol. Bedot recommends that this be done in a large quantity of Flemming's strong solution (§ 36), in which the animals should remain for twenty-four hours at least.

This promises to be a valuable method. Bedot has been able by this means to preserve *Forskalia* and *Halistemma*, which are amongst the most difficult forms known to me.

44 b. Alum.—Alum has been used for fixing purposes, and may therefore be mentioned here. Although quite superseded for general work by other reagents, it may possibly still be found useful for certain special purposes. For instance, for the preservation of *Medusæ* the following process has been recommended (by Pagenstecher). Take two parts of common salt and one of alum, and make a strong solution. Throw the animals into it alive, and leave them there for twenty-four to forty-eight hours. Preserve in weak alcohol. A saturated solution of alum in sea-water preserves very well the forms of *Salpidæ, Medusæ, Ctenophora*, and other pelagic animals. It constitutes a preservative medium in which the objects may remain till wanted.

According to my experience, it is not to be recommended for any but the very coarsest work. It should be noted, however, that Ranvier, *Traité Technique*, p. 279) found that it fixed cartilage-cells better than any other reagent. He employed a solution of 0·5 per cent.

44 c. Permanganate of Potash (DU PLESSIS, *Bull. Soc. Vaud. Sci. Nat.*, 2, sér. xv, pp. 278—280, 1878). — A strong solution in water. I find this reagent has very slight penetrating power, and, besides, macerates some tissues. It is therefore not adapted for general use, but it preserves very well the forms of cells, and has one great virtue, it kills, I fancy, more rapidly than any other agent I have been able to find; even 2 per cent. osmic acid is not equal to it in this respect. I have found it sometimes very valuable for the study of isolated and very contractile cells, such as some spermatozoa.

CHAPTER V.

FIXING AGENTS— CHLORIDES, ORGANIC ACIDS, AND OTHERS.

Chlorides.

45. Bichloride of Mercury (Corrosive Sublimate).—Corrosive sublimate is stated in the books to be soluble in about sixteen parts of cold and three of boiling water. It will probably be found that the aqueous solution contains about 5 per cent. of the sublimate at the temperature of the laboratory. It is more soluble in alchohol than in water, and still more so in ether. Its solubility in all these menstrua is augmented by the addition of hydrochloric acid, ammonious chloride, or camphor. With sodium chloride it forms a more easily soluble double salt; hence sea-water may dissolve as much as 15 per cent., and hence the composition of the liquid of Lang.

For fixing, corrosive sublimate may be, and very frequently is, used pure; but in most cases a finer fixation will be obtained if it be acidified with acetic acid, say about 1 per cent. of the acid. I find that a saturated solution in 5 per cent. acetic acid is a very good formula for marine animals. Van Beneden has recommended a saturated solution in 25 per cent. acetic acid.

In either case, the most concentrated solution obtainable should in general be taken. The cold saturated aqueous solution will suffice in most cases; but for some very contractile forms (coral polypes, Planaria), a concentrated solution in warm or even boiling water should be employed. For Arthropoda the alcoholic solution is frequently indicated. Delicate objects, however, may require treatment with weak solutions. Harting found solutions of 0·2 to 0·5 per cent. suitable for blood-corpuscles, and Pacini's fluids are much of the same strength. For these see the chapter on Examination Media.

Objects should in all cases be removed from the fixing bath

as soon as fixed, that is, in other words, as soon as they are seen to have become opaque throughout, which is practically as soon as they are penetrated by the liquid. Small objects are fixed in a few minutes. I have found that a "salivary" gland of the larva of *Chironomus* is thoroughly fixed in three seconds.

Wash out with water or with alcohol. I consider alcohol almost always preferable. Alcohol of about 70 per cent. may be taken. The extraction of the sublimate is hastened by the addition of a little camphor to the alcohol. Or a little tincture of iodine may be added to the liquid, either alcohol or water, used for washing, and the liquid changed until it no longer becomes discoloured by the objects. It is important that the sublimate be thoroughly removed from the tissues, otherwise they become brittle. They will also become brittle if they are kept long in alcohol.

You may stain in any way you like. Carmine stains are peculiarly brilliant after sublimate, owing to the formation of mercuric carminate. It is not necessary that the objects be thoroughly washed out before staining; the staining processes themselves may be made to constitute a part of the washing-out process.

It must be remembered that the solutions must not be touched with iron or steel, as these produce precipitates that may hurt the preparations. To manipulate the objects, wood, glass, or platinum may be used; for dissecting them, hedgehog spines, or quill-pens.

When properly employed, sublimate is undoubtedly a fixing agent of the very highest order. It is applicable to most classes of objects. It is perhaps less applicable, in the pure form, to Arthropods, as it possesses no great power of penetrating chitin.

Before passing to the sublimate mixtures that have been recommended, it must here be stated that *for general purposes* there seems to be no use in adding anything except a little acetic acid to the pure sublimate.

46. Corrosive Sublimate (LANG's *formula*, 'Zool. Anzeiger,' 1878, i, p. 14). For *Planaria*.—Take—

Distilled water	100 parts by weight.
Chloride of sodium	6 to 10 parts.
Acetic acid	6 to 8 ,,
Bichloride of mercury	3 to 12 ,,
(Alum, in some cases,	½).

The Planaria are to be placed on their backs and the mixture is to be poured over them. They die extended. After the lapse of half an hour they are brought into alcohol, first of 70 per cent., then of 90 per cent., then absolute, and in two days' time are sufficiently hardened.

Second formula (*Ibid.*, 1879, ii, p. 46).—Make a concentrated solution of corrosive sublimate in picro-sulphuric acid, to which has been added 5 per cent. of acetic acid.

47. Platinum Chloride.—An extremely valuable reagent for the study of karyokinesis. RABL, to whom we owe the introduction of this agent, employs an aqueous solution of 1·300. The objects remain in it for twenty-four hours, and are then washed with water, hardened in alcohol, and sectioned. Stain with Delafield's hæmatoxylin, or with safranin.

The action of platinum chloride is similar to that of gold chloride, with the advantage that there is no blackening of the preparations. Rabl finds it give better results (for the study of karyokinesis) than any other reagent except chromoformic acid (§ 32). It causes a slight shrinkage of the chromatin elements, a condition that renders the granules of Pfitzner and the longitudinal division of the elements very distinctly visible (see Rabl's well-known paper in *Morph. Jahrb.*, Bd. x, 1884, p. 216).

Platinum chloride is an extremely deliquescent salt, and for this reason had better be procured in solution. Ten per cent. solutions are found in commerce.

For *Merkel's* solution (chromo-platinic mixture), *see ante*, § 42.

48. Palladium Chloride.—Palladium chloride has been recommended by experienced workers. It is used in solutions of 1·300, 1·600, or 1·800 strength, for from one to two minutes. Cattaneo recommends it as being the best of fixatives for Infusoria. Tissues are impregnated and coloured brown by it. For small objects one or two minutes will suffice for fixation.

This salt is found in commerce in the solid state. To dissolve it, take 10 grammes of the salt, one litre of water, and four to six drops of hydrochloric acid. Solution will be effected in twenty-four hours.

49. Gold Chloride.—When used for fixing (and not for the object of staining by impregnation) gold chloride is generally used in solution of ½ per cent. strength, for a few minutes (30 at most). Weaker solutions (⅕th per cent.) or stronger (1 to 2 per cent.) may also be used. Wash out with water.

Gold chloride is one of the most faithful fixing agents we know of. But it is not fitted for general work on account of the capricious fashion in which it undergoes reduction in the tissues, rendering the impregnated elements unsusceptible of staining.

50. Perchloride of Iron.—(FOL, *Zeit. f. wiss. Zool.*, Bd. xxxviii, 1883, p. 491; and *Lehrb. d. vergl. mik. Anat.*, p. 102). Fol recommends vol. 1 of *Tinct. Ferri Perchlor.* P. B. diluted with 5 to 10 vols. of 70 per cent. alcohol. This gives better results than the weaker (2 per cent.) mixture at first recommended. Aqueous solutions do not give nearly so good results.

Fix for a short time only and wash with alcohol. The preparations are best stained with pyrogallol (see the chapter on Impregnation Methods, § 214). Fol recommends this process chiefly for Infusoria, and other ciliated objects, but also as a general zoological method. I find it fixes cytoplasm well in the cases to which I have applied it (nerve-end organs), but I hear from Naples that it has been tried for the preservation of marine organisms, and found wanting. Its chief value seems to me to lie in the pyrogallol staining method, which may be found very useful for nerve-end organs, which are impregnated somewhat selectively by it.

Perchloride of iron (the tincture diluted with 3 to 4 vols. of either alcohol or water) has lately been recommended for fixing medullated nerve by Platner (*Zeit. f. wiss. Mik.*, vi, 2, 1889, p. 187).

Organic Acids.

51. Acetic Acid.—The place of honour amongst organic acids considered as fixing agents appears rightfully to belong to this old-fashioned reagent. In the first edition of this work it was merely stated that acetic and formic acid "are useful and well-known fixatives of nuclei. Flemming, who has made a special investigation of their action, finds (*Zellsubstanz*, &c., p. 380) that the best strength is from 0·2 to 1 per cent. Strengths of 5 per cent. and more bring out the nuclein structures clearly at first, but after a time cause them to swell and become pale, which is not the case with the weaker strengths (*Ibid.*, p. 103)." It must now be stated that, thanks to V. BENEDEN the *strong* acid has become established as a most precious fixative of the most varied zoological objects. It is particularly applicable to very contractile objects, such as are found in the Vermes and Cœlenterata; it kills with the utmost rapidity, *and has a tendency to leave them fixed in the state of extension*. The *modus operandi* is in general as follows:—Pour glacial acetic acid in liberal quantity over the organisms, leave them until they are penetrated by it,—which should be in five or six minutes, as the strong acid is a highly penetrating reagent, —and wash out in frequent changes of alcohol of gradually increasing strength. Some persons begin with 30 per cent. alcohol, but this appears to me rather weak, and I think 70 per cent. or at least 50 per cent. alcohol should be preferred.

In the *Traité des Méth. Techn.*, 1887, I stated that the reason why glacial acetic acid was not more used was that it did not faithfully preserve delicate histological and cytological detail. I now believe that if the instructions above given be followed, in particular as regards the employment of the *glacial* acid, and the washing out with somewhat strong alcohol, the most delicate detail will generally be found admirably preserved.

52. Acetic Alcohol (CARNOY, *La Cellule*, t. iii, 1, 1886, p. 6; and *Ibid.*, 1887, 2, p. 276; v. BENEDEN et NEYT, *Bul. Ac. roy. d. sci. de Belg.*, t. xiv, 1887, p. 218; ZACHARIAS, *Anat. Anz.*, iii Jahrg., 1, 1888, pp. 24—27; v. GEHUCHTEN, *Ibid.*, 8, p. 237). CARNOY has given two formulæ for this important reagent. The first is—

Glacial acetic acid	1 part.
Absolute alcohol	3 parts.

The second is—

Glacial acetic acid	1 part.
Absolute alcohol	6 parts.
Chloroform	3 ,,

The addition of chloroform is said to render the action of the mixture more rapid.

V. BENEDEN and NEYT take equal volumes of glacial acid and absolute alcohol.

ZACHARIAS takes—

Glacial acetic acid	1 part.
Absolute alcohol	4 parts.
Osmic acid	A few drops.

Acetic alcohol is one of the most penetrating and quickly acting fixatives known. It preserves nuclei admirably, and admits of admirable staining in any way that may be preferred. It was imagined by all of the authors quoted for the study of karyokinesis in the ova of *Ascaris*,—proverbially one of the most difficult objects to fix,—but it is applicable to tissues in general. You may wash them out with alcohol and treat them afterwards in any way that may be preferred. It will be well, however, to avoid treatment with water as much as possible.

53. Formic Acid may be used dilute in the same way as acetic acid (*supra*, § 51). It is probable that it might also take the place of acetic acid in the concentrated form, but I am not aware of any experiments in this direction.

54. Picric Acid.—Picric acid should always be employed in the form of a *strong* solution. (That is to say, strong solutions must always be employed when it is desired to make sections or other preparations of tissues with the elements *in situ*, as weak solutions macerate; but for dissociation preparations, or the fixation of isolated cells, weak solutions may be taken. Flemming finds that the fixation of nuclear figures is equally good with strong or weak solutions.) The saturated solution is the one most employed. Objects should remain in it for from a few seconds to twenty-four hours, according to their size. For Infusoria, one to at most two minutes will suffice; whilst objects of a thickness of several millimètres require from three to six hours' immersion.

Picric acid should *always be washed out with alcohol,* as water is hurtful to tissues that have been prepared in it. For the same reason, during all remaining stages of treatment, water should be avoided; staining should be performed by means of alcoholic solutions, the only exceptions to this rule being in favour of picro-carmine, which, probably on account of the picric acid contained in it, does not appear to exert so injurious an influence as other aqueous stains, and of methyl green, and some few other aqueous stains that are themselves weak hardening agents. It is one of the advantages of picric acid that, by sufficiently prolonged soaking, it can with certainty be entirely removed from any tissue by means of alcohol.

Tissues fixed in picric acid can, after removal of the acid by soaking, be perfectly stained in any stain. Mayer's cochineal, alcoholic borax-carmine, Kleinenberg's hæmatoxylin, Grenacher's alcoholic carmine, may be recommended.

The most important property of picric acid is its great penetration. This renders it peculiarly suitable for the preparation of chitinous structures. For such objects, alcohol of 70 per cent. to 90 per cent. should be taken for washing out, and staining should be done by means of Mayer's cochineal or Kleinenberg's hæmatoxylin.

In very many if not most cases it is advantageous to employ picric acid in the manner suggested by Kleinenberg (*see* below), that is, in combination with sulphuric acid; or with nitric acid, or hydrochloric acid, as suggested by P. Mayer (see Picro-sulphuric Acid, Picro-nitric Acid, Picro-hydrochloric Acid and the directions there given).

55. Picro-sulphuric Acid (Kleinenberg, *Quart. Journ. Mic. Sci.*, April, 1879, p. 208; Mayer, *Journ. Roy. Mic. Soc.* (N.S.), ii (1882), p. 867).—By picro-sulphuric acid, without any qualifying term, I understand a fluid made (following Mayer, l. c.) as follows :—Distilled water, 100 vols. ; sulphuric acid, 2 vols.; picric acid, as much as will dissolve. This may also, in any case in which confusion is likely to arise, be called "concentrated" or "undiluted picro-sulphuric acid."

By "liquid of Kleinenberg" I understand a mixture suggested by Kleinenberg (l. c.), and best made by diluting the concentrated picro-sulphuric acid prepared as above with three times its volume of water. (Kleinenberg also directed the addition of as much creosote as would mix. This was done with the idea of eliminating the swellings produced in some objects by the liquid, but it has been found not to have the effect attributed to it, and has been abandoned. Fol (*Lehrb.*, p. 100) states that the same end may be attained by adding about one third vol. of 1 per cent. chromic acid.)

Of these two formulæ the one commonly employed is that given by Kleinenberg,—the dilute mixture; undiluted picro-sulphuric acid being reserved for objects requiring special treatment, chiefly Arthropods. I may as well say at once that in my opinion this practice should be reversed, for I think it will be found that Kleinenberg's solution is much weaker than is desirable in the majority of cases, and should be reserved for special cases, such perhaps as that for which it was originally proposed, the embryology of the earthworm; and the concentrated solution should be the one taken for general work. This particularly applies to marine organisms.

The treatment is the same in either case. "The object to be preserved should remain in the liquid for three, four, or more hours; then it should be transferred, in order to harden it and remove the acid, into 70 per cent. alcohol, where it is to remain five or six hours. From this it is to be removed into 90 per cent. alcohol, which is to be changed until the yellow tint has either disappeared or greatly diminished."

Warm alcohol extracts the acid much more quickly than cold, with which *weeks* may be required to fully remove the acid from chitinous structures. I call attention here to what was said as to washing out under the head of *picric acid*, viz. that washing out must *never be done with water*. This is a most important point, and one that is not sufficiently attended to. You may stain as directed above for picric acid. You may, of course, stain sections with alcoholic solutions of safranin or the like.

The advantages of picro-sulphuric acid as a fixing agent are, that it kills tissues very rapidly, that it has great penetrating power, that it can be totally soaked out of the structures with alcohol (it is much more easily removed from the tissues than pure picric acid), leaving them in a good condition for staining, and, in the case of marine organisms, that it effectually removes the different salts of sea-water that are present in them.

It has some disadvantages. For vertebrata it should be used with caution, on account of the swelling caused by sulphuric acid in connective tissue. In parasitic Crustacea it also produces swelling and maceration, and should be avoided (as was found by Fraisse, *Entoniscus Cavalini*, n. sp., *Arbeiten Zool. Zoot. Inst. Wurzburg*, 1877–78, iv, p. 383). Notwithstanding this, it is, however, according to Emery, very suitable for fishes, and for embryos of vertebrates generally, provided they are not allowed to remain in it more than three or four hours. For structures that contain much lime it is not to be recommended, for it dissolves the lime and throws it down as crystals of gypsum in the tissues. For such structures the picro-nitric or picro-hydrochloric acid is to be preferred.

56. Picro-nitric Acid (MAYER, *Mitth. Zool. Stat. Neapel*, 1881, p. 5 ; *Journ. Roy. Mic. Soc.* (N.S.), ii, 1882, p. 868).—Prepared in the same way as picro-sulphuric acid except that instead of 2 vols. sulphuric acid you take 5 vols. pure nitric acid (of 25 per cent. N_2O_5). Mayer now dissolves the picric acid in the nitric acid water, so that the formula runs :

    Water . . . . 100 vols.
    Nitric acid (of 25 per cent. N_2O_5) . . . 5 „
    Picric acid, as much as will dissolve.

The fluid is used undiluted.

The properties of this fluid are very similar to those of picro-sulphuric acid, with the advantage of avoiding the formation of gypsum crystals, and the disadvantage that it is much more difficult to soak out of the tissues. "Mayer recommends it strongly, and states that with eggs containing a large amount of yolk material, like those of Palinurus, it gives better results than nitric, picric, or picro-sulphuric acid."

57. Picro-hydrochloric Acid (MAYER, *Ibid.*).—Prepared in the same way as picro-sulphuric acid, except that instead of 2 vols. of sulphuric acid you take "8 vols. of pure hydrochloric acid of 25 per cent. HCl." Mayer now dissolves the picric acid in the hydrochloric acid water, so that the formula runs—

    Water . . . . 100 vols.
    Hydrochloric acid (of 25 per cent. HCl) . . 8 „
    Picrid acid, as much as will dissolve.

The fluid is used undiluted.
The properties of this fluid are similar to those of picro-nitric acid.

58. Picro-chromic Acid. *See ante*, § 41.

59. Picro-osmic Acid.—FLEMMING (*Zells. Kern-u.-Zellth.*, p. 381) has experimented with mixtures made by substituting picric for chromic acid in the chromo-osmic mixtures (*ante*, §§ 34 and 35). The results are identical so far as regards the fixation (of nuclei); but staining is rendered more difficult.

Other Fixing Agents.

60. Alcohol.—For fixing, only two grades of alcohol are found generally useful—very weak alcohol on the one hand, and absolute alcohol on the other hand. Absolute alcohol ranks as a fixing agent because it kills and hardens with such rapidity that structures have not time to get deformed in the process by the energetic dehydration that unavoidably takes place. Dilute alcohol ranks as a fixing agent in virtue of being of such a strength as to possess a sufficiently energetic coagulating action and yet contains enough water to have but a feeble and innocuous dehydrating action. The intermediate grades do not realise these conditions, and therefore should not be employed alone for fixing. But they may be very useful in combination with other fixing agents (such as corrosive sublimate, chromic acid or nitric acid) by greatly enhancing their penetrating power; 70 per cent. is a good grade for this purpose.

61. One-third Alcohol.—The one grade of weak alcohol that is found generally useful for fixing is one-third alcohol, or RANVIER'S ALCOHOL, known in France as "*Alcohol au tiers,*" which is the name given to it by Ranvier himself; in Germany as "*Drittelalcohol*" or "*Ranviersche alcohol dilutus;*" in Italy, as "*alcool al terzo.*" In consists of two parts of water and one part of alcohol *of* 36° *Baumé.* Now, since alcohol of 36° Baumé *contains nearly* 89·6 *per cent. of absolute alcohol,* it follows that *one-third alcohol contains, approximately,* 29·9 *per cent. of absolute alcohol* (or very nearly 30°, and not 33·3°, as was stated by a most regrettable oversight in the 1st edition). (*See* the *Traité Technique* of Ranvier, p. 241, *et passim.*)

Care should be taken that the alcohol is of the strength specified, as the effects of this reagent depend to a remarkable degree on its strength.

Objects may be left for twenty-four hours in this alcohol; not more, unless there be no reason for avoiding maceration, which will generally occur after that time. You may conveniently stain with picro-carmine, alum-carmine, or methyl green.

This classical reagent is a very mild fixative. Its hardening action is so slight that it is seldom indicated for the fixing of objects that are intended to be sectioned. Its chief use is for extemporaneous and dissociation preparations.

62. Absolute Alcohol.—This is also a very valuable reagent. It preserves very well the structure of the nuclei, which is by no means the case with one-third alcohol. It has over the latter also the advantage of superior penetrating power, being indeed one of the most penetrating of known fixing agents. Mayer finds that boiling absolute alcohol is often the only means of killing certain Arthropoda rapidly enough to avoid maceration brought about by the slowness of penetration of common cold alcohol (especially in the case of Tracheata).

It is important to employ for fixing a very large proportion of alcohol. Alum-carmine is a good stain for small specimens so fixed. For preservation, the objects should be put into a weaker alcohol, 90 per cent. or less.

Absolute alcohol is found in commerce. It is a product that it is almost impossible to preserve in use, on account of the rapidity with which it hydrates on exposure to air. Fol recommends that a little quicklime be kept in it. This absorbs part at least of the moisture drawn by the alcohol from the air, and has the further advantage of neutralising the acid that is frequently present in commercial alcohol.

Another plan that I have seen recommended is to suspend strips of gelatin in it. It is stated that by this means ordinary alcohol may be rendered absolute.

Ranvier adopts the following plan for preparing an alcohol absolute enough for all practical purposes. Strong (95 per cent.) alcohol is treated with calcined cupric sulphate, with which it is shaken up and allowed to remain for a day or two. It is then decanted and treated with fresh cupric sulphate, and the operation is repeated until the fresh cupric sulphate no longer becomes conspicuously blue on contact with the alcohol; or until, on a drop of the alcohol being mixed with a drop of turpentine, no particles of water can be seen in it under the microscope. The cupric sulphate is prepared by calcining common blue vitriol in a porcelain capsule over a spirit lamp or gas burner until it becomes white, and then reducing it to powder (*see* *Proc. Acad. Nat. Sci. Philad.*, 1884, p. 27; *Science Record*, ii, 1884, p. 65; *Journ. Roy. Micr. Soc.*, (N.S.), iv (1884) pp. 322 and 984).

63. Acidulated Alcohol (Paul Mayer, *Mitth. Zool. Stat. Neapel*, ii (1881), p. 7.—To 97 vols. of 90 per cent. alcohol, in which is dissolved a small quantity of picric acid, add 3 vols. pure hydrochloric acid. Leave the specimens in the mixture only just long enough to ensure that they are thoroughly penetrated by it. Wash out with 90 per cent. alcohol, the disappearance of the yellow stain of the picric acid being a sign that all the acid is removed.

The use of this mixture is for the preparation of coarse objects it is intended to preserve in alcohol. The object of the acid is to prevent both that glueing together of organs by the perivisceral liquid, which is often brought about by the coagulating action of pure alcohol, and the precipitation on the surface of organs of the salts contained in sea-water, which is a hindrance not only to the penetration of the alcohol, but also to subsequent staining.

Whitman (*Journ. Roy. Mic. Soc.* (N.S.), ii (1882), p. 870) states that "acid alcohol as above prepared loses its original qualities after standing some time, as ether compounds are gradually formed at the expense of the acid." He also states that 70 per cent. alcohol may be taken instead of 90 per cent., for washing out.

64. Chloride and Acetate of Copper (*Ripart et Petit's formula*, Carnoy, *La Biologie Cellulaire*, p. 94).

Camphor water (not saturated)	75 grammes.
Distilled water	75 ,,
Crystallised acetic acid	1 ,,
Acetate of copper	0·30 ,,
Chloride of copper	0·30 ,,

This is a very moderate and delicate fixative. I consider that it has not sufficient hardening power for objects that are intended to be dehydrated and mounted in balsam, but is frequently excellent and sometimes indispensable for objects that are to be studied in as fresh a state as possible in aqueous media. Objects fixed in it stain instantaneously and perfectly with methyl green. Osmic acid may be added to the liquid to increase the fixing action. *For cytological researches* this is a most invaluable medium.

65. Acetate of Uranium (Schenk, *Mitth. a. d. Embryol. Inst. Wien*, 1882, p. 95; cf. Gilson, *La Cellule*, 1, 1885, p. 141). This reagent is very similar in its properties to picric acid. It has a mild fixing action, and a high degree of penetration, which may make it useful for Arthropoda. It may be combined with methyl green, which it does not precipitate.

66. Iodine.—Iodine possesses considerable hardening properties, and a very high degree of penetration; and, in point of fact, iodised serum, which is generally employed as an "indifferent liquid," that is, one which is supposed to exert no action whatever on tissues, is, in reality, a feeble hardening agent, and forms a most admirable fixing agent for delicate tissues. It is so classed by Ranvier (*see* **Iodised Serum**, §§ 349, 496). KENT (*Manual of the Infusoria*, 1881, p. 114, *Journ. Roy. Mic. Soc.* (N.S.), iii (1883), p. 730), has found it to act in a manner almost identical with osmic acid, and in some instances even more efficiently (for fixing Infusoria). His instructions are as follows :—" Prepare a saturated solution of potassic iodide in distilled water, saturate this solution with iodine, filter, and dilute to a brown-sherry colour. A very small portion only of the fluid is to be added to that containing the Infusoria."

Or you may use the solution of LUGOL, of which the formula is as follows:

Water	100 parts.
Iodide of potassium	6 ,,
Iodine	4 ,,

Iodine certainly kills cells very rapidly, without deforming them. Personally I have found it very useful for the examination of spermatozoa. Unfortunately I am not acquainted with any nuclear stain that will work well with it.

CHAPTER VI.

HARDENING AGENTS.

67. The Obligation of Hardening.—Methods of imbedding have now been brought to such a degree of perfection that the thorough hardening of soft tissues that was formerly necessary in order to cut thin sections from them is now, in the majority of cases, no longer necessary; by careful infiltration with paraffin or some other good infiltration mass, most soft objects can be satisfactorily cut with no greater an amount of previous hardening than is furnished by the usual passing of the tissues after fixing through successive alcohols in order to prepare them for the paraffin-bath. Almost the only exceptions to this statements are, I believe, to be found in the cases in which it is desired to cut very large sections, such as sections of the entire human brain. Such an organ as this cannot be duly infiltrated with alcohol in a few hours, and it is doubtful whether it can be duly infiltrated with paraffin or any other imbedding mass in any reasonable time. The processes employed for hardening such specimens as these will be described when treating of the organs in question; in this chapter, which may be considered as parenthetical, I confine myself to such general statements concerning the employment of the usual hardening agents as appear likely to be generally useful.

68. The Practice of Hardening—Hints and Cautions.—Employ in general a relatively large volume of hardening liquid, and change it very frequently. The exact proportions may be made out by experiment for each reagent and each class of objects. If the volume of liquid be insufficient, its composition will soon become seriously altered by the diffusion into it of the soluble substances of the tissues; and the result may be a macerating instead of a hardening liquid. Further, as soon as in consequence of this diffusion the liquid has acquired a composition similar in respect of the proportions of colloids and crystalloids contained in it to that of the liquids of the tissues, osmotic equilibrium will become established, and diffusion will cease. That is to say, the hardening liquid will cease to penetrate. This means, of course, maceration of internal parts. On the other hand, it appears that a certain slight proportion of colloids in the hardening liquid is favorable to the desired reaction, as it gives a better consistency to the tissues by preventing them from becoming brittle. Hence the utility of employing *a certain proportion* of hardening agent.

Hardening had better be done in tall cylindrical vessels, the objects being suspended by a thread at the top of the liquid. This has the advantage of allowing diffusion to take place as freely as possible, whilst any precipitates that may form fall harmlessly to the bottom.

Always begin hardening with a weak reagent, increasing the strength gradually, as fast as the tissues acquire a consistence that enables them to support a more energetic action of the reagent.

Let the objects be removed from the hardening fluid as soon as they have acquired the desired consistency.

69. Choice of a Hardening Reagent.—If you wish, above all, for a rapid and energetic action, take chromic acid. If you wish for a more moderate and more equable action, take a chromic salt, or one of the compounds of which the chromic salts are the principal ingredients.

Mineral Acids.

70. Chromic Acid.—Chromic acid is generally employed in strengths of $\frac{1}{8}$th per cent. to $\frac{1}{2}$ per cent., the immersion lasting a few days or a few weeks, according to the size and nature of the object. Mucous membrane, for instance, will harden satisfactorily in a few days, brain will require some six weeks.

Large quantities of the solution must be taken (at least 200 grammes for a piece of tissue of 1 centimètre cube, Ranvier).

In order to obtain the best results, you should not employ portions of tissue of more than an inch cube. For a human spinal cord, you should take two litres of solution, and change it for fresh after a few days. Six weeks or two months are necessary to complete the hardening.

The solution should be taken weak at first, and the strength increased after a time. The objects should be removed from the solution as soon as they have acquired the desired consistency, as if left too long they will become brittle. (These precautions are peculiarly necessary in the case of chromic acid.) They may be preserved till wanted in alcohol (95 per cent.). It is well to wash them out in water for twenty-four or forty-eight hours before putting them into the alcohol. I think it is frequently useful to add a little glycerin to the hardening solution, there is less brittleness and, I think, less shrinkage.

The reader's attention is called to the statements made in § 30 concerning the action of light on the alcohol containing chromic objects.

Further directions for the employment of chomic acid will be given in the special paragraphs. Chromic acid is a most powerful and rapid hardening agent (by it, you may obtain in a few days a degree of hardening that you would hardly obtain in as many weeks with bichromate, for instance). It has the defect of a great tendency to cause brittleness.

71. Chromic Acid and Spirit (URBAN PRITCHARD, *Quart. Journ. Mic. Sci.*, 1873, p. 427).—Chromic acid, 1 part; water, 20 parts; rectified spirit, 180 parts. Dissolve the chromic acid in the water first, and then add the spirit (violent action will ensue if the dry chromic acid be added directly to the spirit). The colour of the solution soon becomes brown. If, after a few days, it turns semi-gelatinous, it should be changed for fresh. From a week to ten days is required to harden such tissues as retina, cochlea, &c., for which this fluid is particularly well adapted.

72. Chromo-osmic Acid (MAX FLESCH). **Chromo-aceto-osmic Acid** (FLEMMING).—Either of these mixtures may be used for prolonged hardening, and are admirable. (*See* §§ 34 and 35.)

73. Chromic Acid and Platinum Chloride (MERKEL's Solution). —The same remark applies to this excellent mixture. *See* § 42, *ante*.

74. Picro-chromic Acid.—This fixative may be found useful for hardening objects that are only penetrable with difficulty. Some Tunicata, for instance. *See ante*, § 41.

75. Osmic Acid.—Osmic acid is much more useful as a fixing agent than as a hardening agent. Long immersion in osmic acid is sure to cause blackening, and may cause brittleness in the tissues. The strengths employed for hardening vary from ¼th per cent. to 1 per cent., and the tissues are left in the solutions for twelve to twenty-four hours, seldom more. See the further information as to the employment of this reagent given above, §§ 26, 27, 28.

76. Nitric Acid.—Nitric acid is taken of a strength of from 3 per cent. to 10 per cent. or more, and may be allowed to act for two or three weeks. It gives, thus employed (10 per cent. to 12 per cent.), very tough preparations of brain. It is also conveniently used by employing a very short immersion and completing the hardening with alcohol, in which case it is properly considered as a fixing agent. See the information given under this head, § 37, *ante*.

Salts.

77. Bichromate of Potash.—Perhaps the most important of all known hardening agents, *sensu stricto*. It hardens slowly, much more so than chromic acid, but it gives an incomparably better consistency to the tissues, and it has not the same tendency to make them brittle if the reaction be prolonged. They may remain almost indefinitely exposed to its action without much hurt.

The strength of the solutions employed is from 2 to 5 per cent. As with chromic acid it is extremely important to begin with weak solutions and proceed gradually to stronger ones. About three weeks will be necessary for hardening a sheep's eye in solutions gradually raised from 2 to 4 per cent. Spinal cord requires from three to six weeks; a brain, at least as many months.

After hardening, the objects should be well soaked out in water before being put into alcohol. They had better be kept in the dark when in alcohol (*see* above, § 30). *If you wish to have a good stain with carmine, especially ammonia-carmine, which is admirable for portions of nervous system so hardened, you should not put the objects into alcohol at all, even for a second, until they have been stained.*

You may stain either with carmine or hæmatoxylin.

Bichromate objects have an ugly yellow colour which cannot be removed by soaking in water. It is said that it can be removed by washing for a few minutes in a 1 per cent. solution of chloral hydrate. Gierke, however, says that this treatment is prejudicial to the preservation of the tissues.

78. Müller's Solution.—

  Bichromate of potash .   2—2½ parts.
  Sulphate of soda . . . 1 „
  Water. . . . . 100 „

The duration of the reaction is about the same as with the simple solution of chromic salts.

This fluid was very highly in vogue for many years, but seems lately to be much less used. I confess that I do not understand what is the part played by the sodic sulphate, and that I fancy that the superiority of this mixture over the simple bichromate solution is mostly illusory. Fol says that for mammalian embryos, for which it has been recommended, it is worthless.

79. Erlicki's Solution (*Warschauer med. Zeit.*, xxii, Nos. 15 and 18).

  Bichromate of potash . . 2·5 parts.
  Sulphate of copper . . 1·0 „
  Water . . . . . 100·0 „

Here the addition of the cupric sulphate is intelligible. This salt is itself a hardening agent of some energy, and may well serve to reinforce the somewhat slow action of the bichromate. As a matter of fact, "Erlicki" hardens very much more rapidly than either simple bichromate or Müller's solution. A spinal cord may be hardened in it in four days at the temperature of an incubator, and in ten days at the normal temperature (Fol, *Lehrb. d. vergl. mik. Anat.*, p. 106). I believe it to be one of the best hardening agents known for voluminous objects. Human embryos of several months may be conveniently hardened in it.

<small>Nerve-centres that have been hardened in Erlicki's fluid frequently contain dark spots with irregular prolongations, simulating ganglion-cells. These were at one time taken to be pathological formations, but they are now known to consist of precipitates formed by the action of the hardening fluid. They may be removed by washing with hot water, or with water slightly acidified with hydrochloric acid, or by treating the specimens with 0·5 per cent. chromic acid before putting them into alcohol (Tschisch, *Virchow's Arch.*, Bd. xcvii, p. 173; Edinger, *Zeit. f. wiss. Mik.*, ii, 2, p. 245; Loewenthal, *Rev. Méd. de la Suisse romande*, 6me année i, p. 20).</small>

80. Bichromate of Ammonia.—

A review of the literature of the subject shows that this salt is in considerable favour, for what precise motive is not apparent. Its action is very similar to that of the potassium salt. Fol says that it penetrates somewhat more rapidly, and hardens somewhat more slowly. It should be employed in somewhat stronger solutions, up to 5 per cent.

81. Neutral Chromate of Ammonia is preferred by some anatomists. It is used in the same strength as the bichromate. Klein has recommended it for intestine, which it hardens, in 5 per cent. solution, in twenty-four hours.

82. Sulphate of Copper.—This salt is seldom used alone, perhaps because it does not give a sufficiently favorable consistency to the tissues hardened by it. I take from the *Lehrbuch* of Fol (p. 106) the following formula, which was first published by REMAK, then modified by GOETTE, and is said to be useful for hardening the ova of Amphibia:

    2 per cent. solution of sulphate of copper . . 50 cc.
    Alcohol of 25 per cent. . . . . . 50 cc.
    Rectified wood vinegar, 35 drops.

Chlorides and others.

83. Platinum Chloride (MERKEL's Solution).—The formula of this admirable reagent has been given above, § 42. It is an admirable hardening medium for delicate objects. Merkel states that he allowed from three to four days for the action of the fluid for the retina; for annelids Eisig employs an immersion of three to five hours, and transfers to 70 per cent. alcohol; for small leeches Whitman finds " one hour sufficient and transfers to 50 per cent. alcohol."

Whitman recommends, for the hardening of pelagic fish ova, a stronger mixture (due, I believe, to Eisig), viz.:

    0·25 per cent. solution of platinum chloride . 1 vol.
    1 per cent. solution of chromic acid . . . 1 ,,

The ova to remain in it one or two days (WHITMAN, *Methods in Micro. Anat.* p. 153).

84. Palladium Chloride (F. E. SCHULTZE, *Arch. mik. Anat.*, iii (1867), p. 477).—This reagent was recommended by Schultze partly as giving to tissues a better consistency than chromic acid or Müller's solution, and partly on account of a special faculty for penetrating organs rich in connective tissue, that he attributes to it. It is an impregnation reagent, staining certain elements of tissues in various tones of brown. For the somewhat lengthy details of the manner of employing it, the reader is referred to the paper quoted.

85. Chloride of Zinc is only employed for brain, *see post*, part ii (GIACOMINI).

86. Picric Acid is a weak hardening agent, little used. It should be employed in saturated solution.

87. Acetate of Lead.— Both the neutral acetate (sugar of lead) and the basic acetate have been used for hardening nerve tissues. ANNA KOTLA-REWSKY found that nerve-cells hardened in 10 per cent. solution of sugar of lead were admirably preserved. See her "Inaug.-Diss.," in *Mitth. d. naturf. Ges. Bern.*, 1887, and *Zeit. f. wiss. Mik.*, iv, 3, 1887, p. 387.

88. Alcohol.—When used alone, alcohol is inferior as a hardening agent to most of the reagents discussed above; but when judiciously employed to complete the action of a good fixing agent, it renders most valuable services. 90 to 95 per cent. is the most generally useful strength. Weaker alcohol, down to 70 per cent., is often indicated. Absolute alcohol is seldom advisable. You ought to begin with weak, and proceed gradually to stronger, alcohol! Large quantities of alcohol should be taken. The alcohol should be frequently changed, or the tissue should be suspended near the top of the alcohol, in order to have the tissue constantly surrounded with pure spirit (the water and colloid matters extracted from the tissue falling to the bottom of the vessel). Many weeks may be necessary for hardening large specimens. Small pieces of permeable tissue, such as mucous membrane, may be sufficiently hardened in twenty-four hours.

89. Iodine may be used in combination with alcohol, and render service through its great penetrating power. See the method of BETZ, *post*, Part II.

90. Pyridin.—Pyridin has been lately recommended as a hardening agent (by A. DE SOUZA). It is said to harden, dehydrate, and clear tissues at the same time. They may be stained after hardening by anilin dyes dissolved in the pyridin, or passed through water and stained by the usual processes. It is said to harden quickly, and to give particularly good results with brain. See *Comptes Rendus hebd. de la Soc. de Biologie*, 8 sér., t. iv, No. 35, p. 622; *Zeit. f. wiss. Mik.*, v, i, 1888, p. 65; *Journ. Roy. Mic. Soc.*, 1888, p. 1054.

CHAPTER VII.

STAINING.

91. The Kinds of Stains.—The chief end for which colouring reagents are employed in microscopic anatomy, is to obtain a *selective* staining of organs. In a selective stain, certain elements are made prominent by being coloured, the rest either remaining colourless or being coloured of a different intensity or of a different tone.

Two chief kinds of this selection may be distinguished,—histological selection, and cytological selection. In the former an entire tissue or group of tissue-elements is prominently stained, the elements of other sorts present in the preparation remaining colourless or being at all events differently stained, as in a successful impregnation of nerve-endings by means of gold chloride. In the latter, the stain seizes on one of the constituent elements of cells in general, namely, either the nucleus, or the extra-nuclear parts.

Stains that thus exhibit a selective affinity for the substance of nuclei, or *nuclear* stains, form at present by far the most important class of stains—in zootomy at any rate. What the zootomist wants, and the histologist too, in the great majority of cases, is either to differentiate the intimate structures of cells by means of a colour reaction, in order to study them for their own sake, or to have the nuclei of tissues marked out by staining in the midst of the unstained material in such a way that they may form landmarks to catch the eye, which is then able to follow out with ease the contours and relations of the elements to which the nuclei belong; the extra-nuclear parts of these elements being expressly left unstained in order that as little light as possible may be absorbed in passing through the preparation. Possibly this may be an irrational procedure, but it has hitherto been found in practice to be the most efficient for general work.

92. The Methods of Staining.—Colouring matters possessing so great an affinity for certain elements of tissues that they may be left to produce the desired electivity of stain without any special manipulation on the part of the operator, are unfortunately rare. In practice, selective staining is arrived at in two ways. In the one, which may be called the *direct* method, you make use of a colouring reagent that stains the element desired to be selected more quickly than the elements you wish to have unstained; and you stop the process and fix the colour at the moment when the former are just sufficiently stained and the latter not affected to an injurious extent, or not affected at all, by the colour. This is what happens—for instance, when you stain the nuclei of a preparation by treatment with very dilute hæmatoxylin; you get, at a certain moment, a fairly pure nuclear stain; but if you prolonged the treatment, the extra-nuclear elements would take up the colour, and the selectivity of the stain would be lost. It may be noted of this method that it is in general the method of *fast* stains ("echte Farbung"), and that it renders great services in the colouring of specimens *in toto*,—a procedure which is not possible with the chief stains of the other class (the anilins). It is the old method of carmine and hæmatoxylin staining.

The second, or *indirect*, method, is the method of overstaining followed by partial decoloration. You begin by staining all the elements of your preparation indiscriminately, and you then wash out the colour from all the elements, except those which you desire to have stained, these retaining the colour more obstinately than the others in virtue of a certain not yet satisfactorily explained affinity. This is what happens—for instance, when you stain a section of one deep red in all its elements with safranin, and then treating it for a few seconds with alcohol, extract the colour from all but the chromatin and nucleoli of the nuclei. It is in this method that the coal-tar colours find their chief employment. It is in general applicable only to sections, and not to staining objects *in toto* (the case of borax-carmine is probably only a seeming exception to this statement). It is a method, however, of very wide applicability, and gives the most brilliant results that have hitherto been attained.

93. The State of the Tissues to be Stained.—It is generally found that precise stains can only be obtained with carefully fixed (*i. e.* hardened) tissues. Dead, but not artificially hardened tissues stain indeed, but not generally in a precise manner. Living tissue elements in general do not stain at all, but resist the action of colouring reagents till they are killed by them.

Staining " intra vitam."—Some few substances, however, possess the property of staining living cells without greatly impairing their vitality. Such are—in very dilute solutions—cyanin (or quinoleïn), methylen blue, Bismarck brown, and, under certain conditions, dahlia, and gentian violet, with perhaps methyl violet and some others whose action is not yet sufficiently established by experiment.* (The paper of Martinotti, *Zeit. f. wiss. Mik.*, v, 3, 1888, p. 305, may be consulted on this point.)

As to the employment of these reagents, it may be noted that they must be taken in a state of extreme dilution, and in neutral or feebly alkaline solution—acids being of course toxic to cells. Thus employed, they will be found to tinge with colour the cytoplasm of certain cells during life (never, so far as I know, nuclear chromatin during life;—if this stain, it is a sign that death has set in). The stain is sometimes diffused throughout the general substance of the cytoplasm, sometimes limited to certain granules in it (which have been taken, perhaps without sufficient reason, to be identical with the granules of Altmann (Altmann's *Studien über die Zelle*, 1886). Methylen blue has the valuable point that it is perfectly soluble in saline solutions, and may therefore be employed with marine organisms by simply adding it to sea-water. The others are not thus soluble to a practical extent, but I find that gentian and dahlia become so if a trace of chloral hydrate—0·25 per cent. is amply enough—be added to the saline solution. Any of these reagents may be rubbed up with serum, or other " indifferent " liquid.

Methylen blue may be fixed in the tissues, and permanent preparations made, by treating for some hours with saturated

---

* Congo, even in strong solution, is not toxic to some organisms, and stains some structures (*see* Scholtz, *Centralb. f. d. med. Wiss.*, 1886, p. 449; also *Journ. Roy. Mic. Soc.*, 1886, p. 1092). Living Rotifera are in part successfully stained by it during life.

solution of iodine in 1 per cent. solution of iodide of potassium (ARNSTEIN, *Anat. Anzeiger*, ii, 1887, p. 125; *Zeit. f. wiss. Mik.*, iv, 1, 1887, p. 85), or with picro-carmine, or with picrate of ammonia (*Anat. Anzeig.*, 1887, p. 551), and mounting in glycerin. Bismarck brown stains may be fixed with 0·2 per cent. chromic acid, and the preparations may be stained with safranin, care being taken not to expose them too long to the action of alcohol.

I may say that personally I have found gentian, dahlia, and methylen blue, added to indifferent liquids, extremely useful in the examination of tissue-cells. Quinoleïn and Bismarck brown are well-known aids to the study of Infusoria. Methylen blue has a specific affinity for sensitive nerves (see *post*, § 127, and Part II).

94. Choice of a Stain.—The following may be recommended with confidence for general work :—For sections, Flemming's method, with safranin or gentian for a single stain (see Nos. 102, 103), and gentian followed by eosin for a double stain.

For staining *in toto*, Grenacher's alcoholic borax-carmine (No. 163), unless the object be so impermeable as to require a more highly alcoholised stain, in which case take Mayer's alcoholic carmine (No. 164), or Mayer's cochineal tincture (No. 168), or, for chromic-acid objects, Kleinenberg's hæmatoxylin (No. 184).

For fresh dissociated tissues or small entire objects, methyl green, if it is not important to have permanent preparations; if it is, take alum-carmine or picro-carmine.

Picric acid may be used for double-staining after carmine or hæmatoxylin.

Many others of the numerous stains discussed in the following chapters render most valuable services, and will be found recommended in the special paragraphs as occasion dictates.

I would add one word of advice to the beginner : Never use a double stain where a single one will do. To do so is too often to go farther and fare worse.

And a word of caution to beginners and others :

You are not likely to succeed in staining, especially in the beautiful processes of staining with coal-tar colours, unless you see to it that you are working with chemicals of the proper quality. You *cannot* ensure this by going to a generally trustworthy house for chemical products—at all events, not in the case of coal-tar colours. It is not sufficient that these should be analytically what they are described to be, they may be pure, and yet not give good stains. I have a collection of coal-tar colours obtained from one of the best London houses, of the purity of which I have no doubt; a large proportion of them are useless for staining purposes. I therefore feel constrained to advise everybody to get his reagents—at all events his anilins—from the well-known chemists Grübler or Münder. Grübler has all the tried reagents in stock, and supplies only such as have been found by experiment with tissues to furnish the desired stain. He also makes up fixing and staining solutions, injection and imbedding masses, &c., according to the classical formulæ, and sends them out neatly packed and ready for use. From experience I can most highly recommend these preparations, which are, in nine cases out of ten, better than those the observer is likely to make for himself. They may be ordered from the price-list, or by quoting the numbers of the formulæ in this work. The address is—Herrn Dr. G. GRÜBLER, Chemiker, Baiersche Strasse, 12, Leipzig. Grübler can correspond in English.

Münder's address is—Herrn Dr. G. MÜNDER, Mikroskopisch-chemisches Institut, Göttingen.

CHAPTER VIII.

ANILIN* COLOURS GIVING INDIRECT NUCLEAR STAINS (FLEMMING'S METHOD).

95. Very few anilins give a precise nuclear stain by the *direct* method (§ 92). Two of them—methyl green and Bismarck brown—are pre-eminently nuclear stains. Many of the others—for instance, safranin, gentian, and especially dahlia, may be made to give a nuclear stain with fresh tissues by combining them with acetic acid; but in ninety-nine cases out of a hundred are not so suitable for this kind of work as the two colours first named, which practically form a class apart.

Again, very few anilins give a pure plasmatic stain (one leaving nuclei unaffected). The majority give a diffuse stain, which in some few cases becomes by the application of the decolouration—or *indirect* method (92) the most precise and splendid stain as yet obtainable by any means.

The indirect staining method, or Flemming's method, will form the subject of the present chapter, and the remaining anilins will be treated of in the next chapter.

The following list shows the colours treated of in the two chapters.

In Chapter VIII.

COLOURS GIVING INDIRECT NUCLEAR STAINS (FLEMMING'S METHOD).

(The order in this chapter is one dictated merely by convenience of exposition.)

Red.

Safranin . . . § 103	Fuchsin (Roseïn, Rubin,
Magdala red (Naphthalin red) . . . 104	Magenta, Solferino, Corallin) . . . 104

* The word "anilin" is here used in the popular sense, to include all coal-tar colours.

56 ANILIN COLOURS GIVING INDIRECT NUCLEAR STAINS.

Red (continued).

Rocellin (Echtroth, Orseillin, Rubidin)	§ 104
Mauvein	104
Rouge fluorescent	104

Brown and Yellow.

Bismarck brown	§ 104
Orange	104
Tropæolin (Chrysaurein)	104

Green.

Anilin green	§ 101
Solid green	104

Blue.

Victoria	§ 100

Violet.

Gentian	§ 102
Dahlia	103
Methyl violet	104

In Chapter IX.

(A) DIRECT NUCLEAR STAINS.

Methyl green, § 105. Bismarck brown (Vesuvin), § 106. Methyl violet, § 107.

(B) PLASMATIC STAINS (STAINS NOT AFFECTING NUCLEI).

Bleu lumière, § 108. Bleu de Lyon, § 109. Indulin (Nigrosin), § 110. Quinoleïn (Cyanin), § 111.

(C) OTHER COLOURS (GROUND STAINS AND SPECIFIC STAINS).

Red.

Saürefuchsin (Acid fuchsin)	§ 112
Congo	113
Benzo-purpurin, Delta-purpurin	114
Biebricher Scharlach	115
Eosin	116
Bengal rose	117

Orange and Yellow.

Picric acid	§ 118
Metanil yellow	119
Saüregelb (Echtgelb)	120
Tropæolin O.	120
Crocein	120
Gold orange	120

Green.

Iodine green	§ 121
Thiophen green	122
Anilin green	123
Picro-anilin green	124

Blue.

Anilin blue	§ 125
Parma blue	126
Methylen blue	127

Violet.

Violet B.	§ 128

Black.

Anilin black (Nigranilin, Blue black, Noir Colin)	129

General Directions for the Indirect or "Flemming" staining Method.*

96. Staining.—*Sections only* can be stained by this method. The solutions employed are made with alcohol, water, or anilin, according to the solubility of the colour. There seems to be no special object in making them with alcohol if water will suffice, the great object being to get *as strong a solution as possible*. The sections must be *very thoroughly* stained in the solution. As a general rule they cannot be left too long in the staining fluid. With the powerful solutions obtained with anilin a few minutes or half an hour will frequently suffice, but to be on the safe side it is frequently well to leave the sections twelve to twenty-four hours in the fluid. Up to a certain point the more the tissues are stained the better do they resist the washing-out process, which is an advantage. For researches on nuclei the solutions made with anilin had better be employed *only* with preparations well fixed in chromo-aceto-osmic acid, as the basic anilin oil may easily attack chromatin if not specially well fixed.

97. Washing out.—Washing out is generally done with alcohol, sometimes pure, sometimes acidulated (with HCl). The stained sections, if loose (celloidin sections), are brought into a watch-glassful of alcohol; if mounted in series on a slide they are brought into a tube of alcohol (washing out *can* be done by simply pouring alcohol on to the slide, but it is better to use a tube or other bath). It is in either case well to *just rinse* the sections in water before bringing them into alcohol.

The sections in the watch-glass are seen to give up their colour to the alcohol in clouds, which are at first very rapidly formed, afterwards more slowly. The sections on the slide are seen, if the slide be gently lifted above the surface of the alcohol, to be giving off their colour in the shape of rivers running down the glass. In a short time the formation of the clouds or of the rivers is seen to be *on the point of ceasing*; the sections have become *pale* and somewhat *transparent*, and

* Historically the principle of this method is due to HERMANN and BOETTCHER; but it is universally known by the name of Flemming, to whom is due the credit of having greatly improved the method in its practical details.

(in the case of chrom-osmium objects) have *changed colour*, owing to the coming into view of the general ground colour of the tissues, from which the stain has now been removed. (Thus chrom-osmium-safranin sections turn from an opaque red to a delicate purple). At this point the washing out is complete, and *must be stopped instantly*.

It is generally directed that absolute alcohol be taken for washing out. This may be well in some cases, but in general strong (95 per cent.) spirit is found to answer perfectly well.

The hydrochloric acid alcohol process had better only be employed with tissues well fixed with "Flemming," as with tissues imperfectly fixed it may cause swellings. Further, the acid extracts the colour much more quickly from resting nuclei than from kinetic nuclei, which is an advantage or a disadvantage according to the end in view.

The length of time necessary for washing out to the precise degree required varies considerably with the nature of the tissues and the details of the process employed; all that can be said is that it generally lies between thirty seconds and two minutes.

In more than one of the methods presently to be described treatment with chromic acid or with iodine forms part of the washing out process. The *rationale* of this is somewhat obscure; the most probable point of view appears to be that the chromic acid acts as a mordant on the chromatin and helps it to retain the stain. It is known on the one hand that chromic acid precipitates safranin from its solutions, so that by admitting a special affinity of chromic acid on the other hand for chromatin, and especially for chromatin in the kinetic state, the explanation is hypothetically complete.

The iodine in Bizzozero's (Gram's) process also appears to act as a fixative of the colour.

As a rough and ready guide to the beginner, it may be stated that washing out should be done with pure alcohol whenever it is desired to have resting nuclei stained as well as dividing nuclei; the other processes serving chiefly to differentiate "mitoses."

98. Substitution.—It was stated above that washing out is generally done with alcohol. There exists another mode of washing out that is both of practical importance and of great theoretical interest; one anilin stain may be made to wash out another. Thus methylen blue and gentian violet are discharged from tissues by aqueous solution of vesuvin or of eosin; fuchsin is discharged from tissues by aqueous solution of methylen blue. The second stain "substitutes" itself for the first in the general "ground" of the tissues, leaving, if the operation have been successfully carried out, the nuclei stained with the first stain, the second forming a "contrast" stain. It appears from the interesting paper of RESEGOTTI in *Zeit. f. wiss.*

Mik., v. 3, 1888, p. 320, that it may be stated as a very general rule that colours that do *not* give a nuclear stain by the indirect method will wash out those that *do*. Thus he found that

Safranin, Dahlia, Methyl Violet, Gentian Violet, Rubin, Victoria Blue, Magenta, Basic Fuchsin, are washed out by the following:

Congo, Methyl Green, Iodine Green, Nigrosin, Methylen Blue, Orange, Ponceau, Acid Fuchsin, Aurantia, Cyanin, Eosin, Methylic Eosin, Magdala Red, Bordeaux, Vesuvin.

Resegotti obtained the best results by washing out methyl violet or dahlia with eosin or acid fuchsin (Saürefuchsin).

The student will note that the colours in the second of Resegotti's lists may be turned to account for washing out and producing a contrast stain at the same time; he should also take note that this washing out is a true chemical decoloration, and if pushed too far will invade the nuclei as well as the rest of the tissues.

99. Clearing.—The washing out of the colour may be stopped by putting the sections into water; but the general practice is to clear and mount them at once.

You may clear with clove oil, *which will extract some more colour* from the tissues. Or you may clear with an agent that does not attack the stain (cedar oil, bergamot oil, xylol, toluol, naphtha, &c. See the chapter on Clearing Agents). If you have used pure alcohol for washing out, you had perhaps better clear with clove oil, as pure alcohol does not always, if the staining have been very prolonged, extract the colour perfectly from extra-nuclear parts. But if you have not stained very long, and if you have used acidulated alcohol for washing out, clove oil is not necessary, and it may be better not to use it, as it somewhat impairs the brilliancy of the stain. A special property of clove oil is that it helps to differentiate karyokinetic figures, as it decolours resting nuclei more rapidly than those in division.

Some colours are much more sensible to the action of clove-oil than others; and much depends on the quality of this much adulterated essence. New clove oil extracts the colour more quickly than old.

Series of sections on slides are conveniently cleared by pouring the clearing agent over them.

When the clearing is accomplished to your satisfaction, mount in damar or balsam, or stop the extraction of the colour if clove oil have been used by putting the sections into some medium that does not affect the stain (xylol, cedar oil, &c.).

The results depend in great measure on the previous treatment of the tissues. If you have given them a prolonged fixation in Flemming's *strong* chromo-aceto-osmic mixture, and have washed out after staining with acid alcohol and cleared with clove oil, you will get, with some special exceptions, nothing stained but nucleoli, and the chromatin of dividing nuclei, that of resting nuclei remaining unstained. If you have given a lighter fixation, with Flemming's weak mixture, or some other fixing agent not specially inimical to staining, and have washed out after staining with pure alcohol, you will get the chromatin of resting nuclei stained as well. With some of these stains—Victoria, for instance—it is easy to get cytoplasm stained, in a lighter tone than the nuclei, by merely washing out lightly.

100. **Victoria Blue (Victoriablau).** (LUSTGARTEN, *Med. Jahrb. k. Ges. d. Aerzte zu Wien*, 1886, p. 285—91).—Stain (specimens strongly fixed in " Flemming " some hours, lightly fixed specimens a few minutes) in saturated aqueous solution. Wash out in pure alcohol (about one minute, more or less). You may clear with clove oil, but you had perhaps better take cedar or bergamot oil, as clove oil washes out the colour very freely.

A most brilliant and useful nuclear stain, and one that I think should be particularly recommended to the beginner, as it is particularly easy to work with. Chromatin and nucleoli, blue. Cytoplasm, if well washed out, colourless; if less washed out, green or greenish blue. The "spongioplasm" is very finely brought out by this method.

Victoria has a special affinity for elastic fibres. For this object, Lustgarten recommends an alcoholic solution of the dye diluted with two to four parts of water. Fixation in chrom-osmium, or at least in a chromic mixture, is, I believe, a necessary condition to this reaction. And you must stain for a long time.

101. **Anilin Green.**—Use precisely as directed for Victoria blue, *supra*. An extremely delicate and absolutely precise nuclear stain, nucleoli being peculiarly brilliantly stained by it. I am unfortunately unable to trace the history of the colour used by me, which may be identical with the Solidgrün of Flemming (*Arch. f. mik. Anat.*, xix, 1881, pp. 317 and 742). It is well to be even more careful in the use of clove oil than in the case of Victoria blue. This colour seems not so generally useful as Victoria, as it does not give so bold a stain.

102. Gentian Violet.—One of the most important of these stains. It may be used in aqueous solution, or in alcoholic solution diluted with about one half of water (FLEMMING, *Zells., Kern. u. Zellth.*, 1882, p. 384), and the stain may be washed out with pure alcohol or (FLEMMING, *Zeit. f. wiss. Mik.*, 1, 1884, p. 350) with acidulated alcohol, as directed below for safranin. But by far the best way of using it is, in general, that due to BIZZOZERO (*Zeit. f. wiss. Mik.*, iii, 1, 1886, p. 24). The tissues may be hardened either in alcohol or in a chromic mixture, but must in the latter case have been well washed out with water. The staining solution is borrowed from that of EHRLICH for bacteria, and consists of—

Gentian violet	1 part.
Alcohol	15 parts.
Anilin oil	3 ,,
Water	80 ,,

The sections are stained in it for five or ten minutes or longer (for objects from Flemming's solution it will frequently be advisable to stain for as many hours.) After staining, rinse the sections with alcohol, and bring them into a 0·1 per cent. aqueous solution of chromic acid. After from thirty to forty seconds, bring them into alcohol, which begins the washing out of the colour. After thirty or forty seconds in the alcohol, put them back for thirty seconds into the chromic acid (this is done in order to fix the colour more completely in the nuclei). Then bring them back into alcohol for thirty to forty seconds in order to wash out more colour and dehydrate them at the same time. Then treat with clove oil, which will extract more colour, and after a short time must be changed for fresh, in which the sections remain until they are seen to give up no more colour, when they are removed and mounted in damar.

You might give a longer treatment with alcohol, and a shorter treatment with clove oil, but you would get a slightly different result. Alcohol washes out colour freely from kinetic nuclei as well as from resting nuclei, whereas clove oil acts much more energetically on the latter than on the former, and thus serves to differentiate dividing nuclei.

In some cases, especially those of tissues whose nuclei have a tendency up to give the colour too freely, better results are obtained by combining the foregoing method with that of GRAM for the staining of bacteria (*Fortschr. d. Medicin*, ii, 1884

No. 6. *British Med. Journ.*, Sept. 6th, 1884, p. 486. *Journ. Roy. Mic. Soc.* (N.S.), iv, 1884, p. 817).

In Gram's method, the sections are treated, after staining, with a solution composed of

Iodine	. . . . .	1 gramme.
Iodide of potassium	. . .	2 ,,
Water	. . . . .	300 ,,

In Bizzozero's adaptation of this process, the series of operations is as follows: Stain in the gentian, wash for five seconds in alcohol; two minutes in the iodine solution; twenty seconds in alcohol; thirty seconds in the chromic acid solution; fifteen seconds in alcohol; thirty seconds in the chromic acid again; thirty seconds in alcohol; and treatment with changes of clove oil until final decoloration.

NISSEN (*Arch. f. mik. Anat.*, 1886, p. 338) employs this process with omission of the treatment with chromic acid.

In resting nuclei, the nucleoli alone are stained, or the chromatin if stained is pale; in dividing nuclei the chromatin is stained with great intensity, being nearly black in the equatorial stage.

This exceedingly powerful stain is quite as precise as that of safranin, to which it is perhaps even preferable for much work with very thin sections (thick sections with closely packed nuclei may easily come out too dark). It lends itself admirably to double staining with eosin, with which it affords one of the most useful and beautiful double-stains known (*see* § 245).

The stain keeps fairly well in damar, though not so well as that of safranin. Flemming found that after a year it had faded a little, though not so much as hæmatoxylin stains (*v. Zells. Kern. u. Zellth.*, p. 384).

Gentian violet in acid solution stains the nuclei of fresh tissues, and dissolved in indifferent media is sometimes very useful for staining *intra vitam* (*see* above, § 93).

103. Dahlia.—(FLEMMING, *Arch. f. mik. Anat.*, xix, 1881, p. 317). Stain in an aqueous solution, either neutral or acidified with acetic acid, and wash out with pure alcohol. The stain is paler in the nuclei than with gentian or safranin. The cytoplasmic granulations of certain cells are sharply stained.

Dahlia is also a useful nuclear stain for fresh tissues (v. EHRLICH, *Arch.*

f. mik. Anat., xiii, 1876, p. 263). For these the aqueous solution must be acidulated with (7·5 per cent.) acetic acid; or you may stain in a neutral solution, and wash out with acidulated water. Dehydrate with alcohol and mount in turpentine colophonium. It is also useful for staining *intra vitam*. See above, § 93.

For the specific staining of Ehrlich's " plasma cells " (see *post*, Part II).

103. Safranin.—One of the most important of these stains, on account of its great power, brilliancy, and superior permanence in balsam, and also on account of the divers degrees of electivity that it displays for the nuclei and other constituent elements of different tissues.

The great secret of staining with safranin is *to get a good safranin*. It is needful here to insist most urgently on what was said above, § 94, *sub finem*. Before thinking of working with this important reagent, you should go to Grübler or to Münder and order the safranin you want, specifying whether you want it for staining nuclei or for staining elastic fibres, or for what other purpose you may require it.

There are presumably at least a score of sorts of safranin in the market, differing to a considerable extent in colour, weight, solubility and histological action. Some are easily soluble in water and not so in alcohol, some the reverse, and some freely soluble in both. Fourteen brands, supplied by Grübler and by Münder, have been studied by RESEGOTTI (*Zeit. f. wiss. Mik.*, v., 3, 1888, p. 320). They all gave positive results with the chromic acid method, to be detailed below; although Grübler had explained that the brands XX, XXBN, TB, had not given positive results (with the usual methods). Resegotti obtained his best results with the brands " Safranin wasserlöslich," " Safranin spirituslöslich," " XX," " XXBN," " TB," furnished by Grübler, and with the brands " Rein," " O," " FII," and " Conc.," supplied by Münder.

Staining.

The majority of safranins are not sufficiently soluble in water, so that solutions in other menstrua must be employed.

A solution much used some time ago is that of PFITZNER (*Morph. Jahrb.*, vi, p. 478, and vii, p. 291), composed of safranin 1 part, absolute alcohol 100 parts, and water 200 parts, the last to be added only after a few days.

The solution of FLEMMING (*Arch. f. mik. Anat.*, xix, 1881, p. 317) is a concentrated solution in absolute alcohol, diluted with about one half of water.

The solutions of BABES (*Arch. f. mik. Anat.*, 1883, p. 356) are (A) a mixture of equal parts of concentrated alcoholic solution and concentrated aqueous solution (this is very much to be recommended), and (B) a concentrated or supersaturated aqueous solution made with the aid of heat.

Some people still employ simple aqueous solutions.

Lastly, there is the anilin solution of BABES (*Zeit. f. wiss. Mik.*, iv, 4, 1887, p. 470). It consists of water 100 parts, anilin oil 2 parts, and an excess of safranin. The mixture should be warmed to from 60° to 80° C., and filtered through a wet filter. This solution will keep for a month or two.

ZWAARDEMAKER (*Zeit. f. wiss. Mik.*, iv, 2, 1887, p. 212) makes a mixture of about equal parts of alcoholic safranin solution and anilin water (saturated solution of anilin oil in water).*

Any of these stains may be used with any of the following washing-out processes. Of course you will have to stain longer in the weaker solutions. As to the anilin solutions, see *ante*, § 96.

Washing-out and Clearing.

For general directions for washing out and clearing, *see* above, §§ 97 and 99.

FLEMMING's *first method* (l. c. in last §). Wash out with pure alcohol, followed by clove oil. This method stains resting chromatin as well as "mitoses."

FLEMMING's *second method* (*Zeit. f. wiss. Mik.*, i, 3, 1884, p. 350). Wash out until hardly any more colour comes away, in alcohol acidulated with about 0·5 per cent. of hydrochloric acid, followed by pure alcohol and clove oil. (You may use the HCl. in watery solution if you prefer it.) The strength given appears unnecessarily high, and I therefore generally use with good results an alcohol of about 0·2 per cent. of HCl. (Objects are supposed to have been well fixed—twelve hours at least—in the *strong* chromo-aceto-osmic mixture, and stained for some hours). PODWYSSOZKI (*Beitr. z. path. Anat. v. Ziegler u. Neuwerk*, i, 1886; *Zeit. f. wiss. Mik.*, iii, 3, 1886, p. 405) prefers to stain for half an hour only, and wash out with 0·1 per cent. of HCl. in alcohol. In each of these ways you get "mitoses" and nucleoli alone stained (if the fixation have been performed as above directed).

* To make "anilin water" shake up "anilin oil" (which is nothing but pure anilin) with water, and filter.

PODWYSSOZKI (l. c.) gives another method, which consists in washing out (for from a few seconds to two minutes) in a strongly alcoholic solution of picric acid followed by pure alcohol. Same results (except that the stain will be brownish instead of pure red).

BABES employed for washing out after staining in the aqueous or alcoholic solutions above mentioned, pure alcohol followed by oil of turpentine. For sections stained in the anilin solution he recommends treatment with iodine according to the method of Gram (see what is said as to the process of Gram in the paragraph on gentian violet, *ante*, § 102). This process has also been recommended by PRENANT (*Int. Monatsschr. f. Anat.*, &c., iv, 1887, p. 358), who notes that the treatment with the iodine solution should be somewhat longer, and the treatment with alcohol somewhat shorter than with gentian violet sections.

MARTINOTTI and RESEGOTTI (*Zeit. f. wiss. Mik.*, iv, 3, 1887, p. 328) recommend washing out with a freshly prepared mixture of one part of 0·1 per cent. aqueous solution of chromic acid with nine parts of absolute alcohol, followed by pure alcohol and bergamot oil. In my experience this method does not give better results (I think less good) than that of washing-out in the simple aqueous solution of chromic acid of Bizzozero followed by alcohol (*see* the paragraph on gentian violet, *ante*, § 102). The latter is certainly a most useful method. It should be mentioned that Martinotti and Resegotti's results refer to lightly stained alcohol-fixed objects, and not to chromo-aceto-osmic objects, which may make a great difference.

GARBINI (*Zeit. f. wiss. Mik.*, v. 2, 1888, p. 170) has recommended that sections be dehydrated after staining in methylic alcohol (wood spirit) in which safranin is only very slightly soluble, and decoloured in a mixture of two parts of clove oil with one part of cedar oil. I have not been able to obtain good results by this method.

The reader will remember that safranin may be washed out by substitution (see *ante*, § 98). In preparations made with chromo-aceto-osmic acid, safranin stains, besides nuclei, elastic fibres, the cell-bodies of certain horny epithelia, and the contents of certain gland-cells.

104. Other Nuclear Stains by the Indirect Method.—The foregoing paragraphs nearly exhaust the list of colours giving *good* nuclear stains by the indirect process. FLEMMING (*Arch. f. mik. Anat.*, xix, 1881, pp. 317 and 742) mentions the following:

MAGDALA RED (NAPHTHALIN RED, ROSE DE NAPHTHALINE).—Nearly if not quite as good a stain as any of the foregoing, and superior to all except safranin in respect of permanency. This and the following should, as far as is yet made out, be used in alcoholic solution diluted with about one half of water, and be washed out with pure alcohol, followed by clove oil.

Mauvein and **Rouge Fluorescent** are good stains, but colour some nuclei more deeply than others in the same preparation.

Solid Green (perhaps the same as the analin green discussed above) is very elective for nucleoli.

Fuchsin (meaning the basic fuchsins, a series of Rosanilin salts having very similar reactions and found in commerce under the names of FUCHSIN, ANILIN RED, RUBIN, ROSEÏN, MAGENTA, SOLFERINO, CORALLIN).—A good but somewhat weak stain, by the alcohol method. Good results are obtained by substitution in the following manner (GRASER, *Deutsche Zeit. f. Chirurgie*, xxvii, 1888, pp. 538—584; *Zeit. f. wiss. Mik.*, v, 3, 1888, p. 378). You either employ the colour as directed for methyl violet (*post*, § 107), or you stain for twelve to twenty-four hours in a dilute aqueous solution, wash out for a short time in alcohol, stain for a few minutes in aqueous solution of methylen blue, and dehydrate with alcohol. A double stain. Chromatin and nucleoli red, all the rest blue.

ORANGE, precise but weak.

BISMARCK BROWN is not very satisfactory with chromic objects. With alcohol objects it gives a good chromatin stain, but cannot be thoroughly removed from cytoplasm by any means yet discovered.

To these may be added—

METHYL VIOLET, perhaps best used according to the method of Resegotti given in the last §; and (according to GRIESBACH, *Arch. f. mik. Anat.*, xxii, p. 132).

TROPÆOLIN OOO, No. 2 (ORANGE II; CHRYSAUREÏN, β NAPHTOLORANGE), a fine dark orange stain, and

ROCELLIN (ECHTROTH, ORSEILLIN No. 3, RUBIDIN, LA RAUVARIENNE), a cherry-red stain.

BENZOAZURIN has been lately recommended by MARTIN (see *Zeit. f. wiss. Mik.*, vi, 2, 1889, p. 193). Stain for an hour or so in dilute aqueous solution and wash out with HCl alcohol.

CHAPTER IX.

OTHER ANILIN STAINS.

104a. For a list showing the colours treated of in this chapter, see § 95.

As regards the direct nuclear stains, the reader is reminded that, as was stated in § 95, many if not most of the anilins give a nuclear stain of greater or less purity if they are used in solutions acidified with acetic acid. Under the present heading, only those are mentioned which give in all respects, alike as regards precision and permanence, simplicity of manipulation and other qualities, a really valuable stain. The very existence of methyl green and Bismarck brown is a sufficient reason for being silent, in this connection, with regard to the rest.

A. *Direct Nuclear Stains.*

105. Methyl Green.—This is the most common, in commerce, of the "anilin" greens. It appears to go by the synonyms of *Methylanilin green*, *Vert Lumière*, *Lichtgrün*, *Grünpulver*. When first studied by Calberla, in 1874 (*Morphol. Jahrb.*, iii, 1887, p. 625), it went by the name of *Vert en cristaux*. It is commonly met with in commerce under the name of more costly greens, especially under that of Iodine green. It is important not to confuse it with the latter, nor with Aldehyde green (*Vert d'Eusèbe*), nor with the phenylated rosanilins, Paris green, and *Vert d'alcali* or *Véridine*.

The chief use of methyl green is as a nuclear stain for *fresh* or *recently fixed tissues*. For this purpose it should be used in the form of a strong aqueous solution containing a little acetic acid (about 1 per cent. in general). The solutions *must* always be acid. You may wash out with water (best acidulated) and mount in some acid aqueous medium containing a little of the methyl green in solution.

Employed in this way, methyl green is a pure nuclear stain, in the sense of being a precise colour-reagent for nucleïn. For *in the nucleus* it stains nothing but the chromosomes, or nucleïn elements: it does not stain either nucleoli of any sort, nor caryoplasm, nor achromatic filaments. *Outside* the nucleus it stains some kinds of cytoplasm and some kinds of formed material, especially glandular secretions (sericin, for instance). But the nucleïn elements are invariably stained of a bright green (with the exception of the nucleïn of the heads of some spermatozoa), whilst extra-nuclear structures are in general stained in tones of blue or violet.

The following paragraph, translated from the paper by Calberla above quoted, appeared inadvertently without comment in the first edition, and has since been repeated without comment in several places.*

"He then found that 'the nuclei of subcutaneous connective tissue and those of vessels and nerve-sheaths stained rose-red, cells of the corium reddish white, and the cells of epidermis greenish blue to pure blue.'"

It should have been added that Calberla here says what he presumably did not mean to say. Methyl green certainly never stains nuclei red, and Calberla's observation should be taken to refer to the results of a double stain with methyl green and eosin, which is mentioned in the somewhat obscurely expressed passage from which the quotation is taken.

Besides being a perfectly pure chromatin stain, methyl green has other advantages. Staining is instantaneous; overstaining never occurs. The solution is very penetrating, kills cells instantly without swelling or other change of form, and preserves their forms for at least some hours, so that it may be considered as a delicate fixative. Osmic acid (0·1 to 1 per cent.) may be added to it, or it may be combined with solution of Ripart and Petit (this, by the way, is an excellent medium for washing out in and mounting in).

Alcoholic solutions may also be used for staining. They should be acidulated with acetic acid.

The stain does not keep. It is difficult to mount it satisfactorily in balsam, because the colour does not resist alcohol (unless this be sufficiently charged with the colour); and of preparations mounted with excess of colour in aqueous media the most fortunate only survive for a few months.

It was first pointed out, I believe by Heschl (*Wiener med. Wochenschr.*, 2, 1879), that methyl green is a reagent for amyloid degeneration. His observations were confirmed by Curschmann (*Virchow's Arch.*, t. 79, 1880, p. 556), who showed that it colours amyloid substance of an intense violet.

Undoubtedly, methyl green is one of the most valuable

* Calberla's misstatement is repeated, and made one of the grounds of an important theoretical deduction, by Griesbach, in *Zeit. f. wiss. Mik.*, iii, 3, 1886, p. 365.

stains yet known. It is the classical nucleïn stain for fresh tissues.

106. Bismarck Brown (Manchester Brown, Phenylen Brown, Vesuvin, La Phénicienne).—A fairly pure nuclear stain that will work either with fresh tissues or with such as have been hardened in chromic acid.

The colour is not very easily soluble in water. You may boil it in water, and filter after a day or two (WEIGERT, in *Arch. f. mik. Anat.*, xv, 1878, p. 258). You may add a little acetic or osmic acid to the solution. MAYSEL (*Ibid.*, xviii, 1880, pp. 237, 250) dissolves the colour in acetic acid (this solution does not give a permanent stain). Alcoholic solutions may also be used. Paul Mayer recommends a saturated solution in 70 per cent. alcohol; or Calberla's mixture (§ 385), or dilute glycerine (say of 40 per cent. to 50 per cent.) may very advantageously be employed.

The watery solutions must be frequently filtered. The addition to them of carbolic acid has lately been recommended, vide *Journ. Roy. Mic. Soc.*, 1886, p. 908. Bismarck brown stains rapidly, but never overstains. The stain is permanent both in balsam and in glycerin.

As has been noted above (§ 93), Bismarck brown has the property of staining certain cellular elements during life (for this purpose it is necessary to see that the colour employed be pure *and neutral*).

107. Methyl Violet (Methylanilin = anilin-violet = Paris violet = inchiostro di Leonardi).—The following process has been recommended by Orth (*Amer. Mon. Micr. Journ.*, i, 1880, p. 143; *Journ. Roy. Mic. Soc.*, N.S., i, 1881, p. 137): Sections are to be soaked in water, and then brought into the following solution:

Anilin violet . . . . . . 1 part.
Acetic acid . . . . 300 parts.

Mount, without washing out, but simply draining, in acetate of potash (acetate 2 parts, water 1 part).

The stain will probably fade within a year or two.

This process does not appear to be of more than very limited applicability. The following, however, due to GRASER (*Deutsche Zeit. f. Chirurgie*, xxvii, 1888, p. 538—584; *Zeit. f. wiss. Mik.*, v. 3, 1888, p. 378) may be very generally useful.

Sections are stained for from twelve to twenty-four hours in a (presumably aqueous) solution so dilute that at the end of that time the sections will have taken up all the colour from the liquid. They are then washed out for a short time in acidulated alcohol, and then in pure alcohol (followed presumably by clearing and mounting in balsam). Schiefferdecker, whose account is here quoted, says that the results, as regards nuclear figures, are even finer than with safranin. The method is applicable to objects fixed in "Flemming."

A useful stain for fresh tissues is also obtained by using dilute acetic acid in the manner recommended above (by EHRLICH) for Dahlia, § 103.

Amyloid matter appears red in preparations stained with methyl violet. This appears to be an optical effect, see the curious experiments of CAPPARELLI, in *Archivio per le scienze mediche*, iii, No. 21, p. 1.

107a. Fuchsin may also be used in GRASER'S way (last §). See also § 104.

B. *Plasmatic stains (stains not affecting nuclei). (This group is put in small type because although the colours of which it consists are stated to be pure plasma stains, they will probably not be found in practice so generally useful, especially for double staining, as many diffuse stains, such as eosin or picric acid.)*

108. Bleu Lumière is stated to be a plasma stain not affecting nuclei. I have not been able to make out whether it is identical with Parma blue, which is one of the numerous toluidin blues. If it is, Frey recommends a solution in water of 1 : 1000, in which tissues stain in a few minutes, and may be mounted either in glycerin or balsam. "*Lichtblau*" is possibly a synonym of this colour. The principal use of such a colour is for making double stains.

109. Bleu de Lyon (Bleu de Nuit, Grünstichblau).—I quote this colour here, although I am not sure to what extent it is a pure plasmatic stain. It is said to be very useful for double staining with carmine.

110. Indulin (Nigrosin, Bengalin, Anilin Blue-black, Blackley blue, Artificial Indigo). (Introduced by Calberla, see *Morph. Jahrb.*, iii, 1877, p. 627). Indulin dissolves into a dark-blue solution in warm water or in dilute alcohol. For staining, the concentrated aqueous solution should be diluted with six volumes of water. Sections will stain in the dilute solution in five to twenty minutes; they may be washed in water or in alcohol, and examined either in glycerin or oil of cloves.

The peculiarity of this stain is that it never stains nuclei; the remaining cell-contents and intercellular substance are stained blue. In its general effects it resembles quinoléin blue, and is exactly the opposite of methyl green. The stroma of tendinous tissue, for instance, stains of a fine blue, the connective tissue that surrounds the bundle, hardly at all, and the tendon-corpuscles of Ranvier remaining perfectly colourless stand out as white stellate figures on a blue ground.

111. Quinoléin Blue (Cyanin, Chinolinblau ; v. Ranvier, *Traité*,

p. 102).—Quinoléin should be dissolved in alcohol of 36° strength (*i. e.* 90 per cent.), and the solution diluted with an equal volume of water. (If the alcohol were taken dilute in the first instance, the blue would not dissolve.) The solutions employed for staining should be very weak, as quinoléin stains very powerfully.

After staining, wash and mount in glycerin. When first mounted, nuclei will be seen to be stained a fine violet, nerves of a grey-blue, smooth muscle blue, protoplasm blue, fat deep blue. But after twenty-four hours in the glycerin, the aspect of the preparation is changed; the nuclei have become colourless; the protoplasm remains blue, and is seen to contain granulations stained intensely blue; nerves remain grey blue, but frequently contain granulations stained blue. Quinoléin, in a word, has the property of staining fatty matters an intense blue.

If the stained preparations be treated with solution of potash of 40 per cent. strength, the differential reaction is produced immediately; the nuclei are unstained, protoplasm, nerve, and muscle tissue are pale blue, and fatty matters deep blue.

It is, perhaps, not generally known that if tissues be stained slowly (for some hours) in very dilute aqueous solution of quinoléin as directed for methyl violet (*ante*, § 107), and then be mounted in balsam, a fairly precise and permanent nuclear stain may be obtained.

Quinoléin is useful for staining Infusoria, which in dilute solution it stains during life. On this point, see the methods of Certes (*post*, Part II).

Other Anilins.

(The order adopted is—Red, Yellow, Green, Blue, Black).

112. Säurefuchsin (Fuchsin S., Acid Fuchsin), a Fuchsin in which the colouring principle is an acid, instead of being a base as in ordinary fuchsin. It is only made by the "Badische Anilin und Soda Fabrik." It may be obtained from Grübler or the other providers of histological reagents. It is a powerful diffuse stain, having a special affinity for axis cylinders, and is chiefly used for staining nerve centres. See the Chapter on Nerve Centres, in Part II.

113. Congo Red (Congoroth).—See Griesbach, in *Zeit. f. wiss. Mik.*, iii, 3, 1886, p. 379. Also an "acid" colour, in the sense in which that appelative is given to Säurefuchsin. The aqueous solution, however, has a neutral or alkaline reaction. It becomes blue in presence of the least trace of free acid (hence Congo is a valuable reagent for demonstrating the presence of free acid in tissues, see the papers quoted l. c.). A diffuse stain, much of the same nature as that of Säurefuchsin, and like it seems to be at present chiefly useful for staining axis-cylinders. See the Chapter on Nerve Tissue, in Part II. It may also be used for staining some objects during life (see *ante*, § 93).

114. Benzopurpurin, according to Griesbach (*l. c.*), another "acid" colour, very similar in its results to Congo red.

Zschokke (*Ibid.*, v, 4, 1888, p. 466) says that **Benzopurpurin B** is the very best contrast stain to hæmatoxylin known to him. It seems to him preferable to eosin on account of its not being affected by alcohol or the usual clearing agents. Weak aqueous solutions should be used for staining, which is effected in a few minutes, and alcohol for washing out. **Deltapurpurin**, a more purple red, has similar properties, and may be used in the same way. The solution (aqueous) should be moderately concentrated, and allowed to act for a minute or two.

115. Biebricher Scharlach (BIEBRICH SCARLET), a diffuse bright red stain, may be useful as a contrast stain (*see* GRIESBACH, *Arch. f. mik. Anat.*, xxii, p. 132).

116. Eosin, the potassium salt of a bromide of phthaleïn, is found in commerce under the synonyms of **Primerose Soluble, Erythrosin, Pyrosin B., Rose B. à l'Eau**. The preparations indicated by these names are not quite identical in their properties, but vary according to the different modes of manufacture. Most of them are soluble, both in alcohol and in water, but some only in alcohol ("*Primerose à l'Alcohol.*")

Eosin was at one time much employed alone as a general histological stain. Being convinced that this practice is now entirely out of date, I suppress all the numerous formulæ for staining solutions which have been recommended from time to time with great enthusiasm (*see*, however, FISCHER, *Arch. f. mik. Anat.* xii, 1875, p. 349; LAVDOWSKY, *Ibid.* xiii, 1876, p. 359; ÉLOUI, *Rech. Hist. sur le Tissu Conj. de la Cornée*, Paris, 1881, and *Zeit. f. wiss. Mik.*, i, 1884, p. 389). It may be added that experiments have not been made, or at all events have not been published, in the direction of dissolving eosin in anilin water (*see* footnote to § 103), or combining its alcoholic or aqueous solution with anilin or anilin-water—which might very possibly give results very different from those obtained by the ordinary solutions.

But eosin is of very great importance as a secondary or contrast stain. It has already been mentioned that it washes out many of the anilins that give nuclear stains by the indirect method (*see* § 98). Combined with a blue anilin, or with hæmatoxylin, it gives instructive and durable double stains, which are among the most beautiful that can be produced. These will be treated of in the chapter on double-stains.

Here it may merely be mentioned that the majority of the authors who recommend these combinations proceed by staining first with the eosin and then with the nuclear stain, and that the majority also employ alcoholic solutions of eosin. I am by no means convinced that this practice is the best. As regards the anilins, at all events, good results are certainly obtained by staining first with the nuclear stain (gentian violet, for instance) and washing out (sections) for a few seconds or a minute or two in a tolerably strong aqueous solution of eosin, and then dehydrating with alcohol and mounting in balsam. If the eosin is too much extracted by the alcohol or clearing agent, these should be used charged with a little of the colour. Eosin stains may also be kept in glycerin, if this be perfectly neutral or, better, slightly alkaline (which may be brought about by the addition of 1 per cent. of sodium chloride), and also charged with a little eosin.

117. Bengal Rose (GRIESBACH, *Zool. Anz.*, No 135, 1883, p. 172).— Bengal rose, or "Rose bengale," or "Bengal rosa," is an eosin dye. It is the bluest of the eosin dyes as yet known, approaching in hue to fuchsin, but possessing far greater brilliancy and purity of hue. In aqueous solution it is useful for staining chromic-acid objects, especially spinal cord, in which the grey matter stains of a deep bluish-red, and stands out boldly from the less deeply coloured white matter. It is also useful for double and treble stains, as will be explained below.

118. Picric Acid.—Not used alone as a stain, but one of the most useful of all colouring agents as a secondary or groundstain. Nothing is easier than to stain with an alcoholic solution of picric acid tissues of which the nuclei have previously been stained by borax-carmin, alum-carmin, hæmatoxylin, or an anilin stain. It should be borne in mind that picric acid has considerable power of washing out other anilin stains; and that in combination with hydrochloric acid it very greatly enhances the power with which this acid washes out carmine stains. It does not otherwise affect any of the usual stains, and may be most highly recommended as a useful though frequently inelegant stain.

119. Metanil Yellow (Metanilgelb).—This colour has lately been studied with great minuteness by GRIESBACH (*Zeit. f. wiss. Mik.*, iv, 4, 1887, p. 418; see also *Journ. Roy. Mic. Soc.*, 1889, p. 464). It does not appear to be worthy of having so much time spent on it from the practical point of view; the interest of Griesbach's work lying rather in the region of chemical theory. The practical outcome is that metanil yellow is a diffuse stain with a certain affinity for certain elements belonging to the group of the connective tissues; and that with some other colours it gives sharply differentiated double stains of certain preparations. These will be mentioned in the proper places.

120. Säuregelb (Echtgelb) Tropæolin O., Crocein, Gold Orange, are all of them more or less diffuse yellow or orange stains, having certain affinities for certain tissues, and may occasionally be found very useful for double staining, being good stains in their way (*see* GRIESBACH, *Arch. f. mik. Anat.*, xxii, p. 132).

121. Iodine Green ("Hofmann's Grün") (GRIESBACH, *Zool. Anz.*, No. 117, vol. v, 1882, p. 406).—Griesbach employs the following solution:

Crystallized iodine green . . 0·1 gr.
Distilled water . . . . 35·0 „

These proportions may be varied according to the desire of the operator, within limits indicated only by the observation that good results can only be obtained from deep-hued solutions.

The objects are to be put into water for a few seconds before staining. They stain instantaneously in general. They are to be washed out in water, and brought into glycerin, or dehydrated in absolute alcohol and passed through oil of cloves or anise-seed into balsam or dammar. *The stain is not destroyed by immersion in alcohol for days.* The preparations are apparently permanent in balsam.

Alcoholic solutions may be used for staining, but Griesbach finds no advantage in so doing.

A nuclear but frequently diffuse stain, valuable for the exceeding rapidity of its action, and for its striking power of marking-out by staining in various hues the different forms of tissue. For instance, in a section through the uterus of a roe-deer, the epithelia are stained blue, the glands dark green, and the muscle-fibres malachite green, whilst the connective tissue remains unstained.

Chromic acid objects stain well. The colour is useful for double staining (see *post*, Chapter XIII).

This colour is somewhat expensive to prepare, and for this reason is no longer found in commerce, having been superseded by methyl green. But the high price is no impediment to the use of iodine green in histology, on account of the small quantity of the substance required for staining.

The observer will do well to assure himself that he has obtained a genuine iodine green, other coal-tar greens being sometimes palmed off on the unwary purchaser for iodine green. The presence of iodine may be tested in the following way : A little of the solid colouring matter is treated with sulphuric acid, and a few small fragments of bichromate of potash are added ; the iodine, if present, escapes in the form of violet vapours. It may also be demonstrated by means of chloroform or sulphide of carbon.

The colour may be obtained of excellent quality from C. A. F. Kuhlbaum's Chemische Fabrik, Berlin, S.O. (*Zool. Anz.*, No. 130, 1883, p. 56.)

122. Thiophen Green (Thiophengrün,) (KRAUSE, *Intern. Monatsschr. f. Anat.*, &c., iv. 1887, Hft. 2).—Krause finds that the double zinc-salt of Thiophen green prepared by V. Meyer is useful for double staining with carmine. Sections should be stained (after borax carmine) for a few minutes in a concentrated aqueous solution, and be well washed out with absolute alcohol.

123. Anilin Green has a special affinity for mucus gland cells. We shall return to this subject in the chapter on glands.

124. Picro-anilin Green (TAFANI; see *Journ. de Microgr.*, 1878, p. 82).—(Mixture of anilin blue and picric acid solution.)

125. Anilin Blue.—Used alone, without special precautions, this is a diffuse stain hardly to be recommended at the present time. See, however, the method of HEIDENHAIN, *Arch. f. mik. Anat.* vi, 1870, p. 404. It is useful for double-staining.

126. Parma Blue (Toluidin Blue, Lichtblau).—A fast stain either in aqueous or balsam mounts, was much recommended by FREY (*Arch f. mik. Anat.*, 1868, p. 346). An aqueous solution of 0·1 per cent. will stain tissues sufficiently in a few minutes.

127. Methylen Blue.—This is an important colour. It is used for staining organisms during life (see § 93). It has a special affinity for sensitive nerve-endings, which it stains during life. It is used alone or combined with other re-agents, for staining nerve centres (*post*, Part II). And it possesses the property of washing out certain other anilins, with which it gives valuable double-stains (*ante*, § 98, and *post*, Chapter XIII).

128. Violet B (S. Mayer, *Sitzb. d. k. k. Akad. d. Wiss*, *Wien*, iii Abth., February, 1882). Used in solutions of 1 grm. of the colour to 300 grms. of 0·5 per cent. salt solution, and with fresh tissues that have not been treated with any reagent whatever, this colour gives a stain so selective of the elements of the vascular system that favorable objects, such as serous membranes, appear as if injected. The preparations do not keep well; acetate of potash is the least unsatisfactory medium for mounting them in.

129. Anilin Black (Blue-Black, Nigranilin, Noir de Colin).—Has a special affinity for ganglion-cells, and is much used in the study of the central nervous system (see *post*, Part II).

CHAPTER X.

CARMINE STAINS.

130. The Sorts of Carmine.—Commercial carmine, prepared as it is by very various processes, is of very variable quality. None but the very best should be used for histological purposes. The finest sorts are known in the trade as Nakaret Carmine. They cost about forty shillings a pound; inferior sorts costing less than a fourth that sum. But the histologist should procure the very best, even though it cost ten times as much as it does.

131. The Use of Carmine in Staining.—What are carmine stains useful for? Is it for staining fresh tissues? With the exception of aceto-carmine, No! Is it for staining sections? Again, No! For in nine cases out of ten, sections are better stained by some of the anilin stains than they can be in any carmine stain. Is it for staining entire objects?—for staining in the mass? Yes. For in many, if not in most, cases that can be done more satisfactorily by means of carmine than by means of any other known agent. So that until a coal-tar colour shall have been discovered that can beat alum-carmine and borax-carmine on their own ground, these must still hold their sway. As soon as that shall have been done, and it may be done any day, carmine stains will become as extinct as the Dodo.

132. Classification of the Formulæ.—In the treatment of the formulæ given in the following pages I have been guided by the considerations set out in the last paragraph. The best stains for staining in the mass are fully treated, the old-fashioned solutions proposed for staining sections being thrown into the background. The formulæ set out below are arranged according to the nature of the menstruum. This gives us two great groups: aqueous carmine solutions and alcoholic carmine solutions. Taking first the group of aqueous solutions, I have arranged the formulæ comprised in it according to the reaction of the solutions. First come the alkaline ammoniacal solutions, then the neutralised ammoniacal solutions, then other neutral solutions, including alum and picro-carmine, and borax-carmine. These last requiring a treatment of the tissues with an acid to fix the stain, lead naturally to the last group, the acid stains. The alcoholic group is too small to require subdivision.

133. Hints.—Tissues to be stained must be freed from acids before being put into the staining fluid. Overstains may in all cases be washed out with weak HCl. (*e. g.* 0·1 per cent.). All carmine stains, with the exception of aceto-carmine, are permanent in balsam. Aqueous mounts should be acid (except for alum-carmine), and the best plan is to let the mounting medium contain 1 per cent. of formic or acetic acid. Formic acid is to be preferred.

Remember that none of Grenacher's fluids can be used with calcareous structures that it is wished to preserve.

Grenacher's alcoholic borax-carmine may be recommended to the beginner as being the easiest of these stains to work with.

A. Aqueous Carmine Stains.

a. Alkaline.

134. Ammonia-Carmine (Beale, *How to Work, &c.*, p. 109, 4th ed.).

Carmine	10 grains.
Liquor Ammoniæ (fortissimus, B.P.)	½ drachm.
Price's glycerin	2 ounces
Distilled water	2 ,,
Alcohol	½ ounce

The carmine, in small fragments, is to be dissolved in the ammonia, with the aid of heat. Boil for a few seconds, and let cool. Leave uncorked for at least an hour, or until the excess of ammonia has evaporated, as tested by the smell. Then add the glycerin, water, and alcohol, and filter, or allow to settle and decant. If after keeping for some months the carmine begins to precipitate, owing to the escape of ammonia, add one or two drops of liquor ammoniæ.

Another formula, given by Beale for the special purpose of staining by means of injection, will be found at p. 304 of the same work.

135. Simple Aqueous Ammonia-Carmine.—The simple solution of carmine in aqueous solution of ammonia is not stable, but from the very moment of its formation is engaged in a series of chemical changes, of which the nature is still imperfectly understood, though it is now known that they are due, in part at least, to the growth of a special microphyte. They may be to some extent prevented by means of antiseptics, as has been proposed by Hoyer, who recommends the addition of 1 to 2 per cent. of chloral hydrate. But it is now generally recognised that it is better to allow the whole series of changes to take place, not using the solutions until the changes have been entirely accomplished. Such solutions, known as "*ausgefaulte Carmin*," "*carmin pourri*," putrefied carmine, are greatly superior in staining power to solutions in which these changes have not taken place. It has been proved by GIERKE (*Zeit. f. wiss. Mik.*, i, 1884, p. 76) that they owe this enhanced power to the presence of traces of carbonate or bicarbonate of ammonia, which is formed in the solutions by the combination of their free ammonia with the carbonic acid of the air. It has been demonstrated that these carbonates act as mordants in the staining process.

The solutions thus prepared form an exception to the statement made above (§ 131) that carmine is of little good for staining sections, inasmuch as they give one of the best stains known for sections of the central nervous system.

136. **Aqueous Ammonia-Carmine** (BETZ, *Arch. mik. Anat.* ix, 1873, p. 112).—Commercial carmine is rubbed up with a little water in a mortar until a thick syrupy mass is obtained; on to this ammonia is poured, with continual stirring. The solution is diluted with a large quantity of water, and filtered. The filtered solution is exposed to the sun in an uncorked vessel, which must be of *green* glass, until a dirty red, flocculent precipitate appears; it is then filtered. The solution is again left to stand in the same conditions as before, and when the precipitate reappears, it is again filtered, and the solution again exposed. Generally no third precipitate appears; if it does, filter again. In either case, the preparation is now finished, and the solution is to be preserved for use in a corked vessel. It will keep for months. It sometimes happens that the solution whilst exposed to the sun acquires a bad smell, and becomes covered with a white flocculent membrane. This does not hinder the preparation, but, on the contrary, furthers it.

For sections of nerve-centres half an hour, or at most an hour, suffices to stain sections. The first elements that stain are the granular mass of the grey matter, then nerve-cells, epithelium, and lastly, other structures. A nuclear stain. Permanent.

137. Aqueous Ammonia-Carmine (GIERKE, *Zeit f. wiss. Mik.*, 1, 1884, p. 76).—To powdered carmine add sufficient ammonia to dissolve it completely, and leave the solution exposed to the air in a capsule for several days. Filter, and keep the solution in a stoppered bottle. Do not use it for staining before two years at least. By that time it will contain no more free ammonia, but carbonate of ammonia in its place. For staining, add to distilled water as many drops of the solution as will give a liquid of a pale rose tint. Sections should remain in the liquid for twenty-four hours. It is very important to carry out the staining as directed, slowly, in dilute solutions. In this way chromic objects stain well. For sections of encephalon the hardening *ought* to have been done in a chromic liquid, and all treatment with alcohol ought to be carefully avoided until staining is accomplished; alcohol should not even be used for wetting the section knife.

138. Aqueous Ammonia-Carmine (HOYER, *Biol. Centralb.* ii, 1882, p. 17).—"Dissolve 1 gr. of carmine in a mixture of 1—2 c.c. of strong liquor ammoniæ and 6—8 c. c. of water. Heat in a glass vessel on a sand bath until the excess of ammonia has evaporated. (So long as free ammonia is present *large* bubbles are formed in the fluid, and the latter shows the usual dark purple colour of carminate of ammonia. When the free ammonia has evaporated *small* bubbles appear, and the solution takes a brighter red tint.) The solution is left to cool and settle, and by filtering the bright red deposit (which may be used over again) is separated from the neutral dark fluid, which by the addition of chloral hydrate can be kept for a long time " (*see also below*, § 141).

139. Aqueous Ammonia-Carmine, other Formulæ (RANVIER, *Traité Technique*, p. 97. HUXLEY and MARTIN, *Pract. Elemen. Biol.*, p. 268. FREY, *Le Microscope*, p. 167).

β. *Neutral and Acid.*

140. Ranvier's Neutral Carmine (kindly communicated by Dr. MALASSEZ, see *Traité des Méthodes Techniques, &c.*, of Lee et Henneguy, p. 82). Make a simple solution of carmine in water with a slight excess of ammonia and expose it to the air in a deep crystallizing dish until it is entirely dried up. It should be allowed to putrefy if possible. Dissolve the dry deposit in pure water, and filter.

141. Hoyer's Neutral Carmine (*Biol. Centralb.*, ii, 1882, p. 17). If the solution made by the process given *supra*, § 138, be mixed with 4—6 times its volume of strong alcohol a scarlet-red precipitate is formed. This is separated by filtration, washed, and dried, or made into a paste with alcohol in which some glycerin and chloral is dissolved. Both the powder and the paste can be kept several months unchanged; they dissolve easily in water, particularly the paste. The solution passes readily through the filter, whilst the ordinary carmine solution can only be filtered with difficulty; it also keeps a long time unchanged, especially with the addition of 1—2 per cent. of chloral, and it has a much more intense colouring power.

"By dissolving the carmine powder in a concentrated solution of neutral picrate of ammonia a combination is obtained which has all the advantages of ordinary picro-carmine without any of its disadvantages."

142. Böhn's Neutral Carmine (*Arch. Anat. u. Phys.* (*Anat. Abth.*), 1882, p. 4).—Three to 4 grms. carmine are rubbed up in a mortar with 200 grms. water, and ammonia is added drop by drop until the solution acquires a cherry-red colour; acetic acid is then added until the colour becomes of a sealing-wax red; and the solution is filtered. If the colour is not intense enough, add before filtering two drops of ammonia, and leave in an open vessel until the smell of ammonia can no longer be perceived.

Tissues should remain for twenty-four hours in the stain (or longer if they are more than 1 mm. thickness), after which it is desirable, in order to ensure a nuclear stain, to wash out with glycerin and water (equal parts) containing ½ per cent. of hydrochloric acid.

These directions apply to blastoderms.

This formula has been erroneously attributed to "Bohm" by more than one writer.

143. Heidenhain's Neutral Carmine (*Arch. f. mik. Anat.*, vi, 1870, p. 402.).—(Beale's carmine, without the alcohol, and neutralised.)

144. As to Picro-Carmine.—By the term "picro-carmine" there ought to be understood a definite chemical substance, a double salt of picric and carminic acid and ammonia. It is commonly used to denote a whole tribe of solutions in which carmine, ammonia, and picric acid exist *uncombined* in haphazard proportions. In the face of this confusion, all that can be done is to distinguish the true picro-carmine as "Ranvier's" picro-carmine, or *picro-carminate of ammonia*.

True picro-carmine is one of the most important of the carmines. It gives very delicate differentiations, and has the great merit of being as innocuous as possible to most tissues. It is a single or a double-stain according to the manner of using it. If the preparations be washed, after staining, with water, it is a *single* stain, the colour of the carmine alone appearing; if they be washed quickly in alcohol it is a *double*-stain, the yellow colouration of the picric acid not being dissolved by the alcohol as it is by water. Of course the washing with alcohol must not be overdone, or the yellow colouration may be entirely removed.

For slow staining, dilute solutions may advantageously have 1 or 2 per cent. of chloral hydrate added to them.

Over-stains may be washed out with hydrochloric acid, say 0·5 per cent., in water, alcohol, or glycerin.

Preparations should be mounted in balsam, or if in glycerin this should be acidulated with 1 per cent. of acetic or, better, formic acid.

The good qualities (especially that of precision and delicacy of stain), above attributed to picro-carmine, apply in their entirety only to Ranvier's picro-carmine or picro-carminate of ammonia. The other pseudo-picro-carmines are in general so inferior as not to take rank as good stains at all. The reader is in consequence warned against an over-confident faith in them, and is especially warned against the so-called picro-carmine sold by the opticians. The true Ranvier's picro-carminate of ammonia can only be prepared *with certainty* by one process as yet known. This I proceed to give, and amongst the other formulæ set out one or two that *may* by chance give rise to the formation of a picro-carminate, and should at all events afford a useful staining solution (which the majority of these formulæ only do by chance).

145. Ranvier's Picro-Carmine or Picro-carminate of Ammonia.—The method of preparation employed in the Laboratory of Histology of the Collège de France, kindly communicated to myself and Henneguy for our *Traité des Méth. Techn.* (q. v., p. 451) by M. VIGNAL, one of the assistants there, is as follows:

Take—

Water	1000 parts.
Picric acid	20 ,,
Carmine	10 ,,
Ammonia	50 ,,

Put them into a stoppered bottle and leave them for two or three months in a warm place. Then put them into a large crystallising dish and let them putrefy. When the liquid has become reduced by evaporation to four fifths of its original volume, remove the crystals that have formed at the bottom, dry them, and dissolve them in a little warm water. Filter the solution, and examine it with the microscope to see whether the carmine is really dissolved. If not, add water and ammonia, and let the solution putrefy again; evaporate and examine as before. When you have got your carmine combined, evaporate the solution to dryness in a stove, and reduce the picro-carminate to powder.

For staining, dissolve 1 gramme of the powder in 100 grammes of water, and add a crystal of thymol to prevent the development of mould.

Ranvier's Original Formula (*Traité*, p. 100) was as follows: To a saturated solution of picric acid add carmine (dissolved in ammonia) to saturation. Evaporate down to one fifth the original volume in a drying oven; and separate by filtration the precipitate, poor in carmine, that forms in the liquid when cool. Evaporate the mother liquor to dryness, and you will obtain the picro-carminate in the form of a crystalline powder of the colour of red ochre. It ought to dissolve completely in distilled water; a 1 per cent. solution is best for use.

146. Picro-Carmine (*Weigert's formula*, *Virchow's Archiv*, Bd. 84, pp. 275, 315; *Zool. Jahr.*, 1881, p. 40).—Two grammes of carmine are soaked for twenty-four hours (in a vessel protected from evaporation) in 4 grammes of ammonia; 200 grammes of concentrated solution of picric acid are then added, and the whole put away for twenty-four hours more. Small quantities of acetic acid are then added "until the first slight precipitate appears even after stirring." The whole is again put away for twenty-four hours more, when it will be found that there has formed a precipitate that can only partially be removed by filtration; ammonia is then added drop by drop at intervals of twenty-four hours, until the solution becomes clear. If the solution stains too yellow, acetic acid is added; if it overstains red, a little ammonia is again added. All badly staining samples of picro-carmine may be improved in the same way by addition of acetic acid or ammonia.

147. Other Formulæ for Ammonia Picro-Carmine.—GAGE, *Am. M. Mic. Journ.*, i, 1880, p. 22; *Journ. Roy. Mic. Soc.*, vol. iii, p. 501

(very elaborate, and has not afforded me a soluble carmine). FOL, *Lehrb. d. vergl. mic. Anat.*, p. 195. RUTHERFORD, *Pract. Hist.*, p. 173. PAUL MAYER, *Mitth. Zool. Stat. Neapel*, ii, p. 20. BABER, *Mon. Micro. Journ.*, xii, p. 48. PERGENS, Carnoy's *Biologie Cellulaire*, p. 92. HOYER, see above, 141. BIZZOZERO, *Zeit. f. wiss. Mik.*, 1885, p. 539. KLEMENSIEWICS, *Sitzb. Acad. Wiss. Wien*, lxxviii, 1878, iii, Juni; *Zeit. f. wiss. Mik.*, i, 1884, p. 501. (Rub up 1 grm. of carmine with 30 of ammonia, and add 200 of water. To the solution add half a vol. of saturated solution of picric acid; boil for 8 to 10 hours on a water bath, making up at first for evaporation by adding dilute ammonia solution; later, evaporate one half or two thirds. On cooling, hardly any precipitate is formed. The solution is clear, very dark red.) CUCCATI, *Zeit. f. wiss. Mik.*, vi, 1, 1889, p. 42.

148. Soda Picro-Carmine (Picro-carminate of Soda.)—Soda picro-carmine is not infrequently referred to as giving better results than ammonia picro-carmine. The methods of preparation do not seem to have received due publication, and it is not easy to meet with a formula. Here is one, due to LÖWENTHAL (*Anal. Anz.*, ii, 1887, No. 1, p. 22). Dissolve 1 grm. of caustic soda in 100 grms. of water, add 0·4 grm. of carmine and dissolve it by the aid of heat or by allowing it to stand for twenty-four hours or more. Filter, and add 100 c. c. of water. Then add gradually 20 to 25 c. c. of 1 per cent. picric acid solution. This produces at first a precipitate that redissolves, later a slight persistent precipitate. Add a few c. c. of the picric acid in excess, let the solution stand an hour, and filter (if necessary, two or three times through the same filter). The filtrate may be concentrated by evaporation if desired.

The 'Magazine of Pharmacy' gives a formula similar to the foregoing, except that the proportion of soda is reduced to 0·05 per cent. (see *Journ. Roy. Mic. Soc.*, 1888, p. 518).

149. Alum-Carmine (GRENACHER's formula, *Arch. mik. Anat.*, xvi, 1879, p. 465).—An aqueous solution (of 1 to 5 per cent. strength, or any other strength that may be preferred) of common or ammonia alum, is boiled for ten or twenty minutes with ½ to 1 per cent. of powdered carmine. (It is perhaps the safer plan to take the alum solution highly concentrated in the first instance, and after boiling the carmine in it, dilute to the desired strength.) When cool filter.

This stain must be avoided in the case of calcareous structures that it is wished to preserve.

TIZZONI (*Bull. Sc. Med. Bologna*, 1884, p. 259) and PISENTI (*Gazz. degli Ospetali*, No. 24; *Zeit. f. wiss. Mik.*, ii, 1885 p.

378) add a small percentage of sulphate of sodium, with the object of enhancing the energy of the stain. It should not be forgotten that sodium sulphate is a substance that exercises a very peculiar action on nuclei (*see* PFITZNER, *Morph. Jahrb.*, xi, 1, 1885).

It has already been sufficiently repeated that alum-carmine is one of the best stains to be found outside the coal-tar colours. It is particularly to be recommended to the beginner, as it is easy to work with; it is hardly possible to overstain with it (except muscle).

150. Alum-Carmine with Osmic Acid (ZOLTÁN VON ROBOZ, *in litt.*).—To 50 or 60 grms. of water is added alum-carmine until the mixture is of an almost red rose colour; about ten drops of a $\frac{2}{1000}$ solution of osmic acid are then added. (The mixture should have an appreciable smell of osmic acid.) The objects to be stained remain in the mixture for about thirty-six hours in the dark. It is hardly necessary to wash them, as the stain is perfectly precise without that. It is important to perform the staining in a well-closed vessel, in order to prevent the evaporation of the osmium.

I have used this stain with the most diverse objects, and can most highly recommend it. The result is a sharp nuclear double stain (resting chromatin and nucleoli purple, kinetic chromatin red, the rest brown). Very valuable for staining soft tissues in the mass.

151. Acetic Acid Alum-Carmine (HENNEGUY, *Traité des Méth. Techn.*, LEE et HENNEGUY, p. 88).—Excess of carmine is boiled in saturated solution of potash alum. After cooling, add 10 per cent. of glacial acetic acid, and leave to settle for some days. The deposit of carmine and alum that forms during that time is removed by filtration.

For staining, enough of the solution is added to distilled water to give it a deep rose tint. Stain for twenty-four to forty-eight hours, and wash for an hour or two in distilled water. (It is important that the water should be distilled in order to avoid the formation of crystals.) Dehydrate with alcohol and mount in balsam. You can mount in glycerin, but the preparations do not keep so well as in balsam.

The advantage of this carmine is that it has great power of

penetration, and stains deep-seated layers of tissue just as well as the superficial ones. The colour of the stain is a somewhat inelegant violet, but this can be changed to a warmer tone by treating the objects with dilute HCl, as for borax-carmine objects.

152. Alum-Carmine and Picric Acid.—Alum-carmine objects may be double-stained with picric acid. LEGAL *(Morph. Jahrb.*, viii, p. 353) combines the two stains by mixing 10 vols. of alum-carmine with one of saturated picric acid solution.

153. Lithium-Carmine (ORTH, *Berlin. klin. Wochenschr.*, 28, 1883, p. 421). Two parts and a half of carmine are dissolved in ninety-seven parts and a half of saturated solution of carbonate of lithium.

The solution stains with equal readiness alcohol objects and chromic objects. The stain is diffuse, but becomes restricted to nuclei on treatment with hydrochloric acid (1 per cent. in 70 per cent. alcohol). The stain is permanent both in balsam and glycerin. The great quality of this stain is the readiness with which it stains tissues that refuse to stain in any other medium. It is an excellent reagent, that seems not to be sufficiently known.

154. Aceto-Carmine (Acetic Acid Carmine) (SCHNEIDER's formula, *Zool. Anzeig.*, No. 56, 1880, p. 254).—To boiling acetic acid of 45 per cent. strength add carmine until no more will dissolve, and filter. (Forty-five per cent. acetic acid is the strength that dissolves the largest proportion of carmine.)

To use the solution you may either dilute it to 1 per cent. strength, and use the dilute solution for slow staining, which is the method to be preferred for making glycerin preparations ; or a drop of the concentrated solution may be added to a fresh preparation under the cover-glass.

This is a very important reagent, which in certain cases renders services that no other reagent can render. If you use the concentrated solution, it fixes and stains at the same time, and hence is most valuable for the study of fresh objects. It is very penetrating, a quality that enables it to be used where ordinary reagents would totally fail. The stain is a pure nuclear one. Unfortunately, the preparations cannot be preserved.

ZACHARIAS adds wood vinegar (*Acetum pyrolignosum*) to this solution in the proportion of 1 drop to 10 c.c. (see *Zeit. f. wiss. Mik.*, v, 3, 1888, p. 37½).

155. Schweigger-Seidel's Acid Carmine (RANVIER, *Traité*, p. 99).— Ammonia-carmine neutralised and rendered slightly acid with acetic acid. Wash out with 0·5 per cent. of hydrochloric acid.

156. Hamann's Acid Carmine (*Intern. Mon. f. Anat. u. Hist.*, i, 5, 1884; *Zeit. f. wiss. Mik.*, ii, 1885, p. 87).—Thirty gr. of carmine, 200 c.c. of strong ammonia, and acetic acid to neutralisation or slightly acid reaction. This may be used for staining, but it is far better to redissolve in a mixture of ammonia and acetic acid in the same proportions the precipitate that forms when the solution is allowed to stand for from two to four weeks. Treatment with HCl is not necessary.

157. Neutral Borax-Carmine (NIKIFOROW, *Zeit. f. wiss. Mik.*, v, 3, 1888, p. 337).—Boil together 3 parts of carmine, 5 of borax, and 100 of water, adding enough ammonia to get the carmine to dissolve. Evaporate to less than half the original volume. Add dilute acetic acid until the cherry-red colour changes (if you should add too much acetic acid you must re-neutralise with ammonia). Add a little carbolic acid to preserve the solution.

A direct nuclear stain, like that of alum-carmine, but more powerful. Osmic and chromic objects take the stain well. Over-staining does not occur, so that objects may remain for days in the stain. Wash out with water.

158. Neutral Borax-Carmine (GRENACHER, *Arch. f. mik. Anat.*, xvi, 1879, p. 466).

159. Woodward's Borax-Carmine (see *Monthly Micr. Journ.*, vii, 1872, p. 38; *Am. Quart. Micr. Journ.*, 1, 1879, p. 220; *Journ. Roy. Mic. Soc.*, ii, p. 613).

160. H. Gibbes' Borax-Carmine (see *Journ. Roy. Mic. Soc.*, iii, 1883, p. 390).

161. Delage's Osmium-Carmine (*Arch. de Zool. Exp. et gen.*, IV, sér. 2, 1886; *Zeit. f. wiss. Mik.*, iii, 2, 1886, p. 240). Ammonia-carmine neutralised by evaporation over a water-bath, and combined with an equal volume of 1 per cent. osmic acid solution, then filtered under a bell-glass. Stains and fixes at the same time. (The mixture, however, will not preserve its fixative properties for more than a few days.)

162. Other Aqueous Carmines.
Rollet's Carminroth (see *Zeit. f. wiss. Mik.*, i, p. 91).
Perl's Soluble Carmine (see FREY, *Das Mikroskop*, 7 Auf., and *Zeit. f. wiss. Mik.*, i, p. 91).
Carminic Acid, see DIMMOCK (*Amer. Natural.*, xviii, 1884, pp. 324-7; and *Journ. Roy. Mic. Soc.*, 1884, pp. 471-474).
Boric Acid Carmine, ARCANGELI (see *Proc.-verb. Soc. Toscana Sc Nat.*, 1885, p. 283; and *Zeit. f. wiss Mik.*, 1885, p. 377).

Boric Acid Alum-Carmine, Arcangeli, *Ibid.*
Salicylic Acid Alum-carmine, Arcangeli, *Ibid.*
Salicylic Acid Carmine, Arcangeli, *Ibid.*
Picric Acid Carmine, Arcangeli, *Ibid.*
Picric Acid Carmine, Minot (see Whitman's *Methods in Mic. Anat.*, p. 42).

B. Alcoholic Carmine Stains.

163. Alcoholic Borax-Carmine (Grenacher, *Arch. f. mik. Anat.*, xvi, 1879, p. 466, *et seq.*).—Take a *concentrated* solution of carmine in borax solution (2 to 3 per cent. carmine to 4 per cent. borax); dilute it with about an equal volume of 70 per cent. alcohol, allow it to stand some time, and filter. Or, the mixture of carmine and borax solution is *allowed to stand for two or three days* and occasionally stirred; the greater part of the carmine will dissolve. To the solution is added an equal bulk of 70 per cent. alcohol; the mixture is allowed to stand for a week, and then is filtered. If on keeping more carmine is deposited, it must be refiltered.

Preparations should remain in the stain until they are thoroughly penetrated (for days if necessary), and then be brought (*without first washing out*) into alcohol acidulated with 4 to 6 drops of hydrochloric acid to each 100 c.c. of alcohol. They are left in this until they are thoroughly penetrated, and may then be washed or hardened in neutral alcohol. Four drops of HCl is generally enough. Three drops I find not quite sufficient. The stained objects should remain in the acidulated alcohol till they acquire a bright transparent look (three to six hours).

For delicate objects, and for very impermeable objects, it may be well to increase the proportion of alcohol in the stain: it may conveniently be raised to about 70 per cent. The washing out, or decolouration, will be enormously facilitated if picric acid be added to the acidulated alcohol, but in this case the proportion of HCl should be reduced. It should not exceed that of 1 drop of HCl to 100 c.c. of alcohol, and the decolouration should be carefully watched, as the stain may easily be entirely washed out in this mixture. For this reason, the process is not to be recommended in general; I merely mention it because it is well that the student should be acquainted with the reaction.

This stain is probably by far the most popular of any for

staining in the mass, and there can be little doubt that it is deservedly so.

Baumgarten makes a "Borax-picro-carmine" by adding crystals of picric acid to Grenacher's solution until it assumes a blood-red colour (*Journ. Roy. Mic. Soc.*, 1888, p. 676; 1889, p. 149).

164. Alcoholic Hydrochloric Acid-Carmine.—This reagent is important on account of its allowing a very powerful stain to be obtained by means of very highly alcoholic solutions. That is evidently a very precious quality for work with impermeable objects, such as, for instance, the Arthropoda frequently present.

GRENACHER's formula (*Arch. f. mik. Anat.*, xvi, 1879, p. 468). —To 50 cubic centimetres of alcohol (60 to 80 per cent.) add 3 to 4 drops of hydrochloric acid and a knife-pointful of powdered carmine. Boil for ten minutes. When cool, filter.

The solution may or may not be now ready for use; this depends on the proportions of acid and carmine used, and these proportions cannot be exactly prescribed on account of the variability of commercial carmine. If the solution is found to give in five or ten minutes a diffuse stain (like a borax-carmine stain, *see* No. 163), more hydrochloric acid must be cautiously added drop by drop, and the solution tested with fresh sections until the desired effect is produced. If after some days (or at once) the solution gets a yellow hue, it is a sign that too much HCl has been used, and the excess must be neutralised by cautious addition of ammonia, which will restore the purple tint of the solution.

This is, generally speaking, a nuclear stain. Over-staining and diffusion, should either happen, may be corrected by washing out with alcohol *very slightly acidulated* with HCl. Sections must always be washed in alcohol, not water; and alcohol, not water, must be used for diluting it. Dilute solutions often give results different from those given by concentrated solutions.

The foregoing method of preparation will be found troublesome by those who are not expert at neutralising. The following method, due to PAUL MAYER (quoted from GARBINI's *Manuale per la Technica moderna del Microscopio*, first ed., p. 46), is easy and gives excellent results. Take 100 gr. of alcohol (either absolute or of any weaker grade), 1 or 2 drops of HCl and *an excess* of carmine, and boil until you get a clear solution, taking care that there remain *an excess of carmine*. This ought to give a nuclear stain, without the aid of HCl for washing out.

BRASS (*Zeit. f. wiss. Mik.*, ii, 1885, p. 303) takes 100 c.c. of 70 per cent. alcohol, 15 drops of HCl, and an excess of carmine. An old formula of PAUL MAYER's (*M. T. Zool. Stat. Neapel*, iv, 1883, p. 521; *Journ. Roy. Mic. Soc.* (N.S.), iv, 1884, p. 317), which gives a more powerful stain than the preceding, is as follows:

Four gr. carmine are dissolved in 100 c.c. of 80 per cent. alcohol with the addition of 30 drops of concentrated pure hydrochloric acid, and heated for about half an hour in the water-bath; the solution is filtered whilst still hot, and the superfluous acid is carefully removed by the addition of caustic ammonia, added until the carmine begins to be deposited. This solution stains very rapidly (embryos of lobsters are stained in about a minute) and intensely, though diffusely; the preparations must be washed out with HCl alcohol if a nuclear stain is required.

The heating of alcohol of so high a grade as 80 per cent. being troublesome, not to say dangerous, the process may be modified by dissolving the carmine in 15 c.c. of water acidulated with the HCl, adding 95 c.c. of 85 per cent. alcohol, and then neutralising with ammonia.

If it be desired to dilute any of these solutions it should be done with alcohol, not water, and alcohol should be taken for washing out.

165. Alcoholic Boric Acid Carmine (FRANCOTTE, *Bull. Soc. Belge Mic.*, 1886, p. 48). Carmine 0·4 gr.; boric acid, 5 gr.; water, 25 c.c.; 90 per cent. alcohol, 75 c.c. Boil and filter.

166. Dutilleul's Picro-borax Carmine (*Bull. Sci. Dep. Nord*, xvi, 1885, p. 371).—This is quoted here as having the distinction of being the very worst stain I have ever tried.

CHAPTER XI.

COCHINEAL, HÆMATOXYLIN, AND OTHER ORGANIC STAINS.

A. COCHINEAL.

167. The Use of Cochineal.—What is the use of cochineal? In the first place, it gives us the means of getting a *direct* nuclear stain by means of an *alcoholic* solution. For some purposes, this stain is unrivalled. In the second place, it gives us an aqueous stain that takes the place of alum-carmine, with perhaps a greater richness of differentiation.

168. Alcoholic Cochineal (MAYER's *Mitth. Zool. Stat. Neap.*, ii, 1881, p. 14).—Cochineal in coarse powder is macerated for several days in alcohol of 70 per cent. For each gramme of the cochineal there is required 8 to 10 c.c. of the alcohol. Stir frequently. Filter, and the resulting clear, deep red solution is fit for staining.

The objects to be stained must previously be imbibed with alcohol of 70 per cent., and alcohol of the same strength must be used for washing out or for diluting the staining solution, as water, or alcohol of a different strength, gives rise to turbidity and precipitation of colouring matter (the fluid holding in solution matters that are only soluble in alcohol of exactly that degree of concentration). The washing out must be repeated with fresh alcohol until the latter takes up no more colour. Warm alcohol acts more rapidly than cold. Overstaining seldom happens; it may be corrected by means of 70 per cent. alcohol, containing $\frac{1}{10}$th per cent. hydrochloric or 1 per cent. acetic acid.

Small objects and thin sections may be stained in a few minutes, larger animals require hours or days. In the latter case large quantities of the solution must be employed. Very thin sections and delicate objects are best stained in a very dilute solution.

A nuclear stain, slightly affecting protoplasm. The colour varies with the reaction of the tissues, and the presence or absence of certain salts. The salts of the metals and alkaline earths that are present in the tissues, and that are soluble in alcohol, give rise to colourations of a bluish tone, so that when these are present the effect is that of a hæmatoxylin stain. In the presence of acids of course the precipitation of these blue combinations in the nuclei and protoplasm cannot occur, and therefore tissues of an acid reaction, as well as those free from the salts in question, stain red. Crustacea with thick chitinous integuments are generally stained red, most other organisms blue. The stain is also often of different colours in different tissue elements of the same preparation. Glands or their secretion often stain grey-green. In embryos of *Lumbricus* Kleinenberg found the vessels to stain red, their contents of an intense blue.

Acids lighten the stain and make it yellowish-red. Caustic alkalies turn it to a deep purple.

The best stains are obtained in the case of objects that have been prepared with chromic or picric acid combinations, or with absolute alcohol.* The acids must be carefully washed out before staining, or a diffuse stain will result. The stain is permanent in oil of cloves and balsam.

The object for which this stain was imagined is twofold. Firstly, to obtain an *alcoholic* stain which enables us to do away with the necessity of treating with an *aqueous* fluid objects that have been preserved in alcohol and that are intended for mounting in balsam, aqueous fluids being often most deleterious to delicate structures. Secondly, to obtain a fluid whose high penetrating power allows it to be employed in the case of organisms, such as Arthropoda, whose chitinous investments are but very slightly permeable by aqueous solutions of carmine.

I have treated this stain at considerable length, because I am convinced that it ought to be better known. It is very useful in many cases (Annelids, for instance), and indispensable for Arthropoda.

* Osmic acid preparations stain very weakly unless they have been previously *bleached* (No. 542).

169. Alum Cochineal (PARTSCH, *Arch. f. mik. Anat.*, xiv, 1877, p. 180).—Powdered cochineal is boiled for some time in a 5 per cent. solution of alum, the decoction filtered, and a little salicylic acid added to preserve it from mould.

170. Alum Cochineal (CZOKOR, *Arch. f. mik. Anat.*, xviii, 1880, p. 413).—Seven gr. cochineal and 7 gr. calcined alum are rubbed up together into powder in a mortar, add 700 gr. distilled water, and boil down to 400 gr. When cool, add sufficient carbolic acid to be perceptible by the smell, and filter several times. The violet solution is ready for use, and will keep for six months, after which time it must be filtered again, and a fresh trace of carbolic acid added.

This stain possesses considerable elective faculty, and is stated to stain, in a longer or shorter time, all kinds of tissue, *no matter in what way they may have been hardened.* Nuclei are stained hæmatoxylin colour, and other elements different tones of red, so that the effect is that of a double stain with hæmatoxylin and carmine. Alcohol objects require three to five minutes, chromic objects three to five hours.

The formula known as Klein's cochineal fluid (which appears to have been first published in the *Ann. and Mag. Nat. Hist.*, viii, 1881, p. 232) is identical with that of Czokor.

171. Borden (*The Microscope*, 1888, p. 83) has proposed an alcoholic alum-cochineal, which consists of an alum-cochineal similar to that of Partsch, combined with an equal volume of 95 per cent. alcohol. I doubt the success of this modification.

B. HÆMATOXYLIN.

172. The Use of Hæmatoxylin.—We have the coal-tar colours for staining sections, and we have carmine and cochineal for staining in the mass. What, then, do we want hæmatoxylin for? The answer is, that we sometimes want it for staining in the mass on account of the faculty it has of readily staining tissues that have been treated with chromic and osmic mixtures. This it does in general better than any carmine or cochineal. We want it also for some special purposes, such as staining the *Nebenkern* and achromatic figure of nuclei; and for nerve-researches, and other special histological objects.

173. General Remarks.—None of the solutions of hæmatoxylin are perfectly stable; only one or two are fairly so.

In general, freshly prepared solutions stain badly and diffusely; they ought to be allowed to "ripen" before use. This takes, according to the nature of the solution, a few hours, or days, or months. On the other hand, kept solutions easily go bad, by precipitating, or becoming acid, or becoming mouldy.

Most of the solutions, when in good staining order, have a great tendency to over-stain. Over-stains may be corrected by washing out with weak acids, but this is not favorable to the permanence of the stain. If acids be used, it is well to re-neutralise afterwards with ammonia.

A better plan is perhaps to wash out with alum solution; but this frequently requires great patience.

The stain is fairly permanent in balsam; but is sure to fade a little, and may fade a great deal. In aqueous media it cannot be relied on to keep.

Hæmatoxylin stains in different tones of blue or of red, according to the composition of the staining solution.

According to WATNEY (see *Phil. Trans.*, 1882, p. 1075. KRAUSE, *Intern. Zeit. f. Anat. u. Hist.*, i, p. 154; and M. FLESCH, *Zeit. f. wiss. Mik.*, 1885, p. 358, from whom I quote) the colour is an intense blue if the solution has been made *with freshly prepared alum*, whilst a red tone is obtained if the solution has been made with old alum. The reason of this is, that alum that has been long kept almost always contains free acid. This effect is more especially obtained with solutions in which the proportion of alum is less than a third of the extract of logwood employed. The red solutions exhibit a great affinity for connective tissue, and for the granules of "plasma cells;" whilst the blue solutions show a special affinity for mucin and chromatin.

It has been discovered by LANGHANS (*see* MAX FLESCH, *loc. cit.*) that it is possible to obtain these two elective reactions with one and the same solution. All that is necessary is to stain with the solution of Delafield, mount the preparations in balsam, and expose them for some time to the light. The reaction is not obtained with glycerin mounts.

Aqueous solutions.

174. Delafield's Hæmatoxylin.—(*Zeit. f. wiss. Mik.*, ii, 1885, p. 288. The history of this formula is as follows: It had long been in use in the Institute of Pathology at Heidelberg, when it was communicated by Pfitzner to Flemming, who published it and particularly recommended it, in his *Zellstz., &c.*, p. 388, 1882). Flemming then attributed the formula to Grenacher, and in consequence the stain went for years by the name of "Grenacher's hæmatoxylin." Later on, Flemming discovered that this attribution was erroneous, and attributed the formula to Prudden; and in consequence it was thenceforth known for some time as "Prudden's hæmatoxylin." In 1885 matters were set right by Prudden's explaining that the stain was the invention of Delafield, and publishing the correct formula here quoted (in the formula as published by Flemming, the proportions are somewhat different). It is within the bounds of possibility that after a generation this formula may no longer be quoted as Grenacher's ; but up to the present date the Conservatives remain nearly as numerous as the Reformers).

To 400 c.c. of saturated solution of ammonia-alum add 4 gr. of hæmatox. crist. dissolved in 25 c.c. of strong alcohol. Leave it exposed to the light and air, in an unstoppered bottle, for three or four days. Filter, and add 100 c.c. of glycerin and 100 c.c. of methylic alcohol (CH^4O). Allow the solution to stand until the colour is sufficiently dark, then filter and keep in a tightly stoppered bottle.

This solution keeps well—it may be said to keep for years. It is well to allow it to ripen for at least two months before using it.

This famous solution is certainly one of the best hæmatoxylin stains yet published. It is extremely powerful, and when properly used, very precise. For staining, enough of the solution should be added to pure water to make a very dilute stain ; and even then care should be taken not to leave objects too long in the fluid.

175. Böhmer's Hæmatoxylin (*Arch. f. mik. Anat.*, iv, 1868, p. 345 ; *Aerzt. Intelligenzbl., Baiern*, 1865, No. 38).—Make (A) a solution of pure hæmatoxylin (Ɖj) in absolute alcohol (℥ss), and (B) a solution of alumen depuratum (gr. ij) in water (℥j), (or, (A), hæmatox. crist. 1 part, alcohol 12 parts, and (B), alum 1 part, water 320). For staining, add two or three drops of A to a watch-glassful of B.

The staining solution ought not to be made up at the moment of using, but should be made up beforehand and allowed to ripen for some days. The alcoholic solution of hæmatoxylin may be kept in stock ; it becomes brown, but does not lose its properties.

Washing out may be done with a 0·5 per cent. solution of alum in water.

This formula gives a very fine stain, but is not so certain in its results as Delafield's.

176. Ranvier's Hæmatoxylin (*Comptes rend. Ac. Sc.*, 1882, 2 sem., t. 95, p. 1375).—If solutions prepared according to the formula of Böhmer be kept for some weeks they will be found to furnish an abundant precipitate.

Ranvier re-dissolves this precipitate in 1 per cent. solution of alum, and employs the solution for staining sections of epidermis that has been hardened in bichromate of potash. The sections should remain twenty-four hours in the liquid. Nuclei are then found stained light violet, the granules of eleïdin dark violet.

177. Other Aqueous Alum-Hæmatoxylins.—ARNOLD, see *Quart. Journ. Mic. Sci.*, 1878, p. 86. MITCHELL, see *Journ. Roy. Mic. Soc.*, 1884, p. 811—a complicated method of treating logwood, so as to get rid of the tannin, which is inimical to the preservation of the solutions. But why not take *Hæmatox. crist.*? HICKSON, see *Quart. Journ. Mic. Sci.*, 1885. p. 244 —a still more complicated method of arriving at the same mare's nest. COOK, see *Journ. of Anat. and Phys.*, 1879, p. 140,—a sulphate of copper solution.

178. Heidenhain's Hæmatoxylin (*Arch. f. mik. Anat.*, 1884, p. 468, and 1886, p. 383).—Stain for twelve to twenty-four hours in a $\frac{1}{3}$ per cent. solution of hæmatoxylin in pure water (distilled water only should be used). Soak the objects for the same length of time in a 0·5 per cent. solution of neutral chromate of potash. Wash out the excess of chromate with water, and treat further as desired.

The above is a slightly modified form of the original process, in which staining was done in a stronger hæmatoxylin solution (0·5 to 1 per cent.), and bichromate was used for washing out instead of neutral chromate. The more recent process gives a sharper chromatin stain.

The stain succeeds best with alcohol or picric acid objects, but it will succeed with chromic objects if they have been very well washed.

Objects that have been fixed in corrosive sublimate ought to be very carefully washed out with water (many hours in running water), as neutral hæmatoxylin forms a black precipitate with the excess of sublimate that remains after washing out with alcohol (*see* TORNIER, in *Arch. f. mik. Anat.*, 1886, p. 181).

The stain is black to grey (hæmatoxylin forming with chromic salts a black compound). It is a sharp stain, remarkably rich in detail.

The process is one well adapted to staining in the mass. Perhaps its greatest advantage lies in the fact that you can decolour the objects to any extent by prolonging the washing in the chromate.

The method may be varied by washing out after staining

with alum solution (1 per cent.) instead of a chromate. In this case the stain will be blue.

179. Apáthy's Modification of Heidenhain's Process (*Zeit. f. wiss. Mik.*, v, 1, 1888, p. 47). This is an *alcoholic* method. Stain in a 1 per cent. solution of hæmatoxylin in 70 or 80 per cent. alcohol. Wash out (for "thin" sections, *i. e.* sections of 10 to 15 μ, half the time of staining—for "thicker" sections of 25 to 40 μ twice the time of staining) in 1 per cent. solution of bichromate of potash in 70 to 80 per cent. alcohol.

The bichromate solution is conveniently prepared by mixing one part of a 5 per cent. aqueous solution with about four parts of 80 to 90 per cent. alcohol. The mixture should be made immediately before using, and should be kept from the light (light precipitates it) during the process of decolouration, and should also be changed for fresh several times during the process. After the differentiation of the colour has been accomplished, the objects should be thoroughly washed (still in the dark) in several changes of 70 per cent. alcohol.

Preparations made in this manner are more transparent and better preserved than those made by Heidenhain's process.

For staining celloidin series of sections, Apáthy also (*Zeit. f. wiss. Mik.*, vi, 2, 1889, p. 170) recommends the following procedure : Stain in the hæmatoxylin solution as above for ten minutes; then remove the excess of hæmatoxylin fluid from the sections by means of blotting-paper, and bring the series for five to ten minutes into 70 per cent. alcohol containing only a few drops of a strong (15 per cent.) solution of bichromate. This must be done in the dark. If the hæmatoxylin be not removed with blotting-paper as described, the celloidin will take the stain. The sections should appear steel-blue to steel-grey.

180. Weigert's Hæmatoxylin.—This method, which is the inverse of Heidenhain's, is, with some unimportant exceptions, only applicable to nerve-tissues, and therefore will be described in the chapter on Nerve Methods in Part II.

181. Minot's Hæmatoxylin Methods (*Zeit. f. wiss. Mik.*, iii, 2, 1886, p. 177). Minot's account of these is as follows :—" They may be employed with sections of tissues hardened in various ways, and need not be confined to Müller's fluid or chromic acid specimens (as in Weigert's process). The sections are soaked first in a salt solution for ten to fifteen minutes. The following aqueous salt solutions seem to be the most valuable: alum, 2 per cent.; chromic acid, 1 per cent.; bichromate of potassium, 5 per cent.; acetate of copper nearly saturated. After soaking in one of these, the section is passed quickly through distilled water, and placed at once in Weigert's hæmatoxylin (1 part of the crystals in 10 parts alcohol plus 90 parts water), and may be left a short time for direct colouration, then washed and mounted, or a longer time until they become black and are to be washed out by Weigert's iron solution (water 100, borax 2, ferricyanide of potassium 2½). The sections ought to be moved about constantly in the iron solution, otherwise the colour will be extracted irregularly. The copper hæmatoxylin goes out very rapidly, so that with that stain it is better to dilute the iron solution with twice its bulk of water before placing the sections in it. After the iron solution, the sections must be washed very thoroughly in water, to avoid further fading out, from which one is not entirely secure until the sections are actually mounted in balsam.

These methods are all merely modifications of Weigert's, Heidenhain's, and Böhmer's methods."

Glycerin Solutions.

182. Ehrlich's Acid Hæmatoxylin (*Zeit. f. wiss. Mik.*, 1886, p. 150).—The ordinary (alum) hæmatoxylin staining solutions easily decompose, giving rise to a blue precipitate which is formed by the splitting up of the alum into free sulphuric acid and a basic, lake-forming compound of alumina. By adding to a solution an appropriate acid this decomposition may be prevented. The end may be attained by acetic acid. Take—

Water	100 c.c.
Absolute alcohol	100 c.c.
Glycerin	100 c.c.
Glacial acetic acid	10 c.c.
Hæmatoxylin	2 grammes.
Alum in excess.	

Let the mixture ripen in the light until it acquires a dark red colour. It will then keep, with a perfectly constant staining power, for years, if kept in a well-stoppered bottle. Sections are stained in a few minutes. The stain is also very appropriate for staining in the mass, as over-staining does not occur.

In order to get a blue stain with this acid solution, the stained objects should be washed out with common drinking water, which is always slightly alkaline, and not with distilled water.

For double-staining, either "acid colouring matters"

("Farbsäuren") such as eosin, or "basic colouring matters"—("Farbbasen"), may be added to the solution.

This is one of the most important of the alum hæmatoxylins.

183. Renaut's Glycerin Hæmatoxylin ("Glycérine Hématoxylique"; *Arch. de Physiol.*, 1881, p. 640).—Make a saturated solution of alum in strong glycerin. Add drop by drop about a quarter of a volume of concentrated solution of hæmatoxylin in alcohol. (If you add an excess of hæmatoxylin the liquid will become turbid, and you must then add more solution of alum in glycerin, until the turbidity disappears.) Filter, and leave the solution exposed to the light and air for some weeks, until it can be perceived by the smell that it no longer contains any alcohol. Filter.

Sections are stained in this solution in a few minutes. But it can be used in another way, which is an original and really valuable method. If sections, or other objects, be mounted in a drop of the solution, they will, after a few weeks, be found stained, and the glycerin decoloured. Such objects keep the stain for years.

Alcoholic Solutions.

184. Kleinenberg's Hæmatoxylin (*Quart. Journ. Mic. Sci.*, lxxiv, 1879, p. 208).—Prepare a saturated solution of calcium chloride in 70 per cent. alcohol, with the addition of a little alum; after having filtered, mix a volume of this with from 6 to 8 volumes of 70 per cent. alcohol. At the time of using the liquid pour into it as many drops of a concentrated solution of hæmatoxylin in absolute alcohol as are sufficient to give the required colour to the preparation of greater or less intensity, according to desire.

I find it better not to make up the staining solution at the time of using, but to prepare it twenty-four or forty-eight hours beforehand, so that it may "ripen" before using.

At Naples, according to Mayer (*Mitth.*, ii, 1881, p. 13), the solution of calcium chloride is used *saturated* with alum.

Mayer further states that the object of the chloride of calcium is explained by Kleinenberg to be the setting up of diffusion currents between the alcohol in the tissues and the external staining medium, so as to facilitate the penetration of the latter.

Mayer points out that by the reaction of alum and calcium chloride there is formed a precipitate of sulphate of lime, and that it will therefore probably be found better to employ chloride of aluminium in the place of alum.

A powerful, nuclear stain. The stain is permanent, *provided (Mayer) that the tissues have been perfectly freed from acid before staining*. The solution itself is not permanent. A successfully prepared fresh solution should be of a violet colour with a decided touch of blue, and should not be at all reddish. If it becomes reddish after standing for some time that is generally because it has become somewhat acid, and the fault may be corrected by holding over the mouth of the bottle containing the solution the stopper of an ammonia bottle; then on shaking up with the solution the small quantity of vapour of ammonia given off from the stopper, the proper colour is generally regained.

Small objects are best stained slowly with a very dilute solution. If it be required to dilute a solution already prepared for staining this should not be done with alcohol, which may easily cause precipitates to form on the tissues, but with the above-described solution of alum in calcium chloride solution. Over-stains should be washed out with acidulated alcohol. Either oxalic or ($\frac{1}{2}$ per cent.) hydrochloric acid may be used, and the specimens allowed to remain in them until they begin to acquire a reddish hue. The acid is then removed by pure alcohol, which restores the pure blue of the stain.

For large or impermeable objects immersion for days in a very strong solution may be necessary for staining. Osmium and chromic acid objects stain sufficiently.

185. Dippel's Hæmatoxylin (DIPPEL, *Das Mikroskop*, 1882, p. 719; *Zeit. f. wiss. Mik.*, i, 1883, p. 95). In accordance with the recommendation quoted in the last paragraph, Dippel makes a saturated solution of chloride of aluminium in alcohol, dilutes it with six to eight volumes of 70 per cent. alcohol and adds the necessary quantity of alcoholic solution of hæmatoxylin.

186. Acid Alcoholic Hæmatoxylin (Anonymous, *St. Louis Med. and Surg. Journ.*, 1888, p. 165; *Journ. Roy. Mic. Soc.*, 1888, p. 517).—One part of saturated solution of calcium chloride in proof spirit added to eight parts of a similar solution of alum. Extract of logwood added to the mixture until it no longer dissolves freely. Allow the solution to stand for a few days, decant, and to every hundred parts add eighty parts of 1 per cent. acetic acid. Let stand a day and filter.

187. Cuccati's Iodine Hæmatoxylin (*Zeit. f. wiss. Mik.*, v, i, 1888, p. 55).—Dissolve 25 grammes of chemically pure potassic iodide in 25 c.c. of distilled water. Pour this solution gradually and with constant agitation into 75 c.c. of absolute alcohol contained in a stoppered bottle. Close the bottle thoroughly. Rub up in a mortar 75 cg. of crystallised hæmatoxylin with 6 grammes of chemically pure roche alum,* and add 3 c.c. of the iodine solution. Keep the mixture agitated, and add gradually the rest of the iodine solution, then replace the whole in a well-stoppered bottle. Agitate for some time, in order to get the alum to dissolve, and let stand for ten to fifteen hours. Then shake well and filter, taking the usual precautions against evaporation of the alcohol, and preserve in a well-stoppered bottle.

Objects should be left in the liquid for ten hours, then well washed in water, and mounted in glycerin or washed in alcohol and mounted in balsam.

The solution is stated to be perfectly stable, and to give a pure chromatin stain, and to be well adapted for staining in the mass, as it never over-stains.

Other Organic Stains.

188. Alizarin, so far as I am aware, is only used for Nerve Centres. See Part II.

189. Purpurin is useful for cartilage and muscle. It is soluble in a boiling aqueous solution of alum, from which it normally precipitates on cooling, but may be prevented from so doing by the addition of a certain proportion of alcohol. The employment of an alum solution as a vehicle for the colouring matter has the advantage, at least so far as cartilage is concerned, of fixing the cellular elements at the same time that they are

---

* Roche alum, or Roman alum (*allume di rocca, alun de roche, alumen rubrum verum*, and other synonyms), is an alum originally imported from Cività Vecchia, and much esteemed by dyers from being nearly free from iron-alum. That now sold for it in England is ordinary alum coloured with Venetian red, Armenian bole, or rose-pink (*alumen rubrum spurium*). See COOLEY's *Cyclopædia of Pract. Receipts*, s. v. " Alum, Roman."

stained. (Ranvier found that alum in a solution of 5—1000 was the best of all fixing agents for cartilage-cells, *Traité*, p. 279.)

RANVIER's *formula* (*Traité Technique*, p. 280).—200 grammes of water and 1 of alum are boiled in a porcelain capsule; purpurin rubbed up in water is added, and the boiling continued. The purpurin being dissolved to saturation (this is ensured by taking care to have an undissolved excess in the capsule), the solution is filtered hot into a flask containing 60 c.c. of alcohol (36° Cartier, = 90 per cent).

There is thus obtained a solution of an orange-rose colour, presenting a marked degree of fluorescence.

(As regards the quantity of alcohol to be taken, Duval writes that it should always be one fourth in volume of the total mixture, 'Précis de Technique Histologique,' p. 221.)

The solution does not keep well for more than a few weeks.

Sections of fresh cartilage are to be placed in a small quantity (only a few cubic centimetres) of the solution, and after remaining there twenty-four to forty-eight hours, are washed in water and mounted in glycerin. The stain is nuclear, the matrix remaining almost colourless. Duval (l. c.) states that this stain has a special selective action on sections of central nervous system (especially spinal cord) obtained from tissues hardened in bichromate of ammonia (2—1000), and mounted, after staining for forty-eight hours, in Canada balsam. The nerve-cells and processes, axis cylinders, and fibres of connective tissue, are unstained; but the nuclei of connective tissue and of the capillaries are stained red.

GRENACHER's formula (*Arch. f. mik. Anat.*, xvi, 1879, p. 470). —In 50 cubic centimetres of glycerin (pure or diluted with very little water) dissolve from 1 to 3 per cent. powdered alum; added a knife-pointful of purpurin, and boil. (Alcohol must *not* be added.) Let the orange-coloured fluorescent solution stand for two or three days, and then filter.

A nuclear stain; ten to thirty minutes generally suffice to produce good staining. The solution is stable, which Grenacher finds that Ranvier's solution is not, the latter precipitating after a few days.

190. Indigo.—Indigo is employed in histology in the form of solutions of so-called indigo-carmine, or sulphindigotate of soda or potash. The simple aqueous solution gives a diffuse stain, and is therefore not capable of being usefully employed *alone*. It is, however, of great use when employed

to bring about a *double*-stain in conjunction with carmine. Though it has no selective preference for nuclei or protoplasm, it possesses to a high degree the property of imparting different hues and intensity of stain to different tissues; and the nuclei being brought out by carmine, preparations are obtained of a diagrammatic clearness that is not afforded by carmine alone.

Indigo-carmine is found in commerce. The reader who may desire to prepare it himself will find the necessary directions in *Arch. f. mik. Anat.*, x, 1874, p. 32, and in *Journ. Roy. Mic. Soc.*, ii, 1879, p. 614.

191. Thiersch's Oxalic Acid Indigo-carmine (see *Arch. f. mik. Anat.*, i, 1865, p. 150).

192. Tincture of Saffron (H. BLANC, *Zool. Anzeig.*, 129, 1883, p. 23).— Dissolve 5 grammes of saffron in 15 c.c. of absolute alcohol; allow the solution to settle for a few days, filter, and dilute with one half of water. After staining wash out to the desired degree with 80 per cent. alcohol, then dehydrate with absolute alcohol, and mount in balsam.

LEVEN (*The Microscope*, ix, 1889, p. 88; *Journ. Roy. Mic. Soc.*, 1889, p. 467) has lately been using saffron for the study of the regeneration of muscle. He stains in a solution containing—saffron, 1 part; absolute alcohol, 100 parts; water, 200, and washes out in acidulated alcohol (0·5 per cent. HCl). Karyokinetic figures dark red, muscle nuclei pale with dark red nucleoli. Leucocytes stain strongly.

193. Orchella (Orseille) (WEDL, *Arch. f. path. Anat.*, lxxiv, p. 143; *Journ. Roy. Mic. Soc.*, ii, 1879. For an account of this substance *vide* COOLEY's *Cyclopædia, sub voce* " Archil ").—French orchella extract, from which the excess of ammonia has been removed by gentle warming in a sand-bath, is poured into a mixture of absolute alcohol 20 c.c., acetic acid (concentrated, of 1·070 sp. gr.) 5 c.c., and water 40 c.c., until a saturated dark red stain is obtained, which must then be filtered once or twice. Sections are washed with water, drained, and treated with the stain. Mount in levulose. A protoplasmic stain, nuclei remaining colourless. Connective-tissue cells stain deeply, the intercellular substance less deeply. Epithelia, if horny or calcareous, are not stained. The basic substance of bone and teeth take the stain, and so do ganglion-cells and their processes.

This colour ought to be useful for double-staining. FOL (*Lehrb.*, p. 192) advises staining for an hour in Wedl's solution, then rinsing with alcohol, and staining in a complementary stain.

194. Orcin (ISRAËL, *Virchow's Archiv*, cv, 1886, p. 169; *Journ. Roy. Mic. Soc.*, 1887, p. 514).—Orcin is a vegetable dye which unites in itself the staining properties of the basic and acid stains, and also the combination of two contrast colours. Israël stains sections in a saturated acetic acid solution, washes in distilled water, and passes rapidly through absolute alcohol to thick cedar-oil, in which the preparations remain definitively mounted. Nuclei blue, protoplasm red.

195. Kernschwarz (PLATNER, *Zeit. f. wiss. Mik.*, iv, 3, 1887, p. 350; *Journ. Roy. Mic. Soc.*, 1888, p. 675). Kernschwarz is a black liquid of unknown composition, prepared in Russia. It may be obtained from Grübler (address § 93). Sections (sections only, this colour behaving like safranin, for instance) may be stained in a tolerably strong dilution of the concentrated liquid, and washed out (it may be for some hours) in an alkaline aqueous liquid. Dilute ammonia will do, but it is better to take a not quite saturated solution of carbonate of lithium (you may take a saturated solution, and dilute it with three, four, or more volumes of water). The result is a nuclear stain in the cytological sense; nuclear figures of division are stained deeply, resting chromatin less deeply or not at all, cytoplasm unstained or faintly grey. A peculiarity of this stain, on which much stress was laid by Platner in his first announcement of the colour, is that it stains also the *Nebenkern* (which of course safranin and the other anilins do not do).

After some experimentation I feel bound to say that I do not think Kernschwarz has the importance seemingly claimed for it by Platner. It certainly stains the Nebenkern, but it does so less rather than more effectively than hæmatoxylin. Platner seems to have come to this conclusion himself, his latest work on the Nebenkern making no further mention of Kernschwarz, and having been done apparently entirely with hæmatoxylin. And as a nuclear stain, Kernschwarz seems to me inferior to hæmatoxylin, to say nothing of the chromatin-staining anilins. Only in the event of its being found available for staining in the mass (which may be possible, though I have not succeeded in it) will Kernschwarz, as it seems to me, be found to be an important reagent.

196. Other Stains.—Litmus, red cabbage, bilberry juice ("myrtillus"), black currant juice ("ribésine"), and walnut juice ("nucina"), have been recommended by Lawson Tait, Lavdowsky, Fol, and Léon respectively. They do not appear to call for further notice here.

CHAPTER XII.

METALLIC STAINS (IMPREGNATION METHODS).

197. The Characters of Impregnation-stains.—Impregnations are distinguished as *negative* and *positive*. In a negative impregnation, intercellular substances alone are coloured of a deep black or brown or violet, according to the method employed; the cells themselves remaining colourless or very lightly tinted. In a positive impregnation, the cells are stained and the intercellular spaces are unstained. (This explanation is the more needful as a directly contrary statement is made in a recent *Lehrbuch*.)

Negative impregnation is *primary* because it is brought about by the direct reduction of a metal in the intercellular spaces. Positive impregnation is *secondary* (in the case of silver nitrate), because it is brought about by the solution in the liquids of the tissues of the metallic deposit formed by a primary or negative impregnation, and the consequent staining of the cells by the new solution of metallic salt thus formed. These secondary impregnations take place when the reduction of the metal in the primary impregnation is not sufficiently energetic (see on these points: HIS, *Schweizer Zeit. f. Heilk.*, ii, Hft. i, p. 1. GIERKE, *Zeit. f. wiss. Mik.*, i, p 393. RANVIER, *Traité*, p. 107).

There still exists considerable obscurity as to the nature of the black or brown deposit formed in the intercellular spaces in cases of primary impregnation with a silver salt; v. Recklinghausen held that the silver salt combined with a hypothetical intercellular cement-substance (*Kittsubstanz*), forming a compound that blackens under the influence of light. Other authors refuse to believe in the intercellular cement, and hold, either that the coloured lines represent stained cell-membranes, or that the metallic salt combines with the albuminous and saline liquids that surround the cells and is precipitated in simple intercellular *spaces*. SCHWALBE (*Arch. f. mik. Anat.*, vi, 1870, p. 5) thinks that two cases should be distinguished; the *black* lines that are obtained by the action of very weak solutions for a very short time being due to a true precipitate formed by reduction of metal in the inter-cellular liquids; the *brown* lines that are obtained by exposing tissues for a longer period to the action of more concentrated solutions being due to the formation of a compound of metal and cement-substance that becomes brown on exposure to light. (For the history of these questions, *see* GIERKE's *Färberei zu mikroskopischen Zwecken.*)

Silver.

198. Silver Nitrate.—This is the most commonly used salt of silver. The general principles of its employment are so well stated by RANVIER (*Traité*, p. 105) that I cannot do better than abstract his account.

Silver nitrate may be employed either in solution or in the solid state. The latter method is the less frequently employed, but is easy and gives good results. It is useful for the study of the cornea and of fibrous tissue, but is not suitable for epithelia. For the cornea, for instance, proceed as follows. The eye having been removed, a piece of silver nitrate is quickly rubbed over the anterior surface of the cornea, which is then detached and placed in distilled water; it is then brushed with a camel's-hair brush in order to remove the epithelium. The cornea is then exposed to the action of light. On subsequent examination it will be found that the silver nitrate which was dissolved by the liquid that bathes the surface of the cornea has traversed the epithelium and soaked into the fibrous tissue, on the surface of which it is reduced by the action of light. The cells of the tissues will be found unstained.

Silver nitrate is generally employed in solution in the following manner: A 1 per cent. solution is taken, to which 2, 3, or 4 volumes of water are added according to circumstances. The mode of employment varies in its details according to circumstances, a point which it is very important to observe.

In the case of a membrane such as the epiploön, the membrane must be stretched like a drum-head over a porcelain dish,* and washed with distilled water in order to remove the albuminates and white blood-corpuscles that are found on its surface; it is then washed with the solution of silver nitrate. In order to obtain a powerful stain it is necessary that this part of the operation be performed in direct sunlight, or at least in a very brilliant light. As soon as the tissue has become white and has begun to turn of a blackish grey, the membrane is removed, washed in distilled water, and mounted on a slide in some suitable examination medium.

If the membrane were left in the water, the cells would become detached and would not be found in the finished preparation.

If the membrane had not been stretched as directed, the silver would be precipitated not only in the intercellular spaces, but in all the small folds of the surface, and the forms of the cells would be disguised.

If the membrane had not been washed with distilled water before impregnation, there would have been formed a deposit of silver on every spot on which a portion of an albuminate was present, and these deposits might easily be mistaken for a normal structure of the tissue. It is thus that very often impurities in the specimen have been described as stomata of the tissue.

If the solution be taken too weak, for instance, 1·500 or 1·1000, or if the light be not brilliant, a *general* instead of an *interstitial* stain will result; nuclei will be most stained, then protoplasm and the intercellular substance will contain but very little silver.

In general, in a good " impregnation," the contents of cells, and especially nuclei, are quite invisible.

Ranvier notes that when tissues are to be impregnated by immersion, they should be constantly *agitated* in the silver-bath in order to avoid the formation on their surfaces of deposits of chlorides and albuminates of silver which would give rise to deceptive appearances.

Impregnation with silver may be followed by treatment with with picro-carmine (or other carmine stain), which will bring

* The Hoggans' histological rings, for which *see* below, § 201, will be found much more convenient.

out the nuclei, provided the impregnation has not been overdone.

It should be noticed that impregnations only succeed with *fresh* tissues, and cannot be made to succeed with tissues preserved in any way.

199. Silver Nitrate. The Solutions to be Employed (RANVIER).—The solutions generally employed by Ranvier vary in strength from 1·300 to 1·500. Thus 1·300 is used for the epiploön, pulmonary endothelium, cartilage, tendon, whilst a strength of 1·500 is employed for the study of the phrenic centre, and for that of the epithelium of the intestine. For the impregnation of the endothelium of blood-vessels (by injection) solutions of 1·500 to 1·800 are taken.

M. DUVAL (*Précis*, p. 229) recommends solutions of 1, 2, or at most 3 per cent.

V. RECKLINGHAUSEN used, for the cornea, a strength of from 1—400 or 1—500 (*Die Lymphgefässe*, &c., Berlin, 1862, p. 5).

ROBINSKI (*Arch. de Physiol.*, 1869, p. 451) used solutions varying between 0·1 and 0·2 per cent., which he allowed to act for thirty seconds.

REICH (*Sitzb. d. wien. Acad.*, 1873, iii, Abth., April; *Zeit. f. wiss. Mik.*, i, p. 397) takes solutions of from 1—600 to 1—400, for the study of the endothelium of vessels by injection.

ROUGET (*Arch. de Physiol.*, 1873, p. 603) employed solutions as weak as 1—750 or even 1—1000, exposing the tissues to their action several times over, and washing them with water after each bath.

The HERTWIGS take, for marine animals, a 1 per cent. solution (*Jen. Zeit. f. Naturk.*, xvi, pp. 313 and 324).

The HOGGANS (*Journ. of Anat. and Physiol.*, xv, 1881, p. 477) take, for lymphatics, a 1 per cent. solution.

TOURNEUX and HERMANN (ROBIN'S *Journal de l'Anat.*, 1876, p. 200) in their fine studies on the epithelia of invertebrates, employed a solution of 3·1000 strength, and in some cases weaker solutions. The tissues were allowed to remain in the silver-bath for one hour, and were washed out with alcohol of 36° strength.

HOYER (*Arch. f. mik. Anat.*, 1876, p. 649) takes a solution of nitrate of silver of known strength, and adds ammonia to it until the precipitate that is formed just redissolves, then dilutes the solution until it contains from 0·75 to 0·50 per cent. of the salt.

This *ammonio-nitrate* solution is intended principally for the impregnation of the endothelium of vessels by injection, but can also be used for the impregnation of membranes by pouring on. It has the advantage of impregnating absolutely nothing but endothelium or epithelium; connective tissue is not affected by it. It is also said to give a sharper localisation of the stain than the ordinary solutions.

ALFEROW (*Arch. de Physiol.*, 1874; *Laboratoire d'histologie du Collège de France*, 1874, p. 258. DUVAL, *Précis*, p. 230) recommends the soluble silver salts of organic acids, viz. the picrate, lactate, acetate, and citrate, as giving better results than the nitrate. He employs them in solutions of 1·800, and adds to the solution employed for staining a small quantity of the acid of the salt taken (10 to 15 drops of a concentrated solution of the acid to 800 c.c. of the solution of the salt). The object of the free acid is to decompose the precipitates formed by the action of the silver salt on the chlorides, carbonates, and other substances existing in the tissues, leaving only the albuminate, which is a more resistant compound.

200. Silver Nitrate. Reduction.—Reduction may be effected in other media than distilled water.

V. RECKLINGHAUSEN washed his preparations in salt solution before exposing them to the light in distilled water (*Arch. f. path. Anat.*, xix, p. 451). Physiological salt solution (0·75 per cent.) is commonly used for these washings.

MÜLLER (*Arch. f. path. Anat.*, xxxi, p. 110), after impregnation by immersion for two or three minutes in a 1 per cent. solution of nitrate of silver, in the dark, adds to the solution a small quantity of 1 per cent. solution of iodide of silver (dissolved by the aid of a little iodide of potassium). After being agitated in this mixture, the preparations are washed with distilled water, and exposed to the light for two days in a 1 per cent. solution of nitrate of silver (*see also* GIERKE in *Zeit. f. wiss. Mik.*, i, 1884, p. 396).

ROUGET (*Arch. de Physiol.*, 1873, p. 603) reduces in glycerin.

SATTLER (*Arch. f. mik. Anat.*, xxi, p. 672) exposes to the light for a few minutes in water acidulated with acetic or formic acid. THANHOFFER (*Das Mikroskop.*, 1880) recommends this method. He employs a 2 per cent. solution of acetic acid.

KRAUSE brings his preparations, after washing, into a light red solution of permanganate of potash. Reduction takes place very quickly, even in the dark. The method does not always succeed (*see* GIERKE in *Zeit. f. wiss. Mik.*, i, 1884, p. 400).

OPPITZ brings his preparations for two or three minutes into a 0·25 or 0·50 per cent. solution of chloride of tin. Reduction takes place very rapidly (GIERKE, l. c.).

JAKIMOVITCH (*Journ. de l'Anat.*, xxiii, 1888, p. 142; *Journ. Roy. Mic. Soc.*, 1889, p. 297) brings nerve preparations, as soon as they have become of a dark brown colour, into a mixture of formic acid, one part, amyl alcohol, one part, and water 100 parts. The objects exposed to the light in this mixture for two or three days at first become brighter, a part of the reduced silver being dissolved; hence the mixture must be renewed from time to time. When all the silver has dissolved, a darker colour is permanently assumed. The nerve-cells are left in this mixture for five to seven days.

200a. After-blackening.—LEGROS (*Journ. de l'Anat.*, 1868, p. 275) washes his preparations after reduction in hyposulphite of soda which prevents after-blackening. According to DUVAL (*Précis*, p. 230) they should be washed for a few seconds only in 2 per cent. solution, and then in distilled water.

201. The Hoggans' Histological Rings are vulcanite rings made in pairs, of which one ring just fits into the other, so as to clip and stretch pieces of membrane between them. They will be found described and figured in *Journ. Roy. Mic. Soc.*, ii, 1879, p. 357, and in ROBIN's *Journ. de l'Anat.*, 1879, p. 54. They may be obtained, in sets of various sizes (that of seven-eighths of an inch being the most convenient for 3 × 1 slides) of Burge and Warren, 42, Kirby Street, Hatton Garden, London, E.C., price ten shillings the dozen pairs.

This useful little apparatus has lately been reinvented by Eternod (*Zeit. f. wiss. Mik.*, iv, 1, 1887, p. 39), and is made according to his designs by Demaurex, bandagiste, Fusterie, Geneva (Switzerland).

The Hoggans' histological rings were described by me in the 1st edition of this work, p. 375, and in the *Traité d. Méth. Techn.*, LEE et HENNEGUY, p. 138; so that I am in no way responsible for the waste of time involved in this reinvention, on the part of the inventor, and the editors of the *Zeit. f. wiss. Mik.*, and other journals.

202. Silver-impregnation of Marine Animals.—On account of the considerable quantity of chlorides that bathe the tissues of marine animals, these cannot be treated *directly* with nitrate of silver.

HERTWIG (*Jen. Zeit.*, xiv, 1880, p. 324) recommends fixing them with a weak solution of osmic acid, then washing with distilled water until the wash-water gives no more than an insignificant precipitate with silver nitrate, and then treating for six minutes with 1 per cent. solution of silver nitrate.

HARMER (*Mitth. Zool. Stat. Neapel*, v, 1884, pp. 44 to 56) has discovered that many marine animals will live for some time (half an hour) in a 5 per cent. solution of nitrate of potash in distilled water. By washing them in this way, they may be freed from the great part of their chlorides, and may then be treated with silver nitrate in the usual way. This method gave good results with *Loxosoma* and *Pedicellina*, with Medusæ, Hydroids, *Sagitta* and *Appendicularia*.

VOSMAER has been able by this means to demonstrate the epithelium of *Chondrosia* and *Thenea*, which Sollas was unable to see; and MEYER has obtained good results with annelids and ova of Teleostea. Few animals resist the action of nitrate of potash so well as *Loxosoma* and *Pedicellina*, but die in the solution in a few minutes. Their tissues, however, suffer but little change and give good impregnations. Harmer thinks that for these animals other solutions having the same density as sea-water might be substituted for the nitrate of potash, and recommends a 4·5 per cent. solution of sulphate of soda.

203. Double-staining Silver-stained Tissues.—The nuclei of tissues impregnated with silver may be stained with the usual reagents, provided that solutions containing free ammonia be avoided, as this would dissolve out the silver. These stains will only succeed, however, with successful negative impregnations; as nuclei that have been impregnated will not take the second stain.

Impregnation with silver may be followed by impregnation with gold. In this case, the gold generally substitutes itself for the silver in the tissues, and though the results are sharp and precise, the effect of a double-stain is not produced.

Gold.

204. The Characters of Gold Impregnations.—Gold chloride differs from nitrate of silver in that it generally gives *positive* (§ 197) impregnations only. It only gives negative images, so far as I know, when caused to act on tissues that have first received a negative impregnation with silver, the gold substituting itself for the silver. In order to obtain these images you first impregnate, very lightly, with silver; reduce; treat for a few minutes with a 0·5 per cent. solution of gold chloride; and reduce in acidulated distilled water.

This process, however, is in but little use, and except for the staining of cytoplasm for cytological researches, and for certain special studies on the cornea, and on connective tissue, the almost exclusive function of gold chloride is the impregnation of nervous tissue. For this tissue, gold chloride exhibits a remarkable selectivity, in virtue of which it justly ranks as a most valuable reagent for the study of nerve-end-organs and the distribution of nerves.

For all the objects above named, gold chloride is capable of furnishing preparations that for beauty and clearness cannot be surpassed if even they can be equalled by any other means. A successful gold preparation shows at a glance, with diagrammatic clearness, a wealth of minute detail which perhaps can only be painfully glimpsed by other means. But not every gold preparation *is* successful. I think there is no use in blinking the fact that very few are successful (one of the most experienced authorities in the matter told me lately that, as to nerve-end-organs at all events, one preparation in ten thousand is successful). I took up in the first edition of this work the doubtless unpopular position that "with all possible precautions gold chloride is uncertain in its action, and that the results obtained by means of it need to be controlled by the employment of other methods," and illustrated that position at considerable length. Time has only confirmed in me the opinion there expressed. It appears to me superabundantly evident that the very best gold preparations give images that are only worthy of credence as to what they show, and furnish absolutely no evidence whatever as to the non-existence of anything that they do not show; for you can never be sure that the imbibition of the salt has not capriciously failed, or its reduction capriciously stopped at any point. That the images frequently *do* stop capriciously short in the representation of reality there is abundant evidence. One such case has been treated by me *ex professo* in *Recueil Zool. Suisse*, i, 1884, p. 685 (*Les organes chordotonaux des Diptères, et la méthode du chlorure d'or*).

The authors of some of the methods about to be described claim for them that they give permanent preparations. I warn the reader against indulging in the hope that, with all possible precautions, his preparations will retain their beauty for more than a few weeks. A successful gold preparation is certainly a thing of beauty, but is exactly the opposite of a joy for ever. The able histologist whose experience I have taken the liberty of quoting above tells me that " as to permanence, they are

"Like the snowfall on the river."

205. The Two Types of Method.—Gold methods may be divided into two groups. The one, chiefly concerned with the study of peripheral nerves or nerve-end-organs, is characterised by employing either perfectly fresh tissues or tissues that have been subjected to a special treatment by organic acids. The other, concerned with the study of nerve-centres, is characterised by the employment of tissues hardened in the usual way.

The hitherto classical rule, that for researches on nerve-endings the tissues should be taken perfectly fresh, seems not to be valid for all cases. For DRASCH (*Sitzb. k. k. Acad. Wiss. Wien*, 1881, p. 171, and 1884, p. 516; and *Abhand. math.-phys. Cl. d. K. Sach. Ges. d. Wiss.*, xiv, No. 5, 1887; *Zeit. f. wiss. Mik.*, iv, 4, 1887, p. 492) finds that better results are obtained with tissue that have been allowed to lie after death for twelve, twenty-four, or even forty-eight hours in a cool place. He even suspects that the function of the organic acids in the methods inspired by Löwit's method, is to bring the tissues into somewhat the state in which they are naturally found at a certain moment of post-mortem process—a state, namely, in which the nerves have a special susceptibility for impregnation with gold.

206. Cohnheim's Method (*Virchow's Arch.*, Bd. xxxviii, pp. 346—349; *Stricker's Handb.*, p. 1100).—This, the archetype of the gold methods, was as follows: Fresh pieces of cornea (or other tissue to be operated on) are put into solution of chloride of gold of 0·5 per cent. strength until they are thoroughly yellow, and then exposed to the light in water acidulated with acetic acid until the gold is thoroughly reduced, which happens in the course of a few days at latest. They are then mounted in acidulated glycerin.

The method in this, its primitive form, often gave splendid results, but was very uncertain, giving sometimes a nuclear or protoplasmic stain, sometimes an extra-cellular impregnation similar to that of nitrate of silver. And the preparations thus obtained are anything but permanent.

207. Löwit's Method.—The principle of this process is that in order to facilitate the penetration of the gold and its subsequent reduction in the tissues, the tissues are made to swell up by treatment with formic acid before being brought into the gold bath, and formic acid is employed to assist the reduction after impregnation.

The following directions as to this method, which may serve as a type of the modern methods of research on nerve-endings, are taken from Fischer's paper on the corpuscles of Meissner (*Arch. f. mik. Anat.*, xii, 1875, p. 366).

Löwit's method was first published by him in the *Wien. Sitzgsber.*, lxxi Bd., iii Abth., 1875, p. 1.

Small pieces of *fresh* skin are put into dilute formic acid (one volume of water to one of the acid of 1·12 sp. gr.), and remain there until the epidermis peels off. They then are put for fifteen minutes into gold-chloride solution (1½ per cent. to 1 per cent.), then for twenty-four hours into dilute formic acid (1 part of the acid to 1—3 of water), and then for twenty-four hours into undiluted formic acid. (Both of these stages are gone through in the dark.) Thin sections are then made and mounted in dammar or glycerin. Successful preparations show the nerves alone stained, but it is not possible always to control the results.

208. Ranvier's Formic Acid Method (*Quart. Journ. Mic. Sci.* (N.S.), lxxx (1880), p. 456).—The method of Löwit has been modified by many workers by omitting the final treatment with undiluted formic acid, and also in some other details. Ranvier proceeds as follows. Reflecting that the action of the one-third formic acid in which Löwit placed his tissues must be hurtful to the finer ramifications of the nerves, he combines the formic acid with a fixing agent designed to antagonise its altering action, and takes for this purpose the chloride of gold itself. The tissues are placed in a mixture of chloride of gold and formic acid (4 parts of 1 per cent. gold chloride to 1 part of formic acid) which has been boiled and allowed to cool (Ranvier's *Traité*, p. 826). They remain in this until thoroughly impregnated (muscle twenty minutes, epidermis two to four hours); the reduction of the gold is effected either by the action of daylight in acidulated water, or in the dark in dilute formic acid (1 part of the acid to 4 parts of water).

The object of boiling the mixture of gold chloride and formic acid is this, that "by boiling in the presence of the acid the gold acquires a great tendency to reduction, and for this reason its selective action on nervous tissues is enhanced."

209. Ranvier's Lemon-juice Method (*Traité*, p. 813).—Instead of combining the formic acid with gold chloride in order to mitigate its action, recourse may be had to a less injurious acid than formic acid. Ranvier finds that of all acids lemon-juice is the least hurtful to nerve-endings. He therefore soaks pieces of tissue in fresh lemon-juice filtered through flannel, until they become transparent (five or ten minutes in the case of muscle). They are then rapidly washed in water, brought for about twenty minutes into 1 per cent. gold chloride solution, washed again in water, and brought into a bottle containing 50 c.c. of distilled water and 2 drops of acetic acid. They are exposed to the light, and the reduction is complete in twenty-four or forty-eight hours. The preparations thus obtained are good for immediate study, but are not permanent on account of their over-blackening with time, the reduction of the gold being incomplete. In order to obtain perfectly reduced, and therefore permanent, preparations, the reduction should be done in the dark in a few cubic centimetres of dilute formic acid (1 part acid to 4 of water). The reduction is complete in twenty-four hours.

210. Viallanes' Osmic-acid Method (*Hist. et. dév. des Insectes,* 1883, p. 42).—The tissues are treated with osmic acid (1 per cent. solution) until they begin to turn brown, then with one-fourth formic acid for ten minutes; they are then put into solution of chloride of gold of 1·5000 (or even much weaker) for twenty-four hours in the dark, then reduced in the light in one-fourth formic acid. According to my experience, this is a very excellent method, both the fixation by osmic acid, and the great dilution of the gold solution, being features likely to be of advantage in many cases.

211. Other Methods.—The numerous other methods that have been proposed differ from the foregoing partly in respect of the solutions used for impregnation, but chiefly in respect of details imagined for the purpose of facilitating the reduction of the gold, and rendering it as complete as possible.

Thus BASTIAN modified Cohnheim's original method by employing a solution of gold chloride of a strength of 1 to 2000, acidulated with HCl (1 drop to 75 c.c.), and performing the reduction in a mixture of equal parts of formic acid and water, *kept warm;* heat being an agent that furthers reduction.

HÉNOCQUE (*Arch. de l'Anat. et de la Physiol.,* 1870, p. 111) impregnates in a 0·5 per cent. solution of gold chloride, washes in water for twelve to twenty-four hours, and reduces, with the aid of heat, in a nearly saturated solution of tartaric acid. The tartaric acid solution must be contained in a well-stoppered bottle. The best temperature for reduction is 40° to 50° C. Reduction is effected very rapidly, sometimes in a quarter of an hour.

This process has been described as the method of CHRSCHT-SCHONOWIC (*Arch. f. mik. Anat.,* vii, 1872, p. 383).

HOYER (*Arch. mik. Anat.,* ix, 1873, p. 222) proceeds as follows:—(For corneal nerves.) The double chloride of gold and potassium has the following advantages over the simple gold chloride. It is more easy to be obtained of unvarying composition, it is more perfectly neutral, and its solutions are more perfectly stable. It is used in solutions of the same strength as chloride of gold, viz. 0·5 per cent. Corneæ must be very thoroughly imbibed with the solution. Small corneæ (rabbit, guinea-pig) require half to one hour, human corneæ two to five hours (in an acidulated solution). It is better to err on the side of too-prolonged immersion, rather than the contrary. In order to demonstrate the intra-epithelial ramifications of nerves, the gold is partially reduced by exposure for sixteen to twenty-four hours in (1 or 2 ounces of) distilled water, and there is added to the water one or two drops of a pyrogallic-acid developing solution, such as is used in photography (vide GERLACH, *Die Photographie als Hülfsmittel der mikroskopischen Forschung*, Leipzig, 1863). Or instead of treating them with the developing solution, the corneæ may be removed to a warm concentrated solution of tartaric acid and remain there at the temperature of an incubating stove until the gold is fully reduced.

I have myself used the double chloride of gold and sodium, with good results.

CIACCIO (*Journ. de Microgr.*, vii, 1883, p. 38; *Journ. Roy. Mic. Soc.* (N. S), iii, 1883, p. 290) prefers the double chloride of gold and cadmium.

GERLACH, whose preparations of nerve-centres are said not to have been equalled since, proceeded as follows (Stricker's *Handb.*, 1872, p. 678): Spinal cord is hardened for fifteen to twenty days in a 1 to 2 per cent. solution of bichromate of ammonia. Thin sections are made and thrown into a solution of 1 part of double chloride of gold and potassium to 10,000 parts water, which is very slightly acidulated with HCl. They remain there from ten to twelve hours, and having become slightly violet, are washed in hydrochloric acid of 1 to 2 : 3000 strength, then brought for ten minutes into a mixture of 1 part HCl to 1000 parts of 60 per cent. alcohol, then for a few minutes into absolute alcohol, and thence into clove oil, for mounting in balsam.

FLECHSIG (*Die Leitungsbahnen im Gehirn*, 1876; *Arch. f. Anat. u. Phys.*, 1884, p. 453) reduces in a 10 per cent. solution of caustic soda.

NESTEROFFSKY treats impregnated preparations with a drop of sulphhydrate of ammonium, and finishes the reduction in glycerin (quoted from Gierke's *Färberei z. mik. Zwecken*).

BÖHM reduces in Pritchard's solution.

PRITCHARD'S SOLUTION consists of amyl alcohol, 1 per cent.; formic acid, 1 per cent.; and water, 98 per cent.

MANFREDI treats fresh tissues as follows (*Arch. per le sci. med.*, v, No. 15) : Gold chloride, 1 per cent., half an hour; oxalic acid, 0·5 per cent.; they are then warmed in a water-bath to 36°, allowed to cool, and examined. Mount in glycerin. Sunny weather is necessary.

He treats tissues previously hardened in 2 per cent. solution of bichromate of potash, as follows (*ibid.*). They are put for half an hour into solution of arsenic acid, or into 1 per per cent. acetic acid. They are then put into 1 per cent. gold chloride for half an hour, washed in water, and reduced in sunlight in 1 per cent. arsenic acid solution, which is changed for fresh as fast as it becomes brown. Mount in glycerin. Sunny weather is necessary.

BOCCARDI (*Lavori Istit. Fisiol. Napoli*, 1886, i, p. 27; *Journ. Roy. Mic. Soc.*, 1888, p. 155) recommends oxalic acid of 0·1 per cent. or of 0·25 to 0·3 per cent. or a mixture of 5 c.c. pure formic acid, 1 c.c. of 1 per cent. oxalic acid, and 25 c.c. of water. Objects should remain in this fluid in the dark not longer than two to four hours.

KOLOSSOW (*Zeit. f. wiss. Mik.*, v, 1, 1888, p. 52) impregnates for two or three hours in a 1 per cent. solution of gold chloride acidulated with 1 per cent. of HCl, and reduces for two or three days in the dark in a 0·01 per cent. to 0·02 per cent. solution of chromic acid.

For the details of the application of the methods of which the principles have been set forth above, and for those of the important processes of impregnation of central nerve organs, the reader is referred to those chapters of Part II which treat of nerve-tissues and organs.

212. Ulterior Treatment of Impregnated Preparations.—Preparations may be mounted either in balsam or in acidulated glycerin (1 per cent. formic acid).

Theoretically, they ought to be permanent if the reduction of the metal has been completely effected.

In practice, all are doomed to destruction in course of time by after-blackening, and few will be found to survive more than a few months. Ranvier states that this can be avoided by putting the preparations for a few days into alcohol, which possesses the property of stopping the reduction of the gold. But this must be taken to mean that by this means the period of usefulness of the preparations may be prolonged for some time, not indefinitely.

Blackened preparations may be bleached with cyanide or ferrocyanide of potassium. REDDING employs a weak solution of ferrocyanide; CYBULSKY a 0·5 per cent. solution of cyanide. But the results are far from being perfectly satisfactory.

Preparations may be double-stained with the usual stains (safranin, methyl green, and iodine green being very much to be recommended), but nuclei will only take the second stain in the case of negative impregnation.

213. Impregnation of Marine Animals.—For some reason that I am unable to explain, the tissues of marine animals do not readily impregnate with gold in the fresh state. It is stated (FOL) that impregnation succeeds better with spirit specimens.

Other Metallic Stains.

214. Perchloride of Iron.—This reagent, introduced by POLAILLON (*Journ. de l'Anat.*, iii, 1866, p. 43) sometimes gives most useful results, especially in the study of peripheral nerve ganglia, in which it stains the nervous tissue alone, the connective tissue remaining colourless. The method consists in impregnating in perchloride of iron, and reducing in tannic, gallic, or pyrogallic acid.

The HOGGANS, who have done very good work with this reagent, proceed as follows (*Journ. Quekett Club*, 1876; *Journ. Roy. Mic. Soc.*, ii, 1879, p. 358) : The tissue (having been first fixed with silver nitrate, which is somewhat reduced by a short exposure to diffused light) is dehydrated in alcohol, and treated for a few minutes with 2 per cent. solution of perchloride of iron in spirit. It is then treated with a 2 per cent. solution of pyrogallic acid in spirit, and in a few minutes more, according to the depth of tint required, may be washed in water and mounted in glycerin.

FOL (see *ante*, § 50) fixes in perchloride solution, and treats the preparations for twenty-four hours with alcohol containing a trace of gallic acid.

POLAILLON (l. c.) reduces in tannic acid.

This method is not applicable to chromic objects.

I should add that in my own experience I have found it very useful in certain special cases.

215. Pyrogallate of Iron (ROOSEVELT, *Med. Rec.*, ii, 1887, p. 84; *Journ. Roy. Mic. Soc.*, 1888, p. 157).—A stain composed of 20 drops of saturated solution of iron sulphate, 30 grm. water, and 15 to 20 drops pyrogallic acid.

216. Palladium Chloride (F. E. SCHULTZE, see *ante*, 48 and 84).

217. Prussian Blue (*see* LEBER, *Arch. f. Ophthalm*, xiv, p. 300; RANVIER, *Traité*, p. 108).

218. Cupric Sulphate (*see* LEBER, *ibid.*).

219. Lead Chromate (*see* LEBER, *ibid.*).

220. Sulphides (*see* LANDOIS, *Centralb. f. d. med. Wiss.*, 1885, No. 55; and GIERKE, in *Zeit. f. wiss. Mik.*, i, 1884, p. 497).

221. Molybdate of Ammonium (Merkel; Krause) (*see* GIERKE, *Zeit. f. wiss. Mik.*, i, 1884, p. 96).

222. Osmic Acid.—Everybody knows that osmic acid stains tissues. Most people, I should think, would be heartily glad if it did not. Meanwhile, to make the best of this willy-nilly stain, you may sometimes find it useful to treat the tissues with weak pyrogallic acid, which will very quickly turn them of a fine greenish black, sometimes giving useful differentiations.

Or following BRÖSICKE (*Centralb. f. d. med. Wiss.*, 1879, p. 873; *Zeit. f. wiss. Mik.*, i, 1884, p. 409) you may treat them for twenty-four hours with a solution of 1 part of oxalic acid in 15 parts of water. This gives a Burgundy-red stain.

Recommendations of osmium for staining medullated nerve-fibres, as recently advocated by KOLOSSOW (in the place quoted, § 28; *sub finem*) appear to me anachronistic.

223. Impregnation with Fats, Altmann's Method (see *post*, § 530).

CHAPTER XIII.

COMBINATION STAINS.

224. The Classes of Multiple Stains.—I distinguish two classes of multiple stains. In the one a pure nuclear stain, taking effect on *all* the nuclei of *all* the tissues of a preparation, is combined with a stain taking effect on *all the extra-nuclear parts of all the tissues*. Borax-carmine followed by indigo-carmine is a typical example of such a combination. In the second class a stain taking effect on the totality of the elements of any one tissue exclusively is combined with a stain or stains of another colour taking effect on the totality of the elements of the other tissues.

The first class, aiming at enhancing the usefulness of a pure nuclear stain by improving the definition of extra-nuclear parts, has a legitimate scientific end in view, and is capable of rendering service in research. It will therefore here be treated much more fully than the second class, which is composed of much less generally useful, and too frequently merely ornamental, stains.

225. Picric-Acid Combinations.—I follow FLEMMING (*Zeit. f. wiss. Mik.*, i, 1884, p. 360) in pointing out that picric acid is perhaps the most generally useful of all secondary stains. It gives useful plasma-stains with most of the nuclear stains, and particularly with carmine and hæmatoxylin. It may be used with the most delicate of these stains, even the delicate coloration of alum-carmine being in no wise injured by it. The *modus operandi* is as simple as possible; it consists merely in adding picric acid to the alcohols employed for dehydrating the objects after staining with a nuclear stain.

Care must be taken in adding picric acid to alcohol acidulated with HCl (see *ante*, § 163); in fact, this practice had better in general be avoided, and the picric acid only added to the pure alcohol used *after* washing out.

Combinations having Carmine for a Primary Stain.

225a. Picro-Carmine.—Picro-carmine is a double stain, if care be taken not to wash out the picrin beyond the point desired. And it is one of the best of double stains.

226. Borax-Carmine and Picro-Carmine.—A very beautiful and precise double stain may be obtained by means of this combination. I add to a watch-glassful of Grenacher's alcoholic borax-carmine a few drops of picro-carmine.

The mixture will precipitate in the course of a few hours, but the stain will be obtained nevertheless.

BAUMGARTEN's borax-picro-carmine (§ 163).

227. Carmine and Picric Acid.—This combination has been already sufficiently explained above (*see* § 225a).

LEGAL's alum-carmine and picric-acid mixture (§ 152).

228. Orth's Picro-Lithium Carmine.—A mixture of one part of lithium-carmine (*ante*, § 153) and two or three parts of saturated aqueous solution of picric acid. If the mixture should over-stain either in yellow or in red, add a little of the other colour. Wash out with acidulated (HCl) alcohol, as for simple lithium-carmine.

229. Seiler's Carmine followed by Indigo-carmine (*Am. Quart. Mic. Journ.*, i, 1879, p. 220; *Journ. Roy. Mic. Soc.*, ii, 1879, p. 613).—Stain in Woodward's borax-carmine (§ 159), wash out in HCl one part, alcohol four parts, until the sections assume a bright rose colour (which appears in a few seconds). Wash the acid out of the sections, and stain for six to eighteen hours in a mixture of two drops of sulphindigotate of soda solution with one ounce of 95 per cent. alcohol. The mixture should be filtered before using (§ 190).

Nuclei red, formed material slightly tinged with blue.

It is obvious that this method may be modified, in most cases with advantage, by using Grenacher's alcoholic borax-carmine or HCl (alcoholic) carmine, or any similar carmine stain, instead of the aqueous solution of Woodward. Hencage Gibbes uses the borax-carmine quoted under his name (§ 160).

I find the method gives admirable results when applied to sections, but very bad results if it be attempted to stain in the mass with the indigo. The indigo over-stains the superficial layers before it has penetrated to the deeper layers.

230. **Merkel's Carmine and Indigo-Carmine in one Stain** (MERKEL, *Unters. a. d. anat. Anst. Rostock*, 1874; *Month. Mic. Journ.*, 1877, pp. 242 and 317).

(A) Take 2 gr. of carmine; 8 gr. of borax; and 128 c.c. of water (or, half a drachm of carmine, two drachms of borax, and four ounces of water). Rub up in a mortar, allow the fluid to stand some time, decant, filter, and keep in a stoppered bottle.

(B) Take 8 gr. of indigo-carmine; 8 gr. of borax; and 128 c.c. of water (or, two drachms indigo-carmine, two drachms borax, and four ounces water). Mix, decant, filter, and preserve, as before.

Before using, mix A and B in equal proportions.

The objects to be stained must be thin; all traces of chromic acid or chromates must have been carefully washed out from them; and they must be soaked in alcohol before staining. Stain for fifteen or twenty minutes (MAX FLESCH finds it better to stain for several hours, see *Zeit. f. wiss. Mik.*, 1885, p. 350). Wash out with saturated aqueous solution of oxalic acid, for a rather shorter time; wash the acid out with water, and mount as desired.

The oxalic acid is necessary for fixing the indigo-carmine, which, being very soluble in water, would otherwise be washed out. Unfortunately, it precipitates carmine, so that successful preparations are not easily obtained, the carmine being generally either precipitated or turned into a straw colour.

Authors (MERKEL, *loc. cit.*; NORRIS and SHAKESPEARE, *Amer. Journ. Med. Sc.*, January 1877; MERKEL, *Mon. Mic. Journ.*, 1877, p. 242; MARSH, *Section Cutting*, p. 85; BAYERL, *Arch. f. mik. Anat.*, xxiii, 1885, p. 36—37) are unanimous in stating that successful preparations show a most richly differentiated and yet very precise colouring. According to Bayerl, the stain is quite specifically elective for red blood-corpuscles, which are stained of an apple green. The ground substance of cartilage and bone stains blue, their cells red.

The stain is not perfectly permanent. Bayerl recommends that benzin be used for clearing, in lieu of clove oil, which oxydises the stain and injures it.

This method has recently been recommended for Nerve Centres. For Bayerl's application of it to ossifying cartilage (*see* Part II).

I have put this method in small type, because though admirable for certain special purposes, it is not at all to be recommended for general work.

231. **Carmine and Anilin Blue (or Bleu Lumière, or Bleu de Lyon).**—DUVAL (*Précis de technique microscopique*, 1878, p. 225) proceeds as follows: Stain with carmine "in the ordinary way;" dehydrate; and stain for a few minutes (ten minutes for a section of nerve-centres) in an alcoholic solution of anilin blue (ten drops of saturated solution of anilin blue

soluble in alcohol to ten grammes of absolute alcohol, for sections of nerve-centres). Clear with turpentine, without further treatment with alcohol, and mount in balsam.

The sections should appear of a fine dark violet when taken from the anilin: they are extremely transparent under the microscope. Nerve-cells and axis-cylinders, reddish violet; blood-vessels, bluish violet, and so sharply marked out that the preparations have the aspect of injections. The connective elements are stained of a nearly pure blue, so that it is easy to distinguish them from the nervous elements.

Applicable to all kinds of tissues, but especially to sections of nerve-centres.

Recent authors recommend instead of anilin blue, Bleu de Lyon, dissolved in 70 per cent. alcohol acidulated with acetic acid (MAURICE and SCHULGIN), or Bleu lumière, which has hardly any effect on nuclei.

The solutions of both these colours should be extremely dilute. They may be used for staining in the mass.

232. Picro-Carmine and Iodine Green (STIRLING, *Journ. Anat. and Physiol.*, xv, 1881, p. 349, *et seq.*).—Stain picro-carmine, wash in acidulated water (acetic acid), stain iodine green. Iodine green stains very rapidly, and care must be taken not to over-stain. Rinse in water, dehydrate *rapidly*, clear with clove oil, *mount in dammar*. (All preparations stained with iodine green *must* be mounted in dammar.)

Iodine green has a specific action on adenoid tissue and mucous glands, which it stains of a bright green.

233. Picro-Carmine and Methyl Green (MAX FLESCH, *Zool. Anz.*, 123, 1882, p. 554).—Sections of cartilage, skin, and glands made from tissues hardened in Müller's solution and alcohol, were stained with picro-carmine, and subsequently (not "previously," as erroneously stated in *Journ. Roy. Mic. Soc.* (N.S.), ii, 1882, p. 883) with an aqueous solution of commercial methyl green made of such a strength that the sections are just distinguishable in a watch-glassful of the solution when placed on a light ground.

The method is easy, gives good differentiations, but the stain does not appear likely to be permanent. Mount in balsam.

234. Picro-Carmine, Rosein, and Anilin Blue; or Picro-Carmine, Anilin Violet, and Anilin Blue; or Picro-Carmine, Anilin Violet, and Iodine Green; or Picro-Carmine, Rosein, and Iodine Green (HENEAGE GIBBES, *Journ. Roy. Mic. Soc.*, iii, 1880, p. 392).—Make a dilute solution of picro-carmine (about 10 drops to a watch-glass of water), stain in it for about half an hour, wash out for an hour in water acidulated with a few drops of acetic or picric acid, and then double-stain either with rosein and anilin blue, or with anilin violet and anilin blue, or with anilin violet and anilin green, or with rosein and anilin green.

H. Gibbes says of these methods that their great utility consists in their power of differentiating glandular structures according to their secretions. In a section of a dog's tongue "the ordinary mucous glands will be found to have taken on a purple colour, while the serous glands which supply the secretions to the taste-organs stain a totally different colour."

235. Picro-Carmine and Eosin (LANG, *Journ. Roy. Mic. Soc.*, ii, 163 ; *Zool. Anz.*, ii, p. 45 ; *Mitth. d. Zool. Stat. zu Neapel*, Bd. ii, p. 1, *et seq.*).— Take 50 parts 1 per cent. picro-carmine, 50 parts 2 per cent. eosin (aqueous solution). The objects, previously hardened in alcohol, are left in the mixture half to four days. Wash out the picrin by 70 per cent. alcohol, which must be frequently changed, and be followed by 90 per cent. and absolute alcohol until no more eosin is dissolved out.

For *Turbellaria*.—It has not been found useful for other objects.

236. Carmine and the Metallic Stains.—These combinations have been sufficiently spoken of in the passages devoted to gold and silver impregnation-methods. It will suffice here to call renewed attention to ZOLTÁN V. ROBOZ, alum-carmine and osmic acid (§ 150).

Combinations having Hæmatoxylin for a Primary Stain.

237. Hæmatoxylin and Picric Acid.—This combination has been treated of above (§ 225).

238. Hæmatoxylin and Eosin.—This is a well-known combination, and one of the most instructive that have yet been imagined. Objects may be stained with hæmatoxylin (either in the mass or as sections), and the sections stained for a few minutes in eosin. I think it is better to take the eosin weak, though it has been recommended (STÖHR, see *Zeit. f. wiss. Mik.*, i, 1884, p. 583) to take it saturated. Either aqueous or alcoholic solutions of eosin may be used.

This method is most particularly recommendable for embryological sections, as vitellus takes the eosin stain energetically, and so stands out boldly from the other germinal layers in which the blue of the hæmatoxylin dominates.

LIST (*Zeit. f. wiss. Mik.*, ii, 1885, p. 148) stains for twenty-four hours in a solution of three or four drops of Renaut's hæmatoxylic glycerin (§ 183) in 250 c.c. of water, and then for a few minutes in a mixture of one part of 0·5 per cent. aqueous solution of eosin with three parts of absolute alcohol.

BUSCH (*Verh. Berl. Phys. Ges.*, 1877; GIERKE, *Zeit. f. wiss. Mik.*, i, 1884, p. 505) treats sections for some days with 0·5 per cent. solution of chromic acid, or 1 per cent. solution of bichromate of potash, washes, and stains first in an aqueous solution of eosin and then in hæmatoxylin. This process has been recommended for the study of the margin of ossification.

It should be noted that sections should be very well washed before being passed from eosin into hæmatoxylin or the reverse, as eosin very easily precipitates hæmatoxylin.

239. Renaut's Hæmatoxylic Eosin (FOL's *Lehrbuch*, p. 196. Renaut has given from time to time several formulæ for this stain. This one, communicated to Fol by Renaut, is the latest, and I suppress the others).

Take

Concentrated aqueous solution of potassic eosin (*éosine à la potasse*)	30 c.c.
Saturated solution of hæmatox. in alcohol (ought to have been kept some time and to have precipitated)	40 c.c.
Saturated solution of potash alum in glycerin (of a density of about 1·26)	130 c.c.

Mix, and let the mixture stand five or six weeks in a vessel covered with a sheet of paper pierced with holes until the alcohol is evaporated, then filter.

For staining, the solution may be used as it is or diluted. Staining goes on very slowly, and at first the colour is not held by the tissues, but disappears on washing. After some days or weeks, however, it becomes localised and fixed in the tissues. You may then mount in balsam, taking care to employ alcohol charged with a sufficient quantity of eosin. But it is frequently preferable to proceed by mounting the objects in the staining fluid diluted with one to two volumes of glycerin. After a few weeks this mounting medium will have become perfectly colourless through the absorption of the colour by the tissues.

The stain has a specific action on the cells of salivary and gastric glands. Mucous-cells become pale blue. Salivary ferment-cells (crescent-cells of Gianuzzi) intense rose.

There is no doubt as to this being a very fine and useful stain.

See also *Comptes Rendus*, 1879, p. 1039 (1re sér.), and *Arch. de Physiol.*, 1881, p. 640.

240 Hæmatoxylin and Benzo-purpurin (ZSCHOKKE).—For this and the combination with **Delta Purpurin**, see *ante*, § 114.

241. Hæmatoxylin and Iodine Green. (STIRLING, *Journ. of Anat. and Physiol.*, xv, 1881, p. 353).—Stain *not too deeply* with logwood, and then stain with iodine green. (Mucus-glands of tongue green, serous glands hæmatoxylin).

242. Hæmatoxylin and Nitrate of Rosanilin (LIST, *Zeit. f. wiss. Mik.*, ii, 1885, p. 149). Stain for twenty-four hours in the dilute hæmatoxylic glycerin of which the formula was given just now, § 238, wash, and stain for about ten minutes in solution of nitrate of rosanilin, then rinse with water, dehydrate with absolute alcohol, and clear. Recommended especially for mucus-glands, epithelia, and cartilage.

243. Hæmatoxylin and Safranin.—This celebrated combination, which was used to such good effect by RABL in his classical researches on nuclei (*Morph. Jahrb.*, x, 1884, p. 215), does not strictly belong to this subdivision, the safranin being the primary stain, though used after the hæmatoxylin. You stain *very lightly* with hæmatoxylin, so lightly that the stain would not be of any use by itself (Rabl uses very dilute Delafield's solution, for twenty-four hours); wash out first with water, and then with alcohol acidulated with HCl., then stain for some hours in (Pfitzner's) safranin, and wash out with pure alcohol. Rabl certainly was not far wrong when he wrote—" This method is unequalled by any other." For richness of detail in both nucleus and cytoplasm and tissue this method has indeed hardly been equalled.

244. Hæmatoxylin and Metallic Stains.—The conditions under which hæmatoxylin can usefully be employed for staining impregnated tissues have been discussed under the heads of Gold and Silver. It only remains here to remind the reader that hæmatoxylin works very well after osmic acid.

Other Combinations.

245. The Anilin Double-Stains.—These very important combinations are so numerous that only a small proportion of them can be mentioned here.

As regards *sections* stained by the indirect or Flemming's method, considerable latitude is allowable in the manipulations. Some persons stain first in the secondary stain (eosin, for instance), then wash, and stain in the primary stain (by "primary" stain, I constantly mean the nuclear stain), and wash out until the colour of the secondary stain reappears.

Another method, which will, I think, frequently be found preferable as allowing a stronger primary stain, is to stain first with the primary, say gentian violet, and then with the secondary, say eosin. Care must be taken not to wash out the first stain as completely as if you were going to mount it at once, else the operations required by the second stain may result in entirely removing the colour of the first.

Lists of some colours that give good results as primary and as secondary stains are given in § 98. Besides the combinations there recommended I would particularly recommend that of **Gentian Violet** and **Eosin**. Stain and wash out (not too far) by Bizzozero's method (§ 102), and then stain for a few seconds, or as much as two to five minutes, in fairly strong aqueous solution of eosin, dehydrate rapidly with alcohol, clear and mount.

Attention should be paid to the decolorising action of the secondary stain on the primary (*see* § 98). This reaction may be utilised, as proposed by BAUMGARTEN (*infra* § 246). RESEGOTTI (*Zeit. f. wiss. Mik.*, v, 3, 1888, p. 323) recommends staining for five minutes (this is for tissues fixed by alcohol) in strong solution of **Methyl Violet** or **Dahlia**, then washing out for one or two minutes in very weak solution of (alcoholic) **Eosin** or of **Säurefuchsin** in alcohol, then dehydrating and clearing.

In the following paragraphs are given some other methods of proved utility.

246. Baumgarten's Fuchsin and Methylen Blue (*Zeit. f. wiss. Mik.*, i, 1884, p. 415). Stain sections (of chromic objects) for twenty-four hours in a stain made by adding 8 to 10 drops of concentrated alcoholic solution of fuchsin to a watch-glassful of water. Rinse with alcohol, and stain for four or five minutes in concentrated aqueous solution of methylen blue, wash out with alcohol for five to ten minutes, and clear with clove oil. Nuclei red, tissues blue, the fuchsin having been driven out of the tissues by the methylen blue, a result which is *not* attained by washing with alcohol alone, either pure or acidified.

247. Baumgarten's Triple Stain (*Bull. Soc. Belg. Mic.*, 7, 1887, and 14, 1888, p. 146; *Journ. de Microgr.*, 1888, p. 415; *Journ. Roy. Mic. Soc.*, 1887, p. 676, and 1889, p. 149).—Consists of borax-picro-carmine (§ 163), picric acid, gentian violet, iodine, and acidulated alcohol and picric acid in succession. Very complicated, and too delicate to manipulate for ordinary work.

248. Garbini's Safranin and Anilin Blue (*Zool. Anz.*, 1886, p. 26; and *Zeit. f. wiss. Mik.*, v, 2, 1888, p. 170).—Stain (sections) one to four minutes in anilin blue (1 per cent. in water, with 1 per cent. of absolute alcohol). Wash in 1 per cent. solution of carbonate of lithium until the colour has nearly disappeared. Treat for five to ten minutes (until the colour has come back) with 0·5 per cent. hydrochloric acid. Distilled water. Safranin (0·5 gr. in 150 c.c. one-third alcohol) ten to twenty minutes, heating for a minute or two. Dehydrate with methyl alcohol (CH_3), finish the differentiation of the stain in a mixture of one part oil of cedar with two parts clove oil, and pass through xylol into balsam.

Somewhat complicated, but a sharp stain, rich in detail, and valuable in so far as there are not many blues that give very satisfactory results with safranin.

You may dehydrate and clear in the ordinary way with alcohol and clove oil if you like, and, as far as I can see, without inconvenience.

249. Safranin and Indigo-Carmine (KOSSINSKI, *Zeit. f. wiss. Mik.*, vi, 1, 1889, p. 61).—Stain sections ten to twenty minutes in saturated aqueous solution of indigo-carmine, wash with water and with alcohol, and stain with safranin (0·5 per cent. in dilute alcohol), dehydrate and mount.

Safranin and Nigrosin is a combination also recommended by the same author. Stain for three to five minutes, in 0·1 per cent. aqueous solution of nigrosin, and proceed as before.

250. Methyl Green and Eosin (CALBERLA, *Morph. Jahrb.*, iii, 1877, 3 Hft., p. 625).—Mix 1 part of eosin with 60 parts of methyl green, and dissolve the mixture in warm 30 per cent. alcohol.

Sections stain in this solution in five or ten minutes; they should be quickly washed in successive alcohols, and mounted in balsam or glycerin.

251. Methyl Green and Eosin (LIST, *Zeit. f. wiss. Mik.*, ii, 1885, p. 147).—Stain for a few minutes in a mixture of three parts of absolute alcohol with one part of aqueous solution of eosin (0·5 per cent.), wash, and stain for five minutes in 0·5 per cent. aqueous solution of methyl green. Wash, dehydrate, clear, and mount in balsam. The preparations do not keep well in glycerin.

The method may be varied (*l. c.*, p. 150) by diluting the methyl green solution with 50 volumes of water, and staining for twenty-four hours.

It may also be varied by diluting the original methyl green solution with 3 volumes of absolute alcohol.

The preparations should not be left in the alcohol used for dehydrating after the colour of the eosin has begun to reappear.

252. Methyl Green, Orange, and Säurefuchsin (BIONDI and HEIDENHAIN, *Pflüger's Arch.*, xlii, 1888, p. 1; *Zeit. f. wiss. Mik.*, v, 4, 1888, p. 520). To 100 c.c. saturated aqueous solution of orange add with continual agitation 20 c.c. saturated aqueous solution of Säurefuchsin and 50 c.c. of a like solution of methyl green.

Dilute the mixture with 60 to 100 volumes of water. The dilute solution ought to redden if acetic acid be added to it; and if a drop be placed on blotting paper it should form a spot bluish-green in the centre, orange at the periphery. If the orange zone is surrounded by a broader red zone, the mixture contains too much fuchsin.

Stain (sublimate sections) for six to twenty-four hours. Wash out with alcohol, clear with xylol, and mount in xylol balsam.

253. Methyl Green and Bismarck Brown (LIST, *Zeit. f. wiss. Mik.*, ii, 1885, p. 145).—Stain for a few minutes in Weigert's Bismarck brown (§ 106), wash, and stain in 0·5 per cent. aqueous solution of methyl green. Clear with bergamot oil or xylol, and mount in balsam.

Or, dilute the Bismarck brown for staining with 3 volumes of absolute alcohol, wash out with strong alcohol, and stain for a few minutes in the methyl green solution diluted with 3 volumes of absolute alcohol.

Or, stain for twenty-four hours in the Bismarck brown solution diluted with 50 volumes of water, and then for twenty-four hours in the methyl green solution diluted with 50 volumes of water.

254. Methyl Green and Nitrate of Rosanilin (LIST, *l. c.*).— Stain for a few minutes in 0·5 per cent. aqueous solution of methyl green, wash, and stain (*l. c.*, iii, 1886, p. 393) for ten to fifteen minutes in 0·0001 per cent. aqueous solution of nitrate of rosanilin, wash out (rapidly) with absolute alcohol, and clear.

255. Anilin Green and Bismarck Brown (LIST, *l. c.*).—To be used in the same way as methyl green and Bismarck brown *supra*, (§ 253), and giving much the same results.

256. Anilin Green and Eosin (SCHIEFFERDECKER, *Arch f. mik. Anat.*, xv, 1878, p. 30).—Stain in a solution made by adding a few drops of aqueous solution of eosin to a watch-glassful of alcohol, for from half an hour to several hours, wash with water and stain for a few minutes in 1 per cent. aqueous solution of anilin green. Wash with water, wash out with alcohol and clear with clove oil.

This combination possesses a rich selectivity, especially for connective and glandular tissues.

LIST (*l. c.*, § 251) recommends this combination, which he employs in the following manner: Stain for a quarter of an hour in the eosin solution there quoted, rinse with alcohol, and stain for a quarter of an hour in 0·5 per cent. aqueous solution of anilin green diluted with 3 volumes of absolute alcohol. Wash out with absolute alcohol and (as soon as the colour of of the eosin begins to reappear) clear with bergamot oil or xylol.

257. Dahlia and Eosin (SCHIEFFERDECKER, *l. c.*).—Proceed as for anilin green and eosin (§ 256), using a 1 per cent. aqueous solution of dahlia.

258. Methyl Violet and Eosin (SCHIEFFERDECKER, *l. c.*).—Proceed as before, using a 1 per cent. aqueous solution of methyl violet.

259. Iodine Green and Eosin (STIRLING, *Journ. Anat. and Physiol.*, xv, 1881, p. 354).—Stain in alcoholic solution of eosin, wash in acidulated water (1 per cent. acetic or hydrochloric acid), stain with iodine green.

260. Rose Bengale and Iodine Green (GRIESBACH, *Zool. Anzeig.*, 135, 1883, p. 172).—The method consists in staining very quickly in a strong aqueous solution of rose bengale (the section must have been soaked in water before bringing into the stain), washing out with water, and staining for a few seconds in iodine green. The sections may then be mounted, or may be further treated with bleu de Lyon. This is done by treating them for five minutes with absolute alcohol, and staining for two or three seconds in a solution of bleu de Lyon in 40 per cent. alcohol. The sections appear not to take the blue stain, but it becomes visible as soon as they are mounted. They are to be dehydrated in absolute alcohol, cleared in oil of anise-seed of 0·99 sp. gr., and 1·811 refractive index, and mounted in dammar. The method gives in some cases very striking differentiations, but the results are by no means constant, and do not admit of being generalised.

261. Roseïn and Anilin Blue, Roseïn and Anilin Green, Anilin Violet and Anilin Blue, Anilin Violet and Anilin Green (*see* H. GIBBES, *Journ. Roy. Mic. Soc.*, 1880, p. 391).

262. Metallic Stains and Anilin Colours.—Most of the coal-tar colours above quoted may be employed after a metallic impregnation. The combination of safranin with gold chloride, first recommended by Pfitzner, is a classical stain. Eosin may usefully be made to follow nitrate of silver impregnation. And other combinations may be found useful on occasion.

CHAPTER XIV.

IMBEDDING METHODS: INTRODUCTION.

263. A Word on Microtomes.—It is no part of the purpose of this work to discuss instruments, yet a word on this subject may be very helpful to the student. The freezing microtome so generally employed in England is less than any other form adapted to the wants of the zoologist. Very thin sections can be obtained by it more readily than with any other microtome, but they are of little use when obtained. The relations of the parts of the organs are deranged by the freezing and by the thawing, and the aqueous nature of the process prevents it from being readily applicable to the mounting of *series* of sections. The microtome of the zoologist, therefore (I am not writing for pathologists or for dilettanti), must be an *imbedding* microtome. The two most important points to be attended to in the choice of such a microtome are the *object-holder* and the *knife motion*. The object-holder should *never*, as is usual in the English forms, be a well in which *the imbedded object* is raised by a screw; the principle of construction should always be that the *object-holder be raised* in its entirety by the screw, *not* the object alone. The *knife motion* should always be mechanical; cutting of sections with a free knife held in the hand is a primitive process, by which only coarse sections can be obtained, and that at the expense of much time and attention. Remember that the mean thickness to which sections are now cut is 5μ, half a hundredth of a millimètre, or less than $\frac{1}{5000}$th of an inch. Clearly the free hand is not capable of producing *regular* series of sections of such a degree of fineness. The student should therefore be careful to provide himself with an instrument in which the knife is guided by a mechanism giving the required precision of stroke. Cylinder-microtomes with free knife motion should be unhesitatingly condemned by him.

Amongst microtomes fulfilling the conditions I have laid

down various forms will be found almost equally convenient. Zeiss makes a good one, Schanze, of Leipzig, makes a good one. Reichert, of Vienna, makes a good one. All these are relatively cheap, and being at the same time perfectly efficient for easy work, may be recommended. Amongst more precise instruments I recommend the following:—*The Thoma sliding microtome*. If the student will obtain from R. Jung, Mechaniker in Heidelberg, a Thoma microtome, *medium size* (No. 2 a or 4), with the newest Naples object-holder and newest form of knife and knife-holder, he will, in my opinion, be possessed of one of the best of all-round microtomes.

This instrument is described in *Journ. Roy. Mic. Soc.* (N.S.), vol. iii, p. 298; the new Naples object-holder (which I consider *essential* for the zoologist) is described and figured p. 915.

The Becker microtome is considered by many good workers to be an improvement on the Thoma model. It is essentially on the same principle, but possesses a mechanical arrangement for moving the knife-carrier. This, I think, is certainly an advantage. A minor point is that the instrument is somewhat cheaper than the Thoma form. It is made by Aug. Becker, Göttingen. Descriptions of two forms (Spengel and Schieferdecker) will be found in *Journ. Roy. Mic. Soc.*, 1886, pp. 884 and 1084.

Both these are admirable all-round microtomes, that is to say, adapted to the most various kinds of work, and in particular equally adapted for paraffin sections and for colloidin sections. There remains to be mentioned an important form that is adapted for paraffin sections alone. This is the beautiful *Cambridge rocking microtome* (furnished by the Cambridge Scientific Instrument Company, St. Tibb's Row, Cambridge, price £5). This instrument is extremely simple and extremely rapid, and, what is more important, cuts better and more level series of sections than any other microtome I am acquainted with. It should be fitted with the improved moveable object-holder of Henneguy and Vignal (*Compt. Rend. Soc. Biol.*, 1885, p. 647), or some equivalent arrangement allowing the precise orientation of the object. (This, as well as the entire instrument, is manufactured in France by Dumaige, 9, Rue de la Bucherie, Paris.)

264. Imbedding Methods may conveniently be divided into

two classes, distinguished by the end it is intended to compass by their employment. In the one it is merely proposed so to surround an object, too small or too delicate to be firmly held by the fingers or by any instrument, with some plastic substance that will support it on all sides with firmness but without injurious pressure, so that by cutting sections through the composite body thus formed, the included object may be cut into sufficiently thin slices without distortion. This is *simple* imbedding. Its object is very easily attained in a variety of ways of which the simple process of immersing the object to be cut in a molten mass of some such material as wax, which when cold acquires a fit consistency for the cutting of thin slices, may be taken as a type. A further object is proposed in the case of the other class of methods, which may be designated methods of *interstitial* imbedding or *infiltration* methods. In these it is proposed to fill out with the imbedding mass the natural cavities of the object in order that their lining membranes or other structures contained in them may be duly cut *in situ*, or, going a step further, it is proposed to surround with the supporting mass not only each individual organ or part of any organ that may be present in the interior of the object, but each separate cell or other anatomical element, thus giving to the tissues a consistency they could not otherwise possess, and ensuring that in the thin slices cut from the mass all the details of structure will precisely retain their natural relations of position. Such a process of imbedding is at the same time practically a process of hardening in so far as it enables us to give to tissues a degree of firmness that could otherwise only be obtained by the employment of chemical processes such as prolonged treatment with chromic acid and the like.

265. Infiltration Methods.—The principle of the methods of this second class is either, like that of the first, that by immersion of the object to be cut in some material that is liquid while warm and solid when cold, all the parts of the object may be duly surrounded by the supporting mass (the second class differing from the first chiefly in the employment of materials possessing greater power of penetration whilst liquid, in longer immersion in the liquid mass, and in such previous preparation of the object, by soaking in some liquid that is a solvent of the imbedding material, as makes it more readily susceptible of infiltration by the latter); or the processes may be based on another principle, namely, that of the employment of substances which whilst in solution are sufficiently fluid to penetrate the object to be imbedded, whilst at the same time after the evaporation or removal by other means of their solvent, they acquire and impart to the imbedded object sufficient firmness for the purpose of cutting. The collodion process sufficiently exemplifies this principle. If a piece of soft tissue be dehydrated, and soaked first in ether and then in collodion, and if the ether contained in the collodion be allowed slowly to evaporate, the tissue and surrounding mass of collodion will acquire a consistency such as to admit of thin sections being cut from them.

The egg-emulsion process, in which a mass that is liquid whilst cold is coagulated by heat, forms a class by itself.

In any of these cases the material used for imbedding is technically termed an "imbedding-mass." (*Einbettungsmasse:—masse d'inclusion*. Imbedding methods are spoken of by French writers as *méthodes d'inclusion*, or *méthodes d'enrobage*).

266. Imbedding Manipulations.—Before proceeding to describe in detail the more important imbedding methods, it is necessary to give an account of the manipulations of the process of imbedding in general. This will serve at the same time for such account as is needful of the methods of Simple Imbedding—these being now relatively quite unimportant—and of the stage of imbedding *sensu stricto* that forms part of the processes of the Infiltration Methods.

To imbed an object in such a substance as liver or spinal cord (which does not strictly come under the category of an imbedding mass at all, as defined above) nothing more is necessary than to take a piece of fresh liver or cord of convenient dimensions, scoop out in it a hole of the size of the object to be imbedded, place the object in the hole, and immerse the mass in alcohol until such time as the mass is sufficiently shrunken and hardened to hold the object firmly and permit of section cutting.

If pith be employed, a cylinder of pith is halved longitudinally, a cavity corresponding to the object to be imbedded is made by scooping out the inner face of either half-cylinder, the object placed in position between these, and the cylinder pushed into the well of a microtome (which it should fit accurately), and moistened with alcohol (or other suitable liquid) in order that the pith may swell round the object. It should be noted that *it is better to make the cavity in the pith by simple pressure and kneading (e. g.*

with the handle of a scalpel) than *by excavation of material*; the pith-cells that have been flattened and pushed to one side by the kneading tend to regain their normal form and position during the soaking in alcohol, and their resilience causes the imbedded object to be grasped with an often surprising tightness. If the cylinder of pith does not fit the holder of the microtome accurately in the dry state (which it should do if possible) it should be wedged in by means of strips or thin wedges of *kneaded* pith inserted dry, and the whole afterwards soaked. With well-hardened objects this method, if skilfully carried out, allows of very accurate section cutting. These processes are still sometimes employed; in cytological researches, for example (Flemming).

Moist paper may be used for imbedding some objects. They should be swathed in strips of printing paper softened in water, so as to form a roll that will go home in the well of a cylinder microtome with a little pressure (RICHARDSON, *Journ. Roy. Mic. Soc.*, 1882, p. 474).

Simple imbedding in a melted mass such as paraffin is performed in one of the following ways. A little tray or box or thimble is made out of paper, some melted mass is poured into it; at the moment when the mass has cooled so far as to have a consistency that will not allow the object to sink to the bottom, the object is placed on its surface, and more melted mass poured on until the object is enclosed. Or the paper tray being placed on cork, the object may be fixed in position in it whilst empty by means of pins and the tray filled with melted mass at one pour. The pins are removed when the mass is cold.

In either case when the mass is cold, the paper is removed from it before cutting.

To make paper trays, proceed as follows. Take a piece of stout paper or thin cardboard, of the shape of the annexed figure (Fig. 1); thin (foreign) postcards do very well indeed. Fold it along the lines $a\,a'$ and $b\,b'$, then along $c\,c'$ and $d\,d'$, taking care to fold always the same way. Then make the folds $A\,A'$, $B\,B'$, $C\,C'$, $D\,D'$, still folding the same way. To do this you apply $A\,c$ against $A\,a$, and pinch out the line $A\,A'$, and so on for the remaining angles. This done, you have an imperfect tray with dogs' ears at the angles. To finish it, turn the dogs' ears round against the ends of the box, turn down outside the projecting flaps that remain, and pinch them down. A well-made postcard tray will last through several imbeddings, and will generally work better after having been used than when new.

To make paper thimbles, take a good cork, twist a strip of

138   IMBEDDING METHODS.

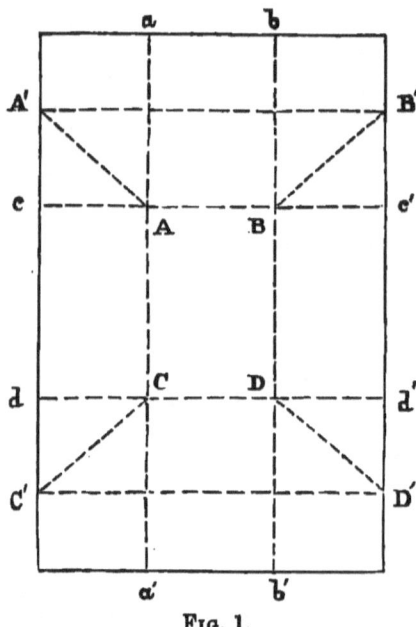

Fig. 1.

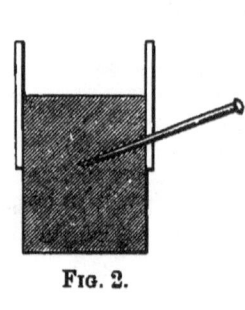

Fig. 2.

paper several times round it so as to make a projecting collar, and stick a pin through the bottom of the paper into the cork. For work with fluid masses, such as celloidin, the cork may be leaded at the bottom to prevent it from floating, when the whole is thrown into spirit or other liquid for hardening (Fig. 2).

"In Professor Leuckhart's laboratory are used boxes made of two pieces of type-metal (Fig. 3). Each of these pieces has the form of a carpenter's "square" with the end of the shorter arm triangularly enlarged outwards. The box is constructed by placing the two pieces together on a plate of glass which has been wetted with glycerin and gently warmed. The area of the box will evidently vary according to the position given to the pieces, but the height can be varied only by using different sets of pieces. In such a box the paraffin may be kept in a liquid state by warming now and then over a spirit-lamp, and small objects be placed

Fig. 3.

in any desired position under the microscope" (*Journ. Roy. Mic. Soc.* (N.S.), ii, p. 880).

SELENKA has recently described and figured a simple but perhaps more efficacious apparatus having the same object. It consists of a glass tube through which a stream of warm water may be passed and changed for cold as desired, the object being placed in a depression in the middle of the tube (see *Zool. Anz.*, 1885, p. 419). A modification of this method is described by ANDREWS in *Amer. Natural.*, 1887, p. 101; cf. *Zeit. f. wiss. Mik.*, iv, 3, 1887, p. 375.

For small paraffin objects the following procedure is very useful. The object is removed from the paraffin solution, the superfluous fluid is removed by means of blotting-paper, and the object placed on a cylinder of paraffin. A piece of stout iron wire is now heated in the flame of a spirit-lamp, and with it a hole is melted in the end of the cylinder, the specimen is pushed into the melted paraffin and placed in any desired position. The advantages of the method lie in the quickness with which it can be performed, and in the fact that by the melting of so small a quantity of paraffin all risk of injury to the tissues by overheating is done away with.

This method may also be used for simple imbedding in the case of solid objects without cavities or irregular outline. They are transferred direct from alcohol to the paraffin-cylinder, and when sections are cut they readily separate from the shaving of paraffin without the application of turpentine.

I strongly recommend the reader not to neglect this simple method, which is capable of sometimes rendering services which no other method can. Those who have to do work with objects so small that their position can only be made out with the aid of a powerful lens, ought to know how to arrange an object with a heated needle under a dissecting microscope, or on the object-carrier of the microtome.

[In the 1st edition, this procedure was attributed to KINGSLEY. It appears to have been first published by BORN, see "Die Plattenmodellirmethode," in *Arch. f. mik. Anat.*, 1883, p. 591.]

There remains the watch-glass method. Melt paraffin in a watch-glass, and throw the object into it; or place the object in the watch-glass, add solid paraffin and warm. After the mass has hardened, cut out a block containing the object (this is of course applicable to other masses, such as celloidin). If paraffin be used, you may, instead of cutting out a block, turn out the whole mass of paraffin by simply warming rapidly the bottom of the glass. To facilitate the removal of the mass, some persons lubricate the watch-glass before pouring in the mass. To do this, a drop of glycerin should be smeared over it and wiped off with a cloth until hardly a trace of it remains.

As regards the merits of the watch-glass process, I wish to say that, as regards small objects at all events, I consider it the very best process of any.

CHAPTER XV.

IMBEDDING METHODS, PARAFFIN AND OTHER FUSION MASSES.

267. As to the Various Infiltration Methods.—Amongst the very various methods of infiltration-imbedding that have been proposed, two are pre-eminently important—the paraffin method, for small objects, and the celloidin or collodion method for large objects.

The subject of the respective merits of paraffin and celloidin still affords matter for discussion to some persons. The case, however, seems to be a very simple one. Celloidin does not afford by a long way the thinnest sections that are obtainable with small objects. For such objects it is therefore not equal to the demands made by modern minute anatomy, and paraffin must be taken. On the other hand, paraffin (as at present employed), will only cut very thin sections with small objects —with objects of 7 millimètres diameter you cannot get with paraffin thinner sections than you can with celloidin; and if you try to cut in paraffin objects of somewhat greater size, 10 mm. and upwards, it will probably happen that you will not get perfect sections at all, blocks of paraffin of this size having a tendency to split under the impact of the knife. So that for large objects, celloidin gives better results.

I have not been able to satisfy myself that the preservation of the tissues is better in celloidin sections than in paraffin sections; so that—convenience apart—the case remains as above stated,—paraffin for small sections, celloidin for large ones.

To this must be added—aqueous masses, such as gum or gelatin, for very special cases.

It is the purpose of this chapter to describe the paraffin method, and to mention some other masses that can be employed in a similar manner.

The celloidin method and the remaining methods will form the subject of the next chapter.

268. Penetration or Clearing.—The first stage of the paraffin method consists in the penetration or infiltration of the object by some substance which is a solvent of paraffin. The process may be called a clearing process, since the chief substances used for infiltration are also "clearing" agents.

The process of penetration or clearing should be carefully performed with well-dehydrated objects in the manner described in a former chapter.

Penetration liquids being merely liquids that are on the one hand miscible with alcohol and on the other hand good solvents of paraffin, are as numerous as could be wished. Amongst them may be mentioned—essence of turpentine, clove oil, creasote, benzol, xylol, toluol, oil of cedar wood, and chloroform.

Turpentine penetrates well and mixes readily with paraffin. I do not, however, recommend it, because in my experience it is of all others the clearing agent that is the most hurtful to delicate structures.

Clove oil penetrates well and preserves delicate structures well; but it mixes very slowly with paraffin, and quickly renders tissues brittle.

Benzin has been recently recommended by BRASS (*Zeit. f. wiss. Mik.*, ii, 1885, p. 301).

Toluol (or toluen) has been recently recommended by HOLL (*Zool. Anz.*, 1885, p. 223).

Chloroform mixes well with paraffin, and after evaporation in a paraffin bath (in the manner described in the next paragraph) leaves behind a pure and very homogeneous paraffin, having but little tendency to crystallise. But it is deficient in penetrating power, so that it requires an excessive length of time for clearing objects of any size; and it must be very thoroughly got rid of by evaporation in the paraffin bath, or by successive baths of paraffin, as if the least trace of it remains in the paraffin used for cutting it will make it soft. Chloroform ought therefore to be reserved for small and easily penetrable objects.

Cedar-wood oil is, according to my continued experience, for the reasons stated by me in *Zool. Anz.*, 1885, p. 563, in general the best clearing agent for paraffin imbedding. It penetrates very rapidly; preserves delicate structure better than any clearing agent known to me; mixes readily with paraffin; and does not make tissues brittle even though they be kept for weeks or months in it.

269. The Paraffin Bath.—The objects having been duly "penetrated" or "cleared," the next step is to substitute melted paraffin for the penetrating or clearing medium.

Some authors lay great stress on the necessity of making the passage from the clearing agent to the paraffin as gradual as possible, by means of successive baths of mixtures of clearing agent and paraffin kept melted at a low temperature, say 35° C. With oil of cedar or toluol at all events, this is not necessary. All that is necessary is to bring the objects into melted paraffin kept just at its melting point, and keep them there till they are thoroughly saturated; the paraffin being changed once or twice for fresh only if the objects are sufficiently voluminous to have brought over with them a notable quantity of clearing agent.

The practice of giving successive baths first of soft and then of hard paraffin appears to me entirely illusory.

It is important to keep the paraffin *dry*, that is, protected from vapour of water during the bath.

It is still more important to keep it as nearly as possible at melting point. If it be heated for some time to a point much over its normal melting point, *the melting point will rise*, and you will end by having a harder paraffin than you set out with. And as regards the preservation of tissues, of course the less they are heated the better.

The duration of the bath must, of course, vary according to the size and nature of the object. An embryo of the size of a pea ought to be thoroughly saturated after an hour's bath, or often less, if cedar oil has been used for clearing.

If chloroform be preferred, choice may be made of two methods; either, as in Giesbrecht's method, the chloroform containing the objects is heated to the melting point of the paraffin, and the paraffin gradually added, and the mass kept at the melting point of the pure paraffin until all the chloroform is driven off; or, as in Bütschli's method, the objects are simply passed direct from chloroform into a solution of paraffin in chloroform, in which they remain until thoroughly impregnated (half to one hour), and which is then evaporated at the melting point of the paraffin. Bütschli recommends a paraffin solution melting at 35°. (Such a solution is made of about equal parts of chloroform and paraffin of 50° melting point.) Or, in the case of larger objects, instead of evaporating the chloroform (which is often a very long process, as the chloroform must be *completely* driven off, or the mass will remain too soft for cutting), Bütschli simply transfers them from the bath of paraffin solution to a bath of pure paraffin.

Giesbrecht's method (*Zool. Anz.*, 1881, p. 484), more fully stated, is as follows:

Objects to be imbedded are saturated with absolute alcohol, and then brought into chloroform (to which a little sulphuric ether has been added if necessary, in order to prevent the objects from floating). As soon as the objects are saturated with the chloroform, the chloroform and objects are gradually warmed up to the melting point of the paraffin employed, and during the warming small pieces of paraffin are by degrees added to the chloroform. So soon as it is seen that no more bubbles are given off from the objects, the addition of paraffin may cease, for that is a sign that the paraffin has entirely displaced the chloroform in the objects. This displacement having been a *gradual* one, the risk of shrinkage of the tissues is reduced to a minimum.

270. Stoves and Water-Baths.—It is important that the paraffin should not be exposed to a moist atmosphere whilst it is in the liquid state. If a water-bath be used for keeping it at the required temperature, provision should be made for protecting the paraffin from the steam of the heated water.

A very convenient apparatus for this purpose is that of Paul Mayer, which will be found described at p. 146 of *Journ. Roy. Mic. Soc.*, 1883. It may be procured from the Zoological Station at Naples (address—" Direzione della Stazione Zoologica, Napoli "), or from M. Paul Rousseau, 17, Rue Soufflot, Paris. See also *Amer. Natural.*, 1886, p. 910 ; and *Journ. Roy. Mic. Soc.*, 1887, p. 167.

Other similar forms of paraffin-heating apparatus are described in several places in the same journal, as also in *Zeit. f. wiss. Mik.*

But whenever the worker has gas at his disposition, it will be found infinitely preferable to employ a regulating stove or thermostat. I recommend the form described in FOL's *Lehrbuch*, p. 121. Other descriptions of similar apparatus will be found also in the above-named journals.

271. Imbedding IN VACUO.—There are objects which, on account of their consistency or their size, cannot be penetrated by paraffin in the ordinary way, even after hours or days in the bath. For such objects the method of imbedding in a vacuum renders the greatest services. It not only ensures complete penetration in a very short time—a few minutes,—but it has the further advantage of preventing any falling-in of the tissues such as may easily happen with objects possessing internal cavities if it be attempted to imbed them in the ordinary way.

The principle of this method is, that the objects are put through the paraffin bath *in vacuo*. In practice this may be realised by means of any arrangement that will allow of maintaining paraffin at the necessary temperature for keeping it fluid under a vacuum.

The apparatus of HOFFMANN will be found described and figured at p. 230 of *Zool. Anz.*, 1884. In this arrangement, the vacuum is produced by means of a pneumatic water aspiration pump, the vessel containing the paraffin being placed in a desiccator heated by a water-bath and furnished with a tube that brings it into communication with the suction apparatus. This arrangement is very efficacious and very simple, if the laboratory possesses a supply of water under sufficient pressure.

In order to obtain the requisite vacuum without the aid of water under pressure, a simple little apparatus has been designed by FRANCOTTE (*Bull. Soc. Belg. Micr.*, 1884, p. 45).

In this the vacuum is produced by the condensation of steam.

FOL (*Lehrb.*, p. 121) employs the vacuum apparatus of Hoffmann, but simplifies the arrangement for containing the paraffin. The paraffin is contained in a stout test-tube furnished with a rubber stopper traversed by a tube that puts it into communication with the pump. The lower end of the test-tube dips into a water-bath. You pump out the air once or twice, wait a few minutes to make sure that no more bubbles rise, then let the air in, turn out the object with the paraffin (which by this time will have become abnormally hard), and re-imbed in fresh paraffin.

272. Imbedding and Cooling.—As soon as the objects are thoroughly saturated with paraffin they should be imbedded by one of the methods given above (§ 266). If the watch-glass method be followed the paraffin bath will naturally have been given in the watch-glass used for imbedding, and no special imbedding manipulation will be necessary. In any case the important point now to be attended to is that *the paraffin be cooled as rapidly as possible*. The object of this is to prevent crystallisation of the paraffin, which may happen if it be allowed to cool slowly, and to get as homogeneous a mass as possible.

Very small objects may be taken out of the paraffin with a needle or small spatula and put to cool on a block of glass, then imbedded in position for cutting on a cone of paraffin by means of a heated needle in the manner described above (§ 266). In the use of the needle it should be noted that it is important *to melt as little paraffin as possible at one time* in order that that which is melted may cool again as rapidly as possible.

If the watch-glass method be adopted, float the watch-glass with the paraffin and objects on to cold water. Do not let it sink till all the paraffin has solidified. When cool, cut out blocks containing the objects; do this with a *slightly* warmed scalpel.

If paper trays be taken, cool them on water, holding them above the surface with only the bottom immersed until all the paraffin has solidified, as if you let them go to the bottom at once you will probably get cavities filled with water formed in your paraffin. Or, you may put them to cool on a block of cold metal or stone.

SELENKA recommends cooling the mass by passing a stream of cold water through the imbedding tube described above (§ 266).

It is well (KOROTNEFF) to let the mass containing the objects lie for a few hours exposed to the sun before cutting; the paraffin thus acquires a more favorable consistency for cutting.

The objects having been mounted on the carrier of the microtome in position for cutting, pare the blocks square to the knife, and sufficiently close down to the objects, and go round them with a lens. If any bubbles or cavities or opaque spots be present, prick with a heated needle till all is smooth and homogeneous. Minutes spent in this way are well invested.

It is well to cut within a few hours of imbedding if the structures be at all delicate, as paraffin may continue to crystallise slowly to a certain extent even after rapid cooling. But this danger is very greatly diminished if the mass have been properly cooled.

273. Cutting and Section-stretching.—Paraffin sections are cut *dry*, that is, with a knife not moistened with alcohol or other liquid. By this means better sections are obtained, but a difficulty generally arises owing to the tendency of sections so cut to curl up on the blade of the knife. It is often impossible by any means to unroll a thin section that has curled. To prevent sections from rolling, the following points should be attended to:

First and foremost, the paraffin must not be too hard, but must be taken of a melting point suitable to the temperature of the laboratory (for the winter season the temperature of the laboratory being between 15° and 17° C., a paraffin melting at about 45° C. should be taken; for hot summer weather, laboratory at 22° C., a paraffin of 48° C. melting point).

The exact degree of hardness necessary must be determined by experiment. If, after cutting has begun, the paraffin be found to be too hard, it may be softened by the following simple expedient (FOL, *Lehrbuch d. vergl. mikr. Anat.*, p. 123) : A lamp provided with a parabolic reflector is set up near the microtome in such a position that the heat-rays of the flame are thrown by the reflector on to the imbedded object. The right temperature is obtained by adjusting the distance of the lamp.

If, on the contrary, the paraffin be found too soft, it may be hardened by exposing it to the cooling influence of a lump of ice placed in the focus of a similar reflector.

It is often sufficient to moderate the temperature of the room by opening or closing the window, stirring the fire, setting up a screen, or the like.

Secondly, the knife should be set square, for the oblique position produces rolling, and the more the knife is oblique the more do the sections roll.

Thirdly, it is better to cut ribbons than disconnected sections; ribbons of sections will often cut perfectly flat even when the same mass will only give rolled sections if cut disconnectedly.

Special masses having less tendency to roll than pure paraffin have been proposed. Thus a mass composed of four parts of hard paraffin and one of vaselin has been recommended. I recommend, however, that all such mixtures be avoided.

Mechanical means may be employed. The simplest of these and perhaps the best is as follows :

During the cutting the edge of the section that begins to curl is caught and held down on the blade of the knife by means of a small camel-hair brush with a flat point, or by a small spatula made by running a piece of paper on to the back of a scalpel. Or, the section is held down by means of an ingenious little instrument called a "section-stretcher." This consists essentially of a little metallic roller suspended over the object to be cut in such a way as to rest on its free surface with a pressure that can be delicately regulated so as to be sufficient to keep the section flat without in any way hindering the knife from gliding beneath it.

(See the descriptions of various forms of section-stretchers, *Zool. Anzeig.*, vol. vi, 1883, p. 100 (SCHULTZE); *Mitth. Zool. Stat. Neapel*, iv, 1883, p. 429 (MAYER, ANDRES, and GIESBRECHT); *Arch. f. mik. Anat.*, xxiii, 1884, p. 537 (DECKER); *Bull. Soc. Belge Mic.*, x, 1883, p. 55 (FRANCOTTE); *The Microscope*, February, 1884, (GAGE and SMITH); WHITMAN's *Meth. in Mic. Anat.*, 1885, p. 91; *Zeit. f. wiss. Mik.*, iv, 2, 1887, p. 218 (STRASSER); as well as *Journ. Roy. Mic. Soc.*, iii, pp. 450, 916, and other places).

Another plan is, to allow the sections to roll but to control the rolling. To this end, the block of paraffin is pared to the shape of a wedge five or six times as long as broad, the object being contained in the head or broad part, and the edge turned towards the knife. The sections are allowed to roll, and come off as coils, the section of the object lying in the outermost coil, which will be found to be a very open one, indeed very nearly flat. Lay the coil on a slide with this end downwards, warm gently, and the part containing the object will unroll completely and lie quite flat.

Alcohol helps sections to unroll, and so does carbolic acid; it is said that if they be brought into concentrated carbolic acid after removal of the paraffin by benzin or turpentine, they will unroll themselves safely and float flat on the surface of the liquid.

274. Chain or "Ribbon" Section-cutting.—It is probably a familiar fact to the majority of workers with the modern microtomes that if a series of paraffin sections be cut in succession and not removed from the knife one by one as cut, but allowed to lie undisturbed on the blade, it not unfrequently happens that they adhere to one another by the edges so as to form a chain which may be taken up and transferred to a slide without breaking up, thus greatly lightening the labour of mounting a series. The following appear to me to be the factors necessary for the production of a chain. Firstly, the paraffin must be of a melting-point having a certain relation to the temperature of the laboratory.

Small sections can always be made to chain when cut from

a good paraffin of 45° C. melting-point in a room in which the thermometer stands at 16° to 17° C. (The temperatures quoted apply to the case of rooms heated by an open fire, and probably would not apply to the case of rooms heated by closed stoves, such as are usual in Germany.) At 15° C. the paraffin will be found a trifle hard. At 22° C. the proper melting-point of the paraffin will probably be found at about 48° C., but my observations at these temperatures are less extended. Secondly, the knife should be set square. Thirdly, the block of paraffin should be pared down very close to the object, and should be cut so as to present a straight edge parallel to the knife edge; and the opposite edge should also be parallel to this. Fourthly, the sections ought to be cut rapidly, with the swiftest strokes than can be produced. It is evident that this condition can only be conveniently realised by means of a sliding microtome; but it is by no means necessary to have recourse to special mechanical contrivances, as in Caldwell's automatic microtome. The Thoma microtome well flooded with oil is sufficient. But the automatic microtomes are certainly most advantageous for this purpose, and amongst them the Cambridge Rocking Microtome may be quoted as giving admirable results.

Various plans, such as coating the edges of the paraffin with softer paraffin, or with Canada balsam, or the employment of specially prepared paraffin, have been recommended, with the idea that they help the sections to stick. None of these devices is necessary. For the prepared paraffin of Spee, Brass, and Fœttinger, *see* below, § 279.

275. Collodionisation of Sections.—Some objects are by nature so brittle that, notwithstanding all precautions taken in imbedding and previous preparation, they break or crumble before the knife, or furnish sections so friable that it is impossible to mount them in the ordinary way without some impairment of their integrity. Ova are frequently in this case. The remedy for this state of things consists in covering the exposed surface of the object just before cutting each section with a thin layer of collodion, which serves to hold together the loose parts of even the most fragile sections in a wonderfully efficacious way.

The primitive form of the process was, to place a drop of

collodion on the free surface of each section just before cutting it. But this practice has two defects; the quantity of collodion employed sensibly softens the paraffin; and the thick layer of collodion when dry causes the sections to roll.

MARK (*Amer. Natural.*, 1885, p. 628; cf. *Journ. Roy. Mic. Soc.*, 1885, p. 738) gives the following directions:

Have ready a little very fluid collodion in a small bottle, through the cork of which passes a small camel-hair brush, which just dips into the collodion with its tip. The collodion should be of such a consistency than when applied in a thin layer to a surface of paraffin it dries in two or three seconds without leaving a shiny surface. Collodion of this consistency does not produce a membrane on the paraffin in drying, and therefore has no tendency to cause sections to roll. It has further the advantage that it penetrates to a certain depth below the surface of the preparation, and fixes the deeper layers of it in their places. The collodion must be diluted with ether as soon as it begins to show signs of leaving a shiny surface on the paraffin.

Take the brush out of the collodion, wipe it against the neck of the bottle, so as to have it merely moist with collodion, and quickly pass it over the free surface of the preparation. Care must be taken not to let the collodion touch the vertical surfaces of the paraffin, especially not the one which is turned towards the operator, as that will probably cause the section to become stuck to the edge or under surface of the knife. As soon as the collodion is dry, which ought to be in two or three seconds, cut the section, withdraw the knife, and pass the collodion brush over the newly-exposed surface of the paraffin. Whilst this last layer of collodion is drying, take up the section from the knife and place it with the collodionised surface downwards on a slide prepared with fixative of Schaellibaum. Then cut the second section, and repeat the manipulations just described in the same order. A skilful operator can cut ribbons of sections, collodionising each section.

HENKING (*Zeit. f. wiss. Mik.*, iii, 4, 1886, p. 478) objects to the above process that the ether of the collodion softens the paraffin, and proposes a solution of paraffin in absolute alcohol. The solution is made by scraping paraffin into absolute alcohol.

For extremely brittle objects, such as ova of Phalangida,

the same author recommends a thin (light yellow) solution of shellac in absolute alcohol.

276. Clearing and Mounting.—The sections having been obtained are generally mounted on a slide in serial order by one of the methods described in the chapter on Serial Section Methods. All that now remains is to get rid of the paraffin and mount or stain as the case may be. The following solvents of paraffin have been recommended for freeing sections from the paraffin with which they are infiltrated :—Turpentine, warm turpentine, a mixture of 4 parts of essence of turpentine with 1 of kreasote, kreasote, a mixture of turpentine and oil of cloves, benzin, xylol, thin solution of Canada balsam in xylol (only applicable to very thin sections), hot absolute alcohol, naphtha, or any other paraffin oil of low boiling point. Any of these may be used, but naphtha and xylol are probably in most respects the best. Toluol may be added to the list, and will be found to work very well. Personally, I prefer naphtha or toluol to anything else; xylol evaporates rather too quickly, and so does chloroform.

If the slide be warmed to the melting point of the paraffin, a few seconds will suffice to remove the paraffin if the slide be plunged into a tube of naphtha or toluol. The sections may be mounted direct from the naphtha, or the slide may be brought into a tube of alcohol to remove the naphtha for staining.

277. Recapitulation of the Paraffin Method, as recommended to be practised. Put into a small test-tube enough oil of cedar to cover your object. On to the oil pour carefully the same quantity of absolute alcohol. Take your (already dehydrated) object and put it carefully into the alcohol. Leave it until it has sunk to the bottom of the cedar oil. Then put it into paraffin kept at melting point in a watch-glass. After a time change the paraffin (*i. e.* put the object into a fresh watch-glass with clean paraffin) once or twice if the object be at all large. As soon as the object is thoroughly soaked with paraffin, float the watch-glass on cold water. When cool, cut out a block of paraffin containing the object and fix it with a heated needle on a cone of paraffin already mounted on the object-carrier of the microtome. Pare it square, and as close down to the object as possible on all sides except the one turned towards the knife; this had better have a wall of a millimetre or two, or more, according to the size of the object, left standing. Set the knife square. Set the block square to the knife-edge. Cut sections in chains or ribbons, collodionising them if necessary. Mount them in serial order on a slide prepared with Schaellibaum's collodion or Mayer's albumin. Warm, and remove the paraffin with naphtha. Stain or mount.

Paraffin Masses.

278. Pure Paraffin.—It is now pretty generally admitted that pure paraffin forms an imbedding mass greatly superior for ordinary work to any of the many fatty mixtures that used to be recommended. I have only to repeat here that a paraffin melting at 45° C. is that which in my experience gives the best results so long as the temperature of the laboratory is between 15°, and 17° C.; whilst for a temperature of 22° C. a paraffin melting at 48° is required. If the temperature of your laboratory have risen much above 22° C. you had better give it up and go dredging, for good section cutting with paraffin under such conditions is next to impossible.

Paraffin of the melting points named is easily found in commerce. Intermediate sorts may be made by mixing hard and soft paraffin. Two parts of paraffin melting at 50° with one of paraffin melting at 36° C. give a mass melting at 48° C.

Many workers of undoubted competence prefer masses somewhat harder than those recommended, viz. of melting points varying between 50° and 55° C. for the normal temperature of the laboratory.

Some authors recommend masses melting at 60° C. or higher. I am convinced that, besides being most hurtful to tissues, such masses have no *raison d'être* whatever in temperate climates.

Paraffin had better be obtained from Grübler, or one of the known dealers in microscopic reagents. Gaule recommends that the bluish-transparent sorts be taken. I should say, transparent by all means, but if possible, rosy rather than bluish. New paraffin is bluish; if kept long, which is well, it generally becomes rosy.

279. Prepared Paraffin (Pure).—GRAF SPEE (*Zeit. f. wiss. Mik.*, ii, 1885, p. 8) recommends the following preparation of commercial paraffin as giving a mass particularly favorable for ribbon-section cutting. Paraffin of about 50° C. melting point is taken and heated in a porcelain capsule by means of a spirit lamp. After a time, disagreeable white vapours are given off, and the mass shrinks a little. This result is arrived at in from one to six hours, according to the quality of the paraffin. The mass then becomes brownish-yellow, and after cooling shows an unctuous or soapy surface on being cut. The melting point will be found to have risen several degrees.

BRASS (*l. c.*, p. 300) recommends the use of paraffin that has been kept for some years, as such has less tendency to crystallise than new paraffin. In this I concur.

FOETTINGER recommends (*Arch. de Biol.*, vi, 1885, p. 124) a somewhat complicated treatment with caustic potash, in which I have no faith (it was tried by one of the writers of the *Traité des Méth. Techn.*, during the preparation of that work).

280. Paraffin Mixtures and other similar Masses.—Of these, the only one that I think can be recommended for a moment is SCHULGIN's paraffin with a little cerisin (this is evidently what Schulgin means by "ceresin"). Or instead of cerisin, white wax (see *Zool. Anz.*, 1883, p. 21).

BRASS (*Zeit. f. wiss. Mik.*, ii, 1885, p. 301) recommends four to six parts of white wax to 100 of paraffin. Sections may be cleared with benzin.

Wax and Oil (STRICKER's formula, ROY's formula, *Handbuch d. Geweblehre*, pp. xxiii and 1202).

Wax and Oil (FOSTER and BALFOUR's formula, *Eléments d'Embryologie*, p. 296, 1877).

Paraffin and Axunge (HUXLEY and MARTIN's formula, FOSTER and BALFOUR's formula), *ibid.*

Paraffin and Tallow (SEILER's formula, *Compendium of Microscopical Technology*, 1881, pp. 47, 48; *Journ. Roy. Mic. Soc.*, N.S., i, p. 840).

Paraffin and Vaselin (FRENZEL's formula, *Zool. Anz.*, 1883, p. 51).

Spermaceti and Castor-Oil (KLEINENBERG's *Eléments d'Embryologie* (FOSTER and BALFOUR), pp. 296-8).

Spermaceti and Cacao-Butter (FOSTER and BALFOUR, *ibid*, p. 296).

Spermaceti, Castor-Oil, and Tallow (STRASSER's *Morph. Jahrb.*, v, 1879, p. 243; *Journ. Roy. Mic. Soc.*, N.S., i, p. 840).

Cacao-Butter, pure, is still used by some histologists. It melts somewhat under 35° C. The objects may be prepared by penetration with clove oil, which may also be used for removing the mass from the sections.

281. Bayberry Tallow (Myrtle Wax, Vegetable Wax, Japan Wax).—There appear to be more than one sort of vegetable wax in the market. That called Bayberry Tallow (see *Journ. Roy. Mic. Soc.*, 1885, p. 735) is a product of the North American bayberry tree, *Myrica cerifera*. Another sort, sold under the name of Myrtle wax or Japan wax (see *op. cit.*, 1888, p. 151), is the product of *Rhus succedanea*. It is this latter sort that, by all accounts, should be used for imbedding.

The wax may be used exactly as paraffin, chloroform being taken as the solvent (MILLER and BLACKBURN, *New York Med. Rec.*, 1885, p. 429; *Amer. Mon. Mic. Journ.*, 1887, p. 164; *Journ. Roy. Mic. Soc.*, 1887, p. 1048). In this case the superiority claimed for it over paraffin appears to the present writer doubtful. Or, it may be used in a similar way, but with *alcohol* as the solvent (*see* FRANCOTTE, *Bull. Soc. Belge de Mic.*, 1887, p. 140; *Zeit. f. wiss. Mik.*, iv, 2, 1887, p. 230; *Journ. Roy. Mic. Soc.*, 1887, p. 681). In this case it has the, for some objects, valuable quality of working without the accompaniment of such solvents as chloroform or benzol, which undoubtedly alter certain tissues more than alcohol. But on the whole I very much doubt the possibility of its rivalling paraffin for general work, and therefore refer the reader to the places quoted for the details of the manner of using it.

Soap Masses.

282. Utility of Soap Masses.—Soap masses certainly have many good points. The solvent is alcohol; the mass is highly transparent, very penetrating, and a good mass cuts far better than even paraffin. The mass may be cut either dry or with alcohol. As to the preservation of tissues, the mass is alkaline, which is against it; yet some workers still prefer soap to paraffin, and it has lately been recommended by so experienced a worker as Chun, for Siphonophora (certainly as delicate a class of objects as any that exist), on the ground of its producing less shrinkage than paraffin.

283. Transparent Soap (PÖLZAM, *Morph. Jahrb.*, iii, 1877, 3tes Heft, p. 558).—The following account is taken from Salensky's paper on the gemmation of Salpa, *l. c.*

Take good white soap ("gewöhnliche Kernseife"), cut it up into thin slices, and put them to dry in the sun for some days, until they become white. The slices are then to be rubbed up to a fine powder, which is mixed with spirit to the consistency of porridge. Now mix the porridge with alcohol and glycerin in such proportions that the whole shall contain for every 10 parts by weight of the soap, 22 parts of glycerin, and 35 parts of alcohol (90 per cent.). Let the whole simmer until there is obtained a perfectly transparent, syrupy, somewhat yellow fluid.

The objects, previously dehydrated in alcohol, are imbedded in this mass in the usual manner.

The mass may be removed from the sections either by means of water or of very dilute alcohol.

284. Transparent Soap (KADYI, *Zool. Anz.*, 37, 1879, vol. ii, p. 477).—Twenty-five grammes of shavings of stearate of soda soap (any stearate of soda soap will do, but the most to be recommended is the sort known in commerce as "weisse Wachskernseife") are heated in a retort with 100 c. c. of 96 per cent. alcohol over a water-bath until the whole is dissolved. Filter if necessary. If a drop of the solution be now poured into a watch-glass it will be seen that it almost immediately solidifies into a *white* mass. This is not what is wanted, and is a sign that the solution does not contain water enough. Small quantities of water are therefore added by degrees to the solution, and the effect tested from time to time by pouring a drop of the mixture into a watch-glass. The mass will be seen to become more and more pellucid until a point is reached at which it is almost perfectly transparent, with merely the slightest blue opalescence. The preparation of the mass is then complete.

It is not possible to state *a priori* the exact proportion of water that should be added, as this naturally depends on the amount of water already present in the sample of soap taken. In very many cases it will be found that for about 120 g. soap solution, 5 to 10 g. of water will be required.

It is necessary to be very cautious in adding the water, as if too much be taken the mass solidifies more slowly or not at all. The greatest amount of elasticity and consistency is possessed by the mass at the moment in which it contains exactly the minimum amount of water necessary to make it transparent.

The reasons for this process are explained as follows : Stearate of soda soap is soluble in divers proportions in warm alcohol. On cooling, the solution either solidifies into a homogeneous and pellucid mass, or into a white granular mass; or, in certain cases, does not solidify at all. The result in each case depends on the proportion of water present in the solution. For instance, if 5 to 6 parts of a tolerably dry soap be dissolved in 100 parts of 96 per cent. alcohol, a solution is generally obtained that solidifies into a transparent mass. But such a mass is too soft, and its melting point too low; it melts by the heat of the finger. If now, in order to get a harder mass, you add more soap, you will get a solution that solidifies on cooling into a *white granular* mass; and it is only after adding to it a *certain (small)* quantity of water that you will obtain a solution that solidifies on cooling into a *transparent* mass. If you add more water than is just absolutely necessary to this end the mass will have too low a melting point, and will solidify more slowly; and if still more water be added the solution will not solidify for hours, or, indeed, not at all. The more soap you have in your alcoholic solution the more water *must* you add in order to get a transparent mass, and the more *may* you add without depriving the solution of the faculty of solidifying. Besides the mass prepared in the proportions given above, useful masses may be made for certain purposes with 10, 20, 30, 40 per cent., or more or less of soap in alcohol. Weisker has employed a mass composed of about equal parts by weight of soap and alcohol. Such a mass is transparent, but yellow and oily, and takes a long time to solidify. When cool it is very tough. It requires a considerable temperature to liquefy it, and has less penetrating power than the more alcoholic masses. It is, however, very suitable for hard, and especially for chitinous, structures.

The mass recommended above boils at about 60° to 70° C. Objects should be imbedded in it in a watch-glass or in paper cases in the usual way. Whilst cutting, wet the knife and the mass with strong alcohol (one advantage of this method is that the knife remains perfectly clean). The sections are brought into 96 per cent. alcohol, which frees them from the mass instantaneously if warmed, and after a time if left cold.

285. Other Soap Masses (FLEMMING, *Arch. f. mik. Anat.*, 1873, p. 123; PFITZER, *Ber. deutsch. bot. Ges.*, 1887, p. lxv; *Journ. Roy. Mic. Soc.*, 1888, p. 316).

Gelatin Masses.

286. Gelatin Imbedding is a method that has the advantage of being applicable to tissues that have not been in the least degree dehydrated; and may render great service in the study of very watery objects.

The *modus operandi* is on the whole the same as for other fusion masses, with the difference that the objects are prepared by penetration with *water* instead of alcohol or a clearing agent. After the cooling of the mass it may sometimes be cut at once; but it is generally necessary to harden it. This may be done by treatment for a few minutes with absolute alcohol (KAISER), or for a few days with 90 per cent. alcohol (KLEBS), or chromic acid (KLEBS), or it may be frozen (SOLLAS).

The mass is removed from the sections by means of warm water.

287. Klebs' Gelatin (Glycerin Jelly) (*Arch. f. mik. Anat.*, v, 1869, p. 165).—A concentrated solution of isinglass is mixed with half its volume of glycerin.

288. Kaiser's Gelatin (*Bot. Centralb.*, i, 1880, p. 25; *Journ.*

Roy. Mic. Soc., iii, 1880, p. 504).—One part by weight of the finest French gelatin is left for about two hours in 6 parts by weight of water; 7 grammes of glycerin are added, and for every 100 grammes of the mixture 1 gramme of concentrated carbolic acid. The whole is warmed for ten to fifteen minutes, stirring all the while, until the whole of the flakes produced by the carbolic acid have disappeared. Filter whilst warm through the finest spun glass which has been previously washed in water and laid whilst wet in the filter.

289. Gerlach's Gelatine (GERLACH, *Unters. a. d. Anat. Inst. Erlangen*, 1884; *Journ. Roy. Mic. Soc.*, 1885, p. 541).—Take gelatin, 40 grammes; saturated solution of arsenious acid, 200 c.c.; glycerin, 120 c.c. Clarify with white of egg. The mass may be kept for years in a well-stoppered bottle. The objects to be prepared for imbedding by a bath of one third glycerin.

CHAPTER XVI.

COLLODION (CELLOIDIN) AND OTHER IMBEDDING METHODS.

290. Collodion or Celloidin.—The collodion method is due to DUVAL (*Journ. de l'Anat.*, 1879, p. 185). Celloidin, recommended later on by MERKEL and SCHIEFFERDECKER (*Arch. f. Anat. u. Phys.*, 1882, p. 200), is merely a patent collodion.

It is stated to be a preparation of pure pyroxylin, and is patented for Germany and England under the name of "Schering's Celloidin." It is manufactured by the *Chemische Fabrik auf Actien* (vorm. E. Schering), Berlin, N. Fenstrasse, 11, 12. It may be obtained through the post by writing to Schering's Grüne Apotheke, Wittick and Benkendorf, Berlin, N. Chaussée-Strasse, No. 19, or from Grübler, or the other dealers in histological reagents.

It is stated to be prepared with the purest pyroxylin, and to be always of a uniform composition. It is sent in the form of tablets of a tough, gelatinous consistency, and slightly milky-white transparency. These tablets have exactly the consistency that is required for section-cutting. They contain 20 per cent. of pure pyroxylin. Celloidin is entirely soluble, in all proportions, in ether and alcohol. It is free from acids. It is not detonant. If ignited it burns like paper; heated in a test-tube it carbonises without exploding.

In order to make a 2 per cent. collodion, take one tablet of celloidin (which contains 40 grammes of the dry pyroxylin) and such a quantity of alcohol and ether that the whole shall weigh 2000 grammes. For a 3 per cent. collodion you take such a quantity of alcohol and ether that the whole shall weigh 1333 grammes; and for a 4 per cent. collodion such a quantity that the whole shall weigh 1000 grammes. The relative portions of alcohol and ether may be taken according to discretion. To prepare a medicinal collodion according to the Prussian Pharmacopœia, you take for each tablet 720 grains of ether and no alcohol, as the celloidin already contains the prescribed proportion of alcohol. The tablets cost three marks (= three shillings) each. A single tablet would, I think, suffice for imbedding many hundreds of embryos.

There is a strife of opinion amongst authorities as to the relative merits of celloidin and common collodion. DUVAL, re-stating the method lately (*Journ. de Micr.*, 1888, p. 197), teaches that celloidin has no real advantage over common collodion, whilst the latter has the advantage of being more transparent. SCHIEFFERDECKER, also re-stating the method, declares that celloidin has "many points of superiority" (*Zeit. f. wiss. Mik.*, v, 4, 1888, p. 504). On carefully reading Schiefferdecker's paper, however, it appears that the only reason alleged in support of this superiority is that it is more "convenient" to make a thick mass by making a strong solution of celloidin than by allowing common collodion to concentrate by evaporation.

Personally I incline to Duval's point of view. The superiority of celloidin, if it exists, is of the nature of a mere matter of convenience. (Unless, indeed, the mass recommended by Apáthy, § 292, of which the description reaches me whilst preparing these sheets for the press, should prove to be really superior.) Otherwise there is hardly a pin to choose between the two, and therefore in this work the terms collodion and celloidin are used indifferently.

291. Preparation of Objects.—The objects must first be thoroughly dehydrated with absolute alcohol. They are then soaked till thoroughly penetrated in ether, or, which is better, in a mixture of ether and absolute alcohol. DUVAL (*l. c.*) takes for this purpose a mixture of ten parts of ether to one of alcohol; SCHIEFFERDECKER (and the majority of workers) a mixture of equal parts of ether and alcohol.

This stage may be omitted if the objects are of a sufficiently permeable nature, and they may be brought direct from alcohol into the collodion bath.

292. The Collodion Bath.—The next step is to get the objects penetrated with thick collodion. The secret of success here is to penetrate them first with a thin solution, then with the definitive thick one.

If celloidin be taken the solutions are made in a mixture of ether and absolute alcohol in equal parts. If collodion be taken the thin solution may be made by diluting it with ether. Apáthy recommends that celloidin be allowed to dry in the air until it becomes yellow, transparent, and of a horny consistency, and that it be then dissolved in the alcohol and ether (sulphuric, free from acid). The solutions thus prepared are free from the excess of water that is present in the undried celloidin, and give after hardening a mass that is more transparent and of a better consistency for cutting (*Zeit. f. wiss. Mik.*, vi, 2, 1889, p. 164).

The objects ought to remain in the first bath until very thoroughly penetrated;—days, even for small objects,—weeks, or months, for large ones (human embryos of from six to twelve weeks for instance). If the object contain cavities, these should be opened to ensure their being filled by the mass.

When the object is duly penetrated by the solution, it should be brought into the thick one. SCHIEFFERDECKER (*l. c.*) recommends that this be done by allowing the thin solution to concentrate slowly (the stopper of the containing vessel being raised, for instance, by means of a piece of paper placed under it), and making up the loss from evaporation with thick solution.

293. Imbedding.—At this stage, if it has not been done before, the objects must be imbedded, that is, arranged in position in the receptacle in which they are to be hardened. For the usual manipulations, *see* § 266. I recommend the paper thimbles or cylindrical trays, fig. 2, as being very convenient for collodion imbedding. The bottoms, however, should be made of soft wood in preference to cork; cork is elastic, and bends in the object-holder of the microtome, deforming the mass and object. The box should be prepared for the reception of the object by pouring into it a drop of collodion, which is allowed to dry. The object of this is to prevent bubbles coming up through the wood or cork and lodging in the mass.

Objects may also be imbedded on a piece of pith or leather, which should also be prepared with a layer of dry collodion.

Watch-glasses, square porcelain water-colour moulds, and the like, also make convenient imbedding receptacles. Care should be taken to have them perfectly *dry*. Any of these receptacles or supports may be set with the mass under a glass shade, allowing of just enough communication with the air to set up a slow evaporation. Or porcelain moulds or small dishes may be covered with a lightly-fitting cover.

294. Hardening.—This is logically the next step, but as a matter of fact is frequently begun before. · For the different processes of the collodion method so run into one another that it is difficult to assign natural lines of demarcation between them.

The objects being imbedded, and in the stage at which we left them at the end of § 292, the treatment should be as follows. As soon as the added thick collodion (of which only just enough to cover the object should have been taken) has so far sunk down that the object begins to lie dry, fresh thick solution is added, and the whole is left as before. Provision should be made for slow evaporation, either in one of the ways above indicated, or, which is perhaps better, by setting the objects under a *hermetically* fitting bell-jar, which is lifted for a few seconds only once or twice a day. I have sometimes found it advantageous to set the objects under a bell-jar together with a dish containing alcohol; so that the evaporation is gone through in an atmosphere of alcohol. This is especially indicated for very large objects.

When the mass has attained a consistency such that the ball of a finger no longer leaves an impress on it, it should be scooped out of the dish or mould, or have the paper removed if it has been imbedded in paper, and be submitted to the next stage of the hardening process. (If the mass is found to be not quite hard enough to come away safely, it should be put for a day or two into weak alcohol, 30 to 70 per cent.)

Several methods are available for the definitive hardening process. One of these is the *chloroform* method, due to VIALLANES (*Rech. sur l'Hist. et le Dev. des Insectes*, 1883, p. 129). I recommend this method for *small* objects, because I find it more certain and more rapid than the alcohol method and preferable on account of a superior consistency it gives to the mass. (SCHIEFFERDECKER does not find this, v. *Zeit. f. wiss. Mik.*, v, 4, 1888, p. 506). For *large* objects the method is inferior to the alcohol method, because the rapid hardening of the external layers is an obstacle to the diffusion necessary to the hardening of the inner layers.

The method consists in bringing the objects into chloroform. " Under the influence of this reagent, the collodion coagulates into a mass having the consistence of wax, but having also an elasticity that renders it unbreakable, and having besides the precious quality of being admirably transparent, and possessing exactly the index of refraction of glass."

In some cases, a few *hours'* immersion is sufficient to give the requisite consistence. In no case did my specimens require more than three days. But the length of time required varies in a very inexplicable way, so that no rule can be given. The collodion frequently becomes opaque on being put into the chloroform, but regains its transparency after a time.

Small objects may be hardened by chloroform without preliminary hardening by evaporation. All that is necessary is to expose the mass to the air for a few seconds until a membrane has formed on it, and then bring it into chloroform. If the mass is in a test-tube this may be filled up with chloroform, and left for two or three days. By this time the collodion mass will be considerably hardened, and also somewhat shrunk, so that it can be shaken out of the tube. It is then brought into fresh chloroform in a larger vessel, where it remains for about six days, after which time it is generally ready for cutting.

Good chloroform is a necessity, as the reaction cannot be obtained with samples of chloroform that are not free from water.

The more commonly employed hardening method is the *alcohol* method. The objects are thrown into alcohol and left there until they have attained the right consistency (one day to several weeks). The bottle or other vessel containing the alcohol *ought not to be tightly closed, but should be left at least partly open.*

The strength of the alcohol is a point on which the practice of different writers differs greatly. Some take very weak alcohol; so ROLLETT, one-third alcohol (*see* ROLLETT'S *Unters. üb. d. Bau d. quergestr. Muskelf.*, 1885; or *Zeit. f. wiss Mik.*, iii, 1, 1886, p. 93). Others take equal volumes of absolute alcohol and water. SCHIEFFERDECKER (*Zeit. f. wiss. Mik.*, v, 4, 1888) recommends alcohol of 50 to 60 per cent.; THOMA alcohol of 82 per cent. (0.842 sp. gr., see *Journ. Roy. Mic. Soc.*, 1883, p. 305); DUVAL, alcohol of 36° (= 90 per cent., *Journ. de Microgr.*, 1888, p. 197).

Lastly, the mass may be frozen. After preliminary hardening by alcohol, it is soaked for a few hours in water, in order to get rid of the greater part of the alcohol (the alcohol should not be removed entirely, or the mass may freeze too hard). It is then dipped for a few moments into gum mucilage, in order to make it adhere to the freezing plate, and is frozen. The sections are brought into warm water. If the mass have frozen too hard, cut with a knife warmed with warm water.

A paper has been lately written by FLORMAN (*Zeit. f. wiss. Mik.*, vi, 2, 1889, p. 184), to recommend that the definitive hardening should be done without the aid of alcohol or chloroform, by simply cutting out the blocks, turning them over, and carefully continuing the evaporation process in the way described above. No doubt the author is right in claiming for this process a superior degree of hardening of the mass; but I doubt whether it is possible to carry the hardening much beyond the point attained by the chloroform or alcohol method without incurring a degree of shrinkage that must be destructive of the scientific value of the preparations.

295. Preservation.—The hardened blocks of collodion may be preserved till wanted in weak alcohol (70 per cent.). They may also be preserved dry by dipping them into melted paraffin (APÁTHY, *Zeit. f. wiss. Mik.*, v, 1, 1888, p. 45).

Reference numbers may be written with a soft lead pencil on the bottom of the paper trays, or with a yellow oil pencil on the bottom of the watch-glasses in which the objects are imbedded. On removal of the paper from the collodion after hardening, the numbers will be found impressed on the collodion.

296. Cutting.—If the object has not been stained in the mass, it will form so transparent a mass with the collodion that the arrangement of the object and sections in the right position may be rendered very difficult. It is, therefore, well to stain the collodion lightly, just enough to make its outlines visible in the sections. This may be done by adding picric acid or carmine dissolved in alcohol to the collodion used for imbedding, or to the bergamot oil used for clearing.

To fix a collodion block to the microtome, proceed as follows. Take a piece of soft wood, or, for very small objects, pith, of a size and shape adapted to fit the holder of the microtome. Cover it with a layer of collodion, which you allow to dry. Take the block of collodion, cut a slice off the bottom, so as to get a clean surface; wet this surface first with absolute alcohol, then with ether (or allow it to dry), place one drop of *very thick* collodion on the prepared wood or pith, and press down *tightly* on to it the wetted or dried surface of the block of collodion. Then throw the whole into weak (70 per cent.) alcohol for a few hours (or even less) in order that the joint may harden.

For objects of any considerable size, it is important not to use cork for mounting on the microtome, especially if the object-holder be a vice, for cork bends under the pressure of the holder, and the elastic collodion bends with it, deforming the object. I have seen large embryos so deformed in this way that the sections obtained were true calottes, segments of a sphere. If the object-holder be of the cylinder type, as in the later forms of the Thoma microtome, the above-described accidents will be less likely to happen, and a good cork may be used, but even then, I think, wood is safer.

Sections are cut with a knife kept abundantly wetted with alcohol (of 50 to 85 or even 95 per cent.). Some kind of drip arrangement will be found very useful here. Apáthy recommends that the knife be smeared with yellow vaselin; it cuts better, is protected from the alcohol, and the mobility of the alcohol on the blade is lessened.

The knife is set in as oblique a position as possible.

Very brittle sections may be collodionised as explained above (§ 275).

The sections are either brought into alcohol (of 50 to 85 or 95 per cent.) as fast as they are made, or if it be desired to mount them in series they are treated according to one of the methods described below, in the chapter on Serial Section Methods.

297. Staining.—The sections may now be stained in almost any stain that may be desired, either loose, or mounted in series on slides or on paper as described below. It is *not* necessary, nor indeed desirable, to remove the mass before staining.

298. Clearing and Mounting.—You may mount in glycerin without removing the mass, which remains as clear as glass in that medium.

You may mount in balsam, also without removing the mass, which does no harm, and serves the useful purpose of holding the parts of the sections together during the manipulations. Dehydrate in alcohol of 95 or 96 per cent. (not absolute, as this attacks the collodion). Clear with a substance that does not dissolve collodion. The clearing agents most recommended are origanum oil (*Ol. Origan. cretici* should be taken, not *Ol. Orig. gallici*), bergamot oil (said to make sections shrink somewhat); oil of sandal wood; lavender oil; oil of cedar wood (safe and gives excellent results, but acts rather slowly); chloroform; xylol, or benzin (may make sections shrink if not well dehydrated); or Dunham's mixture of three or four parts of white oil of thyme with one part of oil of cloves. (Is not white oil of thyme identical with origanum oil?)

Some specimens of clove oil dissolve collodion very slowly, and may be used, but I would not be understood to recommend it. The action of origanum oil varies much, according to the samples; some sorts do not clear the collodion, others dissolve it, others pucker it. MINOT (*Zeit. f. wiss. Mik.*, iii, 2, 1886, p. 175) says that Dunham's mixture "clarifies the sections very readily and softens the celloidin just enough to prevent the puckering, which is so annoying with thyme alone."

Carbolic acid has been recommended. WEIGERT (*Zeit. f. wiss. Mik.*, iii, 4, 1886, p. 480) finds that a mixture of 3 parts of xylol with 1 part of carbolic acid (anhydrous) clears well. But it must not be used with the basic anilin stains, as it discolours them. For these, anilin oil may be used with the xylol, in the place of carbolic acid.

Anilin oil clears well, but unless thoroughly removed the preparation becomes yellowish-brown (*see* VAN GIESON, *Amer. Mon. Mic. Journ.*, 1887, p. 49, or *Journ. Roy. Mic. Soc.*, 1887, p. 519, for a review of these clearing agents).

Beechwood kreasote has been recommended (by M. Flesch).

299. Double Imbedding in Collodion and Paraffin.—The best-hardened collodion masses are of a more elastic consistency than is desirable for fine section cutting. This defect may be to a certain extent remedied as follows: Dip the block of hardened collodion into chloroform, and imbed it simply in paraffin.

Kultschizky's Celloidin-Paraffin Method.—A more thoroughgoing procedure is the following *infiltration* method (*Zeit. f. wiss. Mik.*, iv, 1, 1887, p. 48) :—After the collodion bath, the object is soaked in oil of origanum (*Oleum Origani vulg.*). It is then brought into a mixture of origanum oil and paraffin, heated to not more than 40° C., and lastly into a bath of pure paraffin.

The mass may be preserved in the dry state, and may be cut dry.

RYDER (*Queen's Micr. Bull.*, 1887, p. 43; *Journ. Roy. Mic. Soc.*, 1888, p. 512) has modified the process by substituting chloroform for the origanum oil. He states that sections chain more easily than even in ordinary paraffin imbedding. They may be cleared for mounting either with chloroform or with Weigert's xylol and carbolic acid mixture (*supra*, § 298. Ryder says, "equal parts," but Weigert's formula is as above given).

Other Evaporation Masses.

300. Joliet's Gum and Glycerin Method (*Arch. Zool. exp. et gén.*, x, 1882, p. xliii; *Journ. Roy. Mic. Soc.* (N.S.) ii, 1882, p. 890).—Pure gum arabic dissolved in water to the consistency of a thick syrup. (Solutions of gum sold under the name of strong white liquid glue ("*colle forte blanche liquide à froid*") may also be used; they have the advantage of having a uniform consistency.)* Pour a little of the solution into a watch-glass, so as not quite to fill it, add from 6 to 10 drops of pure glycerin, stir until thoroughly mixed.

Between the limits of 6 to 10 drops of glycerin the proportions most suitable to the nature of the object and to the season of the year must be found by experimental trials. In the winter or in rainy weather less glycerin should be taken than in the summer or dry weather.

It is often well to soak the object in glycerin before putting it into the mass. In this case less glycerin should be added to the gum, in proportion to the amount of glycerin contained in the object.

The object is imbedded in the mass in the watch-glass, and the whole left to dry for from one to four days. When it has assumed a cartilaginous consistency, a block containing the object is cut out, turned over, and allowed to dry again until

* It is highly probable that these commercial preparations contain gelatin, and perhaps some other gum besides gum arabic.

wanted for use. A stove, or the sun, may be employed for drying, but it is best to dry slowly at the normal temperature. The block may be preserved in good condition almost indefinitely, the gum, when mixed with a sufficient quantity of glycerin, never becoming hard or brittle. It is generally better to wait till the blocks have assumed such a consistency that they cannot be easily bent. It is after having waited almost a week that the author always obtained the best sections. The gum is dissolved out from the sections by means of a drop of water on the slide. The sections are then covered, and a drop of glycerin being added, the preparation is complete as soon as the water has evaporated.

An infiltration mass. It has the advantage of being transparent. Joliet employs it for *Pyrosoma*. A similar mass was employed by Hertwig for *Ctenophora* (*Jen. Zeitsch.*, xiv (1880), pp. 313, 314 ; *Journ. Roy. Mic. Soc.* (N.S.), ii, p. 278).

It would probably be advantageous to add some preservative substance to this mass.

This mass can be cut dry.

301. **Stricker's Gum Method** (*Hdb. d. Gewebel.*, p. xxiv).—A concentrated solution of gum arabic. The object may be prepared in alcohol and imbedded in the gum in a paper case. The whole is thrown into alcohol, and after two or three days may be cut.

I have seen masses of admirable consistency prepared by this simple method.

302. **Hyatt's Shellac Method** (*Am. M. Micr. Journ.*, i, 1880, p. 8; *Journ. Roy. Mic. Soc.*, iii, 1880, p. 320).—Prepare the object by soaking in alcohol, and then put it for a day or two into a clear alcoholic solution of shellac. Take a cylinder of soft wood, split it, and make a groove in one or both of the half cylinders sufficiently large to admit the object without pressure. Imbed in the groove with plenty of thick shellac solution, and tie together the two halves of the cylinder with thread. In a day or two the shellac will be quite hard; the cylinder is then fixed in a microtome, is soaked with warm water, and sections made. Should the shellac prove so opaque as to interfere with a proper examination of the sections, a drop of borax solution will immediately remove this difficulty.

This process is intended for the purpose of making sections through hard chitinous organs consisting of several pieces, such as stings and ovipositors, retaining all the parts in their natural positions.

303. **Von Koch's Copal Method** (*Zool. Anz.* 2, vol. i, 1878, p. 36).—Small pieces of the object are stained in the mass and dehydrated with alcohol. A thin solution of copal in chloroform is prepared by triturating small fragments of

copal in a mortar with fine sand, pouring on chloroform to the powder thus obtained, and filtering. The objects are brought into a capsule filled with the copal solution. The solution is now slowly evaporated by gently heating the capsule on a tile by means of a common night-light placed beneath it. As soon as the solution is so far concentrated as to draw out into threads that are brittle after cooling, the objects are removed from the capsule and placed to dry for a few days on the tile, in order that they may more quickly become hard. When they have attained such a degree of hardness that they cannot be indented by a finger-nail, sections are cut from them by means of a fine saw. The sections are rubbed down even and smooth on one side with a hone, and cemented, with this side downwards, to a slide, by means either of Canada balsam or copal solution. The slide is put aside for a few days more on the warmed tile. As soon as the cement is perfectly hard, the sections are rubbed down on a grindstone, and then on a hone, to the requisite thinness and polish, washed with water, and mounted in balsam.

The process may be varied by imbedding the objects unstained, removing the copal from the sections by soaking in chloroform, decalcifying them if necessary, and then staining.

It is sometimes a good plan, after removing the copal, to cement a section to a slide by means of hard Canada balsam, then decalcify cautiously the exposed half of the specimen, wash, and stain it. In this way von Koch was able to demonstrate the most delicate lamellæ of connective tissue in *Isis elongata*.

This method was imagined in order to enable the hard and soft parts of corals to be studied in their natural relations. It is evidently applicable to the study of any structures in which hard and soft parts are intimately combined. It is certainly a method of the very greatest value.

304. Ehrenbaum's Colophonium and Wax Method (*Zeit. f. wiss. Mik.*, 1884, p. 414).—Ehrenbaum recommends that the objects be penetrated by a mass consisting of ten parts of colophonium to one of wax. The addition of wax makes the mass less brittle. Sections are obtained by grinding in the usual way. The mass is removed from them by means of turpentine followed by chloroform.

305. Weil's Canada Balsam Method (*Zeit. f. wiss. Mik.*, v., 2, 1888, p. 200).—Balsam heated till brittle when cold, then dissolved in chloroform. Heat the objects in the mass on a water-bath. For further details, see *Journ. Roy. Mic. Soc.*, 1888, p. 1042.

White of egg Masses.

306. The White of Egg Method consists in imbedding in a white of egg emulsion, and hardening by alcohol or by the combined action of alcohol and heat. The method had certainly a "*raison d'être*" at one time, as giving results which could not then be obtained by other means. But the method was extremely cumbrous, had many defects, and must now be considered to be entirely superseded. See *Zeit. f. wiss. Mik.*, 1884, p. 223 (RUNGE); *Morph. Jahrb.*, Bd. ii, 3tes Heft, 1876, p. 445 (CALBERLA); *Zool. Anz.*, 6, vol. i, 1878, p. 130 (SELENKA); *Journ. Roy. Mic. Soc.*, 1883, p. 304 (THOMA), and BECKER's *Zur. Anat. d. ges. u. kranken Linse* (RUGE).

Congelation Masses.

307. The Freezing Method.—Fresh tissues may be, and are, frequently frozen without being included in any mass, and in certain cases very satisfactory sections can be obtained in this manner. But the formation of ice crystals frequently causes tearing of delicate elements, and it is better to infiltrate the tissues with a mass that does not crystallise in the freezing mixture, but becomes hard and tough. Gum arabic affords such a mass. Some workers use common gum water, which is either poured into the well of the microtome or round the object on the object plate, according to the form of microtome used.

308. Syrup and Gum Congelation Mass (HAMILTON, *Journ. of Anat. and Phys.*, xii, 1878, p. 254).—Hamilton cuts sections (of hardened brain) in a Rutherford's freezing microtome. The hardening reagent having been soaked out by water the tissues are prepared for freezing in the following manner, which it is important to observe, otherwise it will be found that the crystals of ice so break up the delicate nervous tissue as to render it totally useless for minute examination. The tissues are to be well soaked in syrup. The sugar somewhat retards the freezing, and besides, seems to alter the manner of crystallisation, so that instead of the ice being spicular in form it becomes granular and does no injury to the parts.

The syrup requires to be of a particular strength, viz. double refined sugar, 2 ounces; water, 1 fluid ounce.

Wash the superfluous syrup from the surface, and put into the ordinary mucilage for an hour or so before cutting. Imbed in the freezing microtome with mucilage in the usual way. Float the sections into water.

309. Gum and Syrup Congelation Mass (COLE, *Methods of Microscopical Research*, 1884, p. xxxix; *Journ. Roy. Mic. Soc.* (N.S.) iv, 1884, p. 318).—Gum mucilage (B. P.), 5 parts; syrup, 3 parts. (For brain and spinal cord, retinæ, and all tissues liable to come in pieces put 4 parts of syrup to 5 of gum). Add 5 grains of pure carbolic acid to each ounce of the medium.

(Gum mucilage (B. P.) is made by dissolving 4 ounces of picked gum acacia in 6 ounces of water).

The syrup is made by dissolving 1 pound of loaf sugar in 1 pint of water and boiling.

This medium is employed for soaking tissues previous to freezing. They may remain in it for "any length of time; all the year round" if desired.

The freezing is conducted as follows: the gum and syrup is removed from the *outside* of the object by means of a cloth; the spray is set going and a little gum mucilage painted on the freezing-plate; the object is placed on this and surrounded with gum mucilage; it is thus saturated with gum and syrup, but surrounded when being frozen with mucilage only. This combination prevents the sections from curling up on the one hand, or splintering from being too hard frozen on the other. The mass ought to cut like cheese. Should freezing have been carried too far, wait for a few seconds.

310. Gelatin Congelation Mass (SOLLAS, *Quart. Journ. Mic. Soc.*, xxiv, 1884, pp. 163, 164; *Journ. Roy. Mic. Soc.* (N.S.) iv, 1884, p. 316).—"Instead of gum one uses gelatin jelly. This is prepared and clarified in the usual manner. It should set into a stiff mass when cold. . . . . The tissue to be cut is transferred from water to the melted jelly and should remain in it till well permeated."

The sections are transferred to a slide as soon as cut. On touching the glass they adhere to it. When enough sections have been thus arranged they are covered with a drop of glycerin; a cover is put on, and the mount closed with any suitable cement. In process of time the glycerin will permeate the gelatin and convert it into glycerin jelly; this may be hastened by placing the slide in an oven kept at about 20° to 30° C.

311. Gum-Gelatin Congelation Mass (JACOBS, *Amer. Natural.*, 1885, p. 734; *Journ. Roy. Mic. Soc.*, 1885, p. 900).—Gum arabic, 5 parts; gum tragacanth, 1 part; gelatin, 1 part. Dissolve in enough warm water (containing one sixth of glycerin) to give a mass of the consistency of thin jelly when cold.

312. White of Egg Congelation Mass (ROLLETT, *Denkschr. math. naturw. Kl. k. Acad. Wiss., Wien*, 1885; *Zeit. f. wiss. Mik.*, 1886, p. 92). —Small portions of tissue brought in the white of a freshly-laid egg on to the freezing stage, frozen, and cut. The knife must be well cooled.

CHAPTER XVII.

SERIAL SECTION MOUNTING.

313. Choice of a Method.—All the following methods are excellent, if properly carried out. I recommend for general work the following:—For paraffin sections that have been already stained, Schällibaum's collodion. For paraffin sections that are to be stained on the slide, Mayer's albumen. For collodion objects, one of the forms of Summers' ether-vapour process. For very large collodion sections, Weigert's process.

Methods for Paraffin Sections.

314. Gaule's Xylol Method. (*Arch. f. Anat. u. Phys.*) (*Phys. Abth.*), 1881, p. 156).—A slide is moistened with alcohol, the sections are arranged on it by means of a camel-hair brush, also moistened with alcohol; the slide is slightly warmed so as to cause the sections to stick to the slide; a cover is put on, and a solution of Canada balsam in xylol (equal parts of each) run underneath it. If the sections are not thicker than $\frac{1}{70}$th mm., they will be clear at once, and nothing remains but to refill the cell day by day as the xylol evaporates, in order to have a perfect mount. If, however, the sections are thicker than $\frac{1}{70}$th mm., they will contain more paraffin than the xylol balsam can dissolve. In that case, the excess of paraffin must be removed by means of a drop of pure xylol (the sections being first melted on to the slide as before), and the mount is completed by means of xylol balsam.

Both the moistening with alcohol, and the heating, are necessary for the attachment of the sections to the slide; the effect is not obtainable by means of one of these manœuvres alone.

This simple process is very useful for the preliminary examination of trial sections whilst cutting. It is often sufficient to put on a cover, warm, withdraw the melted paraffin by means of a cigarette paper, and run in a drop of clearing agent.

315. Schällibaum's Collodion (*Arch. f. mik. Anat.*, 1883, p. 565).—One part of collodion is shaken up with three to four volumes (according to the consistency of the collodion) of clove oil or lavender oil. This should give a clear solution. A little is spread thinly on a slide with a small brush. After arranging the sections on the prepared surface, warm over a water-bath, gently, until the clove oil has evaporated (five to ten minutes). The sections are then found to be fixed, and can be treated for days with turpentine, chloroform, alcohol, and watery fluids, without becoming detached. The advantage of this method is, that it allows of staining on the slide. If after staining any cloudiness should appear between the sections, dehydrate the slide and treat it several times with absolute alcohol and turpentine, warming it gently the while; or brush the space between the sections repeatedly with a brush moistened with clove oil. This cloudiness only arises from the collodion solution having been taken too concentrated, or having been laid on too thick on the slide.

I find it is not necessary to evaporate over a water-bath. It is sufficient to hold the slide over a spirit-lamp until the paraffin has melted and the clove oil has collected in drops between the sections. Schällibaum has stated elsewhere that long evaporation of the slide is necessary if the sections are to be secured firmly enough to allow of staining on the slide. That is not so. What is necessary is that the paraffin and clove oil be thoroughly removed from contact with the sections; and that can be done in a second (as was shown me by Professor v. Korotneff). Warm the slide over a flame, and whilst the paraffin is still melted hold it close before your lips and blow down on it vigorously. The paraffin and clove oil are scattered right and left over the slide, leaving the sections high and dry.

Personally, I do not consider Schällibaum's method so safe as Mayer's albumen (and some other methods) for objects that are to be stained on the slide. I recommend it for already stained objects, because it is found to work more pleasantly than any other method. I recommend xylol or naphtha for clearing, in preference to turpentine.

Good collodion is essential in this process.

STRASSER (*Zeit. f. wiss. Mik.*, iv, 1, 1887, p. 45) recommends a mixture of 2 parts collodion, 2 parts ether, and 3 parts castor oil; or (*Ibid.*, vi, 2, 1889, p. 153) 2 parts of collodion with one of castor oil, the sections being painted over with a thicker solution, viz. collodium concentratum duplex, 2 to 3 parts, castor oil, 2 parts, and the slide being plunged at once, without warming, into a bath of turpentine, in which it remains till the paraffin is dissolved (2 to 10 hours, somewhat less if the whole be put in a stove). The turpentine suffices to harden the collodion (benzin, benzol, and chloroform have the same effect).

GAGE prefers preparing slides with a layer of pure collodion, which is allowed to dry, and is rendered adhesive at the instant of using by brushing with clove oil.

SUMMERS (*Amer. Mon. Mic. Journ.*, 1887, p. 73; *Zeit. f. wiss. Mik.*, iv, 4, 1887, p. 482) also employs a dry layer of collodion, which he renders adhesive after the sections are arranged on it, by wetting with a mixture of equal parts of alcohol and ether. As soon as the mixture has evaporated, the sections are found to be fixed.

316. **Strasser's Collodion-Paper Method** (*Zeit. f. wiss. Mik.*, iii, 3, 1886, p. 346). This is an extremely complicated modification of Weigert's collodion method for celloidin sections (*post*, § 327).

In a later communication (*Ibid.*, vi, 2, 1889, p. 154) STRASSER describes the following modified form of the process. For sections of *already stained objects* paper is prepared by saturating it evenly and thinly with wax, its surface is prepared with the thin collodion mixture given in the last paragraph, the sections are arranged and painted over with the thick mixture as there described, and the whole is brought into turpentine, which dissolves the wax and sets free the plate of collodion containing the sections. The plate is then either mounted on glass in the usual way, or is treated as follows. The paper is lifted out of the turpentine, with the collodion plate lying free on it; the collodion plate is painted over with a thickish layer of thick solution of resin, and covered (as with a cover-glass) with tracing paper. The whole is then allowed to dry a little on filter-paper, and put away, lightly wrapped between folds of filter paper, so that the air may have access to it, in an album (avoiding pressure). For the microscopic study of such preparations they are mounted between glass with some clearing fluid, such as oil or kreasote. If it be required to definitely mount them in the usual way, they are put into turpentine, which dissolves the resin, and sets free the plate of collodion with the sections.

For sections that are to be stained the procedure is as follows. Well-sized paper is prepared with a thick solution of gum containing 10 per cent. of glycerin, and dried. It is then prepared for use with a surface of the thin collodion mixture, on which the sections are arranged and painted over with the thick mixture. The whole is then brought into rectified turpentine until the paraffin is dissolved and the collodion hardened, and then (in order to remove the last traces of castor oil) into a second bath of clear turpentine. The preparation is then pressed in a press between sheets of filter paper in order to get rid of the turpentine mechanically (alcohol is not applicable here), and brought for a quarter to half an hour into a mixture of equal parts of chloroform and 95 per cent. alcohol. It is then passed through successive alcohols into the staining fluid.

For ordinary work this process is evidently too elaborate to have any chance of supplanting the classical methods of preparing the series on the slide. Strasser recommends it for very large sections of the central nervous system (entire *pons* of the adult). But it seems to me that he will hardly have many followers in his belief that as good sections of such large objects can be obtained by the paraffin method as by the celloidin method; for which reasons the method appears to me superfluous for ordinary purposes.

317. The Shellac Method (GIESBRECHT, *Zool. Anz.*, 1881, p. 484).—Prepare a stock of slides covered with a thin and even film of shellac. This is done as follows: Make a not too strong solution of brown shellac in absolute alcohol, filter it thoroughly; warm the slides, and spread over them a layer of shellac by means of a glass rod dipped in the solution and drawn once over each slide. Let the slides dry.

Just before beginning to cut your sections take a prepared slide and brush it over *very thinly* with kreasote applied by means of a brush; this forms a sticky surface on which the sections are now arranged one by one as cut, care being taken to bring them on to the slide with as little surrounding paraffin as possible.

When all the sections are arranged the slide is heated on a water-bath for about a quarter of an hour at the melting point of the paraffin; this causes the paraffin to run down into a thin layer, and allows the sections to fall through it and come into close contact with the shellac film, whilst at the same time it evaporates the kreasote.

The slide is allowed to cool, and the sections are now found to be firmly fixed in the shellac. The paraffin is dissolved away by dropping turpentine on to the sections, which are then mounted in Canada balsam. There is no danger of the sections being floated away by the turpentine, because turpentine does not dissolve shellac.

In the note in the *Zool. Anz.* above quoted, the shellac solution is stated to be prepared with common brown shellac (choosing, of course, by preference the paler sorts), on account of the insolubility of white shellac in alcohol. In the *Mitth. d. Zool. Stat.* of Naples, of the same year, "bleached white shellac" is recommended to be dissolved as before, in absolute alcohol.

In the *Journ. Roy. Mic. Soc.* (N.S.), vol. ii, 1882, p. 888, it is stated (on whose authority is not clear) that the solution is made by mixing 1 part of bleached shellac with 10 parts absolute alcohol, and filtering. In the same place it is added that " Dr. Mark uses the bleached shellac in the form in which it is prepared for artists as a 'fixative' for charcoal pictures. It is perfectly transparent, and a film of it cannot be detected unless the surface is scratched. He attaches a small label to the corner of the slide, which serves for the number of the slide and the order of the sections, and at the same time marks the shellac side otherwise not distinguishable." (The latter object is better attained by gumming a paper square, or spinning a ring with ink, in the centre of the unprepared surface of the slide. The disc or ring then serves at the same time for centering the group of sections.

The account given in the *Mitth. d. Zool. Stat.* further varies in one other detail from that given in the *Zool. Anz.* It directs that the shellac slides be brushed before cutting with *oil of cloves*, instead of kreasote, the slide being slightly warmed before brushing.

The white shellac of commerce is sometimes not easily soluble in alcohol. KINGSLEY (*see* WHITMAN's *Methods in Microscopical Anat.*, p. 117) recommends that brown shellac be taken and bleached by exposure to the sun.

CALDWELL (*Quart. Journ. Mic. Soc.* (N.S.), lxxxvii, 1882, p. 336) simplifies the method by merely brushing over the slide (thinly) at the moment of using with a strong solution of shellac in anhydrous kreasote. (To make the solution, warm the kreasote.)

In both the foregoing methods it often happens that the shellac becomes granular or cloudy on the slide. P. MAYER attributes this to the kreasote or clove oil, and proposes to remedy it by employing carbolic acid instead (*Amer. Natural.*, 1882, p. 733; *Zeit. f. wiss. Mik.*, iv, 1, 1887, p. 77); *Journ. Roy. Mic. Soc.*, 1885, p. 910). Powdered white shellac is heated with crystallised carbolic acid till it dissolves, and the solution filtered warm.

But more recently (*Intern. Monatschr. f. Anat. &c.*, 1887, H. 2; *Zeit. f. wiss. Mik.*, iv, 1, 1887, p. 77) the same author, on the ground that hot carbolic acid attacks some tissues, recommends another method. Slides are prepared with alcoholic shellac according to Giesbrecht's plan. The sections are arranged on the dry film and gently pressed down on to it, then exposed for half a minute to vapour of ether.

Chloroform softens shellac; therefore, chloroform balsam is not a safe mounting medium for sections fixed by these methods.

These methods do not allow of staining on the slide.

I feel bound to say that I am at a loss to understand by what virtue it is that the shellac method continues to survive, as it certainly seems to do, in the face of far more convenient and efficient processes.

318. Mayer's Albumen (*Mitth. Zool. Stat. Neapel*, iv, 1883; *Journ. Roy. Mic. Soc.* (N.S.), iv, 1884, p. 317; *Internat. Monatschr. f. Anat.*, 1887, Hft. 2; *Journ. Roy. Mic. Soc.*, 1888, p. 160).—White of egg. 50 c.c., glycerin, 50 c.c., salicylate of soda, 1 gramme. Shake them well together, and filter into a clean bottle.

Fol (*Lehrb.*, p. 134) takes whipped white of egg, filters it through a Bunsen filter, and adds the glycerin and a little camphor or carbolic acid.

According to my experience, carbolic acid is perfectly efficient as a preservative, but is not to be recommended because it precipitates a great deal of the albumen.

A thin layer of the mixture is spread on a cold slide with a fine brush and the sections laid on it, and warmed for some minutes on a water-bath. As the paraffin in the sections melts it carries the albumen away from them, and this is one of the advantages of the method. The sections may be treated with turpentine, alcohol, and aqueous or other stains without any danger of their moving.

The function of the glycerin is merely to keep the layer of albumen moist.

This method allows of the staining of sections on the slide with anilin stains, which is seldom practicable with Schällibaum's method, as the collodion stains with most anilin stains, and does not yield up the colour to alcohol.

The slide should be very thoroughly treated with alcohol after removal of the paraffin, in order to get rid of the glycerin, which will cause cloudiness if not perfectly removed.

According to my experience, this method is *absolutely safe*, and is the one that should in general be preferred for staining on the slide, more especially for staining with anilins by Flemming's method.

319. Flögel's Gum Method (*Zool. Anz.*, 1883, p. 565).—Make a solution of one part gum arabic in twenty parts water, filter, and add a little alcohol to prevent the formation of mould. Slides are prepared by pouring the solution over them, and draining. (It is important that the slides be so perfectly clean as to be evenly wetted all over by the gum solution.) Sections may now be cut and laid on the gum surface before it has become dry, and floated into the proper position; this is the best plan for sections of $\frac{1}{100}$ mm. thickness, and for large sections. For thinner and small sections it is best to take slides that have completely dried, arrange the sections dry on the gum film, and then breathe on it until the gum has become sticky.

A very neat method for cases in which it is not required to treat the slide with watery fluids.

WADDINGTON (*Journ. Quek. M. Club*, vi, 1881, p. 199 ; *Journ. Roy. Mic. Soc.* (N.S.), i, 1881, p. 704) gives the following process for preparing "Arabin," a purified gum arabic which has the advantage of not presenting a granular appearance under the microscope as ordinary gum arabic does.

Dissolve clear and white gum arabic in distilled water to the consistency of thin mucilage. Filter. Pour the filtrate into rectified alcohol, and shake well; the arabin separates as a white pasty mass. Place it on filter paper and wash with pure alcohol until the washings are free from water. Dry.

The white powder thus obtained should be dissolved in distilled water and filtered twice. It may then be placed on slides, which are drained, dried, and put away till wanted. In this condition it may be preserved indefinitely.

320. Frenzel's Gum Method (*Arch. f. mik. Anat.*, Bd. xxv, 1885, p. 51).—Gum arabic is dissolved in water to the consistency of a thin mucilage, and to this is added aqueous solution of chrome-alum. An excess of the latter does no harm. Finally, add a little glycerin and a trace of alcohol (*l. c.*, p. 142). The slide is prepared with this in the usual way, the sections (either cut dry or in the wet way) are gently pressed on to it with a brush and slightly melted on, and heated for at most a quarter of an hour at a temperature of 30° to 45° C., which suffices to render the gum insoluble. This layer has the advantage of not staining with the majority of staining-fluids; fuchsin and safranin are the only ones that stain it to a harmful degree. In the other anilins, and in carmine or hæmatoxylin, it does not stain. Watery stains (it is stated) may be used with it.

321. Born and Wieger's Quince-Mucilage (*Zeit. f. wiss. Mik.*, 1885, p. 346).—To two volumes of the ordinary pharmaceutical quince-mucilage add one volume of glycerin and a trace of carbolic acid. Spread in a thin layer on a carefully cleaned slide, and arrange the sections on the moist surface. Heat for twenty minutes at a temperature of 30° to 40° C. After removal of the paraffin by turpentine the slide is brought for half an hour into *absolute* alcohol. You may then mount, or pass through successive alcohols, and stain. Alkaline staining fluids must be avoided, as they soften the mucilage and cause the sections to become detached.

Methods for Watery Sections.

322. Fol's Gelatin (FOL, *Lehrb.*, p. 132).—Four grammes of gelatin are dissolved in 20 c.c. of glacial acetic acid by heating on a water-bath and agitation. To 5 c.c. of the solution add 70 c.c. of 70 per cent. alcohol and 1 to 2 c.c. of 5 per cent. aqueous solution of chrome alum. Pour the mixture on to the slide and allow it to dry. In a few hours the gelatin passes into the insoluble state. It retains, however, the property of swelling and becoming somewhat sticky in presence of water. The slide may then be immersed in water containing the sections, these can be slid into their places, and the whole lifted out; the sections will be found to be fixed in their places.

This method is especially useful for sections made under water, large celloidin sections amongst others.

323. Poli (MALPIGHIA, ii, 1888, 2, 3; *Zeit. f. wiss. Mik.*, v, 3, 1888, p. 361) arranges sections on a layer of melted Kaiser's gelatin (*supra*, § 288), adds glycerin, and covers.

324. Frenzel and Threlfall's Gutta-percha (or Caoutchouc) Method (*Zool. Anz.*, 1883, pp. 51, 301, and 423).—This extremely elegant method is not perfectly safe, the gutta-percha film being liable to tear; and is now, I believe, very generally abandoned.

Methods for Celloidin Sections.

325. Summers' Ether Method (*Amer. Mon. Mic. Journ.*, 1887, p. 73; *Zeit. f. wiss. Mik.*, iv, 4, 1887, p. 482; *Journ. Roy. Mic. Soc.*, 1887, p. 523).—Besides the method given above (§ 315), which is applicable to celloidin sections, but is needlessly complicated, Summers recommends the following simpler method.—Place the sections in 95 per cent. alcohol for a minute or two, arrange on the slide, and then pour over the sections sulphuric ether *vapour*, from a bottle partly full of liquid ether. The celloidin will immediately soften and become perfectly transparent. Place the slide in 80 per cent. alcohol, or even directly in 95 per cent. if desired. The sections will be found to be firmly fixed, and may be stained if desired.

SCHIEFFERDECKER (*Zeit. f. wiss. Mik.*, v, 4, 1888, p. 507) recommends that the slide be one that has been previously prepared with a layer of collodion, if it is desired to stain on the slide; but if not, a clean slide is perfectly sufficient. The slide may of course be treated with ether vapour in a preparation glass or similar arrangement.

326. Apáthy's Oil of Bergamot Method (*Mitth. Zool. Stat. Neapel*, 1887, p. 742; *Zeit. f. wiss. Mik.*, v, 1, 1888, p. 46, and v, 3, 1888, p. 360; *Journ. Roy. Mic. Soc.*, 1888, p. 670).
—Cut with a knife smeared with vaselin (§ 296) and wetted with 95 per cent. alcohol. Float the sections, as cut, on bergamot oil (must be green, must mix perfectly with 90 per cent. alcohol, and must not smell of turpentine). The sections spread themselves out on the surface of the oil; before they sink, each one is pushed by means of a needle into its place on a slip of tracing paper dipped into the oil. (A good size for the paper is, about as broad as the slide, and three times as long as the cover.) When the requisite number of sections has been arranged on the paper, you drain the paper, dry the under side of it with blotting-paper, turn it over, and gently press it down with blotting-paper on to a carefully dried slide. Remove the paper by rolling it up from one end. The sections remain adhering to the slide, and may have the remaining bergamot oil removed from them by means of a cigarette paper. If they are already stained, nothing remains but to add balsam and a cover.

In the case of unstained or very small objects, it is well to add a little alcoholic solution of safranin to the bergamot oil. The celloidin of the sections becomes coloured in it in a few seconds, and makes them readily visible. The colour disappears after mounting in a few days.

If the sections are to be stained, the slide after removal of the bergamot oil is exposed for a few minutes to the vapour of a mixture of ether and alcohol, then brought into 90 per cent. alcohol, and after a quarter of an hour therein may be stained in any fluid that contains 70 per cent. alcohol or more.

If it be desired to stain in a watery fluid, care must have been taken when arranging the sections to let the celloidin of each section overlap that of its neighbours at the edges, so that the ether vapour may fuse them all into one continuous plate. This will become detached from the slide in watery fluids, and may then be treated as a single section.

326 a. Apáthy's Series-on-the-Knife Method (*Zeit. f. wiss. Mik.*, vi, 2, 1888, p. 168).—The following is in some respects more convenient than the oil of bergamot method. The knife is well smeared with yellow vaselin rubbed evenly on with the finger, and is wetted with alcohol of 70 to 90 per cent. As fast as the sections are cut they are drawn with a needle or small brush to a dry part of the blade, and there arranged in rows, the celloidin of each section overlapping or at least touching that of its neighbours. The rows are of the length of the cover-glass, and are arranged one under the other so as to form a square of the size of the cover-glass. When a series (or several series, if you like) has been thus completed, the sections are dried by laying blotting-paper on them (there is no risk of their becoming attached to it, as they are held down by the vaselin). The series is then painted over with some of the thickest celloidin solution used for imbedding, is allowed to evaporate for five minutes in the air, and is then either wetted with 70 per cent. alcohol, and allowed to remain whilst cutting is proceeded with, or (if no more sections are to be cut, or if the knife is now full) the knife is removed and brought for half an hour into 70 per cent. alcohol. This hardens the celloidin around the sections into a continuous lamella, which can be easily detached by means of a scalpel, and stained, or further treated as desired. It is well to bring it at once on to a slide, moisten the edges of the celloidin plate with ether and alcohol mixture, so that it may not become detached, and bring the whole into the staining solution.

327. Weigert's Collodion Method (*Zeit. f. wiss. Mik.*, 1885, p. 490).—Sections are cut wet, with alcohol. Care should be taken not to have so much alcohol on the knife as to cause the sections to float. Prepare a slip of porous but tough paper (Weigert recommends "closet paper"), of about twice the width of the sections. Soak it in alcohol, take it by both ends, stretch it slightly, and lower it on to the section that is on the knife. The section will adhere to the paper, and is taken up by moving the slip horizontally or slightly upwards, away from the edge of the knife. Take up the first section towards the end of the paper that you hold in your left hand, and let the remaining sections follow in order from left to right. After each section has been taken up, the slip is placed, whilst the next section is being cut, with the sections upwards on a moist surface prepared by arranging several layers of blotting paper, covered with one layer of closet paper, in a plate, and saturating the whole with alcohol. When all the sections have been arranged on the slip, you pass to the next stage of the process, the collodionisation of the series.

This is done in two steps. The first of these consists in transporting the series on to a plate of glass prepared with collodion. The plate is prepared beforehand by pouring on to it collodion and causing it to spread out into a thin layer, as photographers do, and allowing it to dry. (A number of the plates may be prepared and kept indefinitely in stock; microscope slides will do for series of small sections.) Take one of these plates; lay the slip of paper with the sections on the plate, the sections downwards; press it down gently and evenly, and the sections will adhere to the collodion, then carefully remove the paper. (Do not place more than one or at most two lines of sections on the same plate, for those first placed run the risk of becoming dry whilst you are placing the others.) This finishes the first stage of the collodionising process.

Now remove with blotting-paper any excess of alcohol that may remain on or around the sections, pour collodion over them, and get it to spread in an even layer. As soon as this layer is dry at the surface you may write any necessary indications on it with a small brush charged with methylen blue (the colour will remain fast thoughout all subsequent manipulations).

The plate may now either be put away till wanted in 80 per cent. alcohol, or may be brought into a staining fluid. Weigert recommends his hæmatoxylin process (*see* § 180), but other watery stains may be used. The watery fluid causes the double sheet of collodion to become detached from the glass, holding the sections fast between its folds. It is then easy to stain, wash, dehydrate, and mount in the usual way, merely taking care not to use alcohol of more than 90 to 96 per cent. for dehydration. Weigert recommends for clearing the above-described mixture of xylol and carbolic acid (§ 298). Both the dehydration and the clearing take rather longer with the collodionised series than with free sections.

The series should be cut into the desired lengths for

mounting whilst in the alcohol. It is perhaps safer to lay them out for cutting on a strip of closet paper saturated with alcohol.

It is hardly necessary to comment on the great value of this beautiful method.

It is suggested by STRASSER that gummed paper (see *ante*, § 316) might be an improvement on the glass plates used in this process—especially for very large sections.

Other Methods.

328. Giacomini's Collodion-Gelatin Method.—*See* the Chapter on Nerve-Centres in Part II.

CHAPTER XVIII.

CLEARING AGENTS.

329. Introductory Remarks.—Clearing agents are liquids, one of whose functions it is to make microscopic preparations transparent by penetrating amongst the highly refracting elements of which the tissues are composed, the clearing liquids themselves having an index of refraction not greatly inferior to that of the tissues to be cleared. Hence all clearing agents are liquids of high index of refraction. The same substances have also a second function, which consists in getting rid of the alcohol in which preparations are generally preserved, and facilitating the penetration of the balsam or other resinous medium in which preparations are, in most cases, finally mounted. Hence, all of the group of bodies here called " clearing agents " must be capable of expelling alcohol from tissues and must be at the same time solvents of Canada balsam and the other resinous mounting media.

It is important to note again, notwithstanding some repetition, the manner of employing these agents. The old plan was to take the object out of the alcohol and float it on the surface of the clearing medium in a watch-glass. This plan was faulty, because the alcohol escapes from the surface of the object into the air quicker (in most instances) than the clearing agent can get into it; hence the object must shrink. To avoid or lessen this cause of shrinkage, clearing is now generally done by the method suggested by Giesbrecht, which consists in putting the clearing medium *under* the alcohol containing the object. This is done in the following manner. Take a test-tube, and put into it enough alcohol to contain the objects (a watch-glass will often do well, but a test-tube is safer). With a pipette carefully put under the alcohol a sufficient quantity of clearing medium (or, carefully pour the alcohol on to the clearing medium). Then put the objects into the alcohol. They will sink down to the level of separation of the two liquids at once; and after some time they will be found to have sunk to the bottom of the clearing medium. They may then be removed by means of a pipette; or the supernatant alcohol drawn off and the preparations allowed to remain until wanted.

The chief clearing agents are essential oils. A classification of these is given below (No. 330, Stieda and Schiefferdecker).

The penetration of all clearing media may be hastened by using them warm. Directions for clearing are given when necessary under the heads of the different organs and tissues. It will suffice here to advise the beginner to keep on his table the following: Oil of cedar, for general use; clove oil, for making minute dissections in cases in which it is desirable to take advantage of the property of that essence of forming very convex drops on the slide, and of imparting a remarkable brittleness to soft tissues; carbolic acid, for rapidly clearing imperfectly dehydrated objects.

330. Classification of Clearing Agents (STIEDA).—Stieda's experiments with essential oils led him to establish the following classification:

A. The *turpentine* group, capable of clearing in a short time perfectly dehydrated sections, but clearing watery sections only after many hours or not at all.

    Ol. Terebinthinæ.
    Ol. Absynthii.
    Ol. Balsam. Copaivæ.
    Ol. Cortic. Aurantiorum.
    Ol. Cubebarum.
    Ol. Fœniculi.
    Ol. Millefolii florum.
    Ol. Sassafras.
    Ol. Juniperi.
    Ol. Menthæ crispæ.
    Ol. Origani vulgaris.
    Ol. Lavandulæ.
    Ol. Cumini.
    Ol. Cajeputi.

Ol. Cascarillæ cortic.
Ol. Sabinæ.
Ol. Citri.

This, then, for Stieda, is the *Index Expurgatorius* of clearing media.

B. The *oil-of-cloves* group, clearing *very rapidly* sections that have been dehydrated, and clearing watery sections "somewhat more slowly" and with a certain amount of shrinkage.

Ol. Gaultheriæ.
Ol. Cassiæ.
Ol. Cinnamomi.
Ol. Anisi stellati.
Ol. Bergamotti.
Ol. Cardamomi.
Ol. Coriandri.
Ol. Carui.
Ol. Roris marini.

But Stieda found kreasote preferable to any of these.

He relates that kreasote was suggested to him by a paper of KUTSCHIN, *Über den Bau des Rückenmarks des Neunanges*, Kasan, 1863. Kutschin rinsed his sections *in water*, brought them on to slides, drew off the water by means of blotting-paper, and added a drop of kreasote at the side. When clear, he covered, and closed the mounts with a border of dammar.

Stieda modified this process by mounting in dammar instead of kreasote.

He then tried experiments to ascertain whether oil of cloves could be applied in the same manner, that is, to the clearing of non-dehydrated sections. He found that it could, though its employment requires longer time. Sections brought from water into kreasote clear in a few minutes, whilst in oil of cloves they require from half an hour to an hour or more; and this slowness of the process exposes them to the risk of shrinkage.

To the group of good clearing agents should be added—cedar-wood oil, sandal-wood oil, carbolic acid.

NEELSEN and SCHIEFFERDECKER (*Arch. f. Anat. u. Phys.*, 1882, p. 206) examined a large series of ethereal oils (prepared by Schimmel and Co., Leipzig), with the object of finding a not too expensive substance that should combine the properties of clearing quickly alcohol preparations, *not* dissolving out anilin colours, clearing celloidin without dissolving it, not evaporating too quickly, and not having a too disagreeable smell.

The following is a list of twenty-four products examined by them. It seems worth while to give it, although the authors only found three amongst the number that fulfil the conditions; as to know that they have been found wanting in some of these respects may perhaps save somebody a wild-goose chase.

Oils of—Anise, Amber, Birch-tar, Cajeput, Calmus, Cassia, Cedar wood, Citrons, Dill, Field thyme, Fir needles, Mint, Cumin, Niobe, Origanum, Palmarosa, Peppermint, Pennyroyal, Rosemary, Sassafras, Spikenard, Thuja, Sandal wood, Caraway.

Of these, the following three fulfil the conditions and can be *recommended*:—*Cedar wood, Origanum, Sandal wood.*

It would be important to possess a list of the exact indices of refraction of the substances used for clearing. I have, unfortunately, not been able to obtain sufficient information of a trustworthy nature for the compilation of such a list. Cedar oil has nearly the index of crown glass (this is true of the oil in the thick state to which it is brought by exposure to the air, not of the new, thin oil, which is less highly refractive), it therefore clears to the same extent as Canada balsam. Clove oil has a much higher index, and therefore clears more than balsam. Turpentine, bergamot oil, and kreasote, have much lower indices, and therefore clear less.

331. Cedar Oil (NEELSEN and SCHIEFFERDECKER, *l. c.*, last §). —Finest cedar-wood oil, price per kilo varies from fifteen to twenty shillings, say about sevenpence halfpenny per ounce, for small quantities, or about the price of clove oil. Very thin, colour light yellow, odour slight (of cedar wood), evaporates slowly, is not changed by light, is miscible with chloroform balsam, and with castor oil. Clears readily tissues in 95 per cent. alcohol, without shrinkage, does not extract anilin colours. Celloidin sections are cleared in five to six hours.

Cheap, but requires an inconvenient length of time for the clearing of celloidin sections.

Note.—I have examined the clearing properties of a sample of cedar-wood oil obtained from the celebrated firm of

Rousseau, Paris. This sample was absolutely colourless. It *totally* failed to clear absolute alcohol objects after many days.

The authors think that a laboratory supplied with cedar oil and origanum oil is fully equipped for all possible cases (the origanum oil being used merely to take the place of cedar-wood oil for the special case of celloidin sections). *See* below, § 335.

Cedar oil is very penetrating, and for this and other reasons is one of the best media, if not the very best, for preparing objects for paraffin imbedding.

332. Clove Oil.—Samples of clove oil of very different shades of colour are met with in commerce. It is frequently recommended that only the paler sorts should be employed in histology. A word of explanation is here necessary. Doubtless it is, in general, best to use a pale oil, provided it be pure, but it is not always easy to obtain a light-coloured oil that is pure. Clove oil passes very readily from yellow to brown with age, so that in choosing a colourless sample you run great risk of obtaining an adulterated sample, for clove oil is one of the most adulterated substances in commerce.

Two important properties of clove oil should be noticed here. It does not easily spread itself over the surface of a slide, but has a tendency to form very convex drops. This property makes it a very convenient medium for making minute dissections in. The second property I wish to call attention to is that of making tissues that have lain in it for some time very brittle. This brittleness is also sometimes very helpful in minute dissections.

These qualities may be counteracted if desired by mixing the clove oil with bergamot oil.

Clove oil has, I fancy, the highest index of refraction of all the usual clearing agents; it clears objects *more* than balsam. It dissolves celloidin (or collodion), and therefore should not be used for clearing sections cut in that medium, without special precautions. Notwithstanding the opinion of Schiefferdecker, I consider this to be one of the best of clearing agents, and very valuable on account of the properties, to which attention has been called above. New clove oil washes out anilin colours more quickly than old. It is well to possess trustworthy samples of both new and old oil.

333. Cannel Oil.—Greatly resembles clove oil, but is in general thinner. An excellent medium, which I particularly recommend.

334. Oil of Bergamot.—SCHIEFFERDECKER (*Arch. Anat. u. Phys.*, 1882 (Anat. Abth.), p. 206) finds that this oil has many good qualities; it clears 95 per cent. alcohol preparations and celloidin preparations quickly, does not attack anilin colours, but the strong odour is disagreeable; it is as dear as oil of cloves, twice as dear as oil of origanum, and three times as dear as oil of cedar. He considers its action preferable to that of oil of cloves, but, all things considered, gives the palm to cedar and origanum. I think that this is a very valuable medium, and though I do not agree with Schiefferdecker in thinking its action superior to oil of cloves, I think it should always be kept at hand.

Bergamot oil is, I believe, the least refractive of these essences, having a lower index than even oil of turpentine.

335. Oil of Origanum (NEELSEN and SCHIEFFERDECKER, *Arch. Anat. u. Phys.*, 1882, p. 204).—Price per kilo 15 mark (= 15*s*.). Thin, light brown colour, odour not too strong, agreeable, does not evaporate too quickly, is not changed by light, is miscible with chloroform balsam and with castor oil. Ninety-five per cent. alcohol preparations are cleared quickly, and so are celloidin sections, without solution of the celloidin. Anilin colours are somewhat extracted.

For work with celloidin sections care should be taken to obtain *Ol. Origani cretici* (" Spanisches Hopfenöl "), not *Ol. Orig. gallici* (v. GIESON; see *Zeit. f. wiss. Mik.*, iv, 4, 1887, p. 482). Specimens of origanum oil vary greatly in their action on celloidin sections, and care should be taken to obtain a good sample.

336. Sandal-wood Oil (NEELSEN and SCHIEFFERDECKER, *Ibid.*). —" Finest East Indian sandal-wood oil," price per kilo 50 mark (= £2 10*s*. 0*d*.). Somewhat thicker than the last two, light yellow, odour faint, agreeable, evaporation hardly perceptible, unchangeable by light, miscible with chloroform balsam and with castor oil. Ninety-five per cent. alcohol preparations cleared quickly, celloidin more slowly, anilin colours unaffected.

Very useful; its worst fault is its high price.

337. Turpentine.—Generally used for treating sections that have been cut in paraffin, as it has the property of dissolving out the paraffin and clearing the sections at the same time, but many other reagents (naphtha, for instance) are preferable for this purpose (see *ante*, § 276). If used for alcohol objects it causes considerable shrinkage and alters the structure of cells more than any other clearing agent known to me, unless used in the thickened state, a method which is much liked for some purposes in Germany. Thickened turpentine ("Verhartzes Terpentinöl" of German writers) is prepared by exposing rectified turpentine in thin layers for some days to the air. All that is necessary is to pour some turpentine into a plate, cover it lightly so as to protect it from dust without excluding the air, and leave it until it has attained a syrupy consistency. Turpentine has, I believe, the lowest index of refraction of all the usual clearing agents, except bergamot oil; it clears objects *less* than balsam.

338. Carbolic Acid.—Best used in concentrated solution in alcohol. Clears instantaneously, even very watery preparations. This is a very good medium, but it is generally better avoided for preparations of soft parts which it is intended to mount in balsam, as they generally shrink by exosmosis when placed in the latter medium. It is, however, a good medium for celloidin sections (*see* above, § 298).

339. Kreasote.—Much the same properties as carbolic acid. *Beech-wood* kreasote is the sort that should be preferred for many purposes,—for clearing celloidin sections (for which it is a very good medium) amongst others.

340. Xylol, Benzol, Toluol, Naphtha, Chloroform.—Too volatile to be recommendable as general clearing agents, but may be used for celloidin sections or for paraffin sections. For these, I greatly recommend naphtha.

341. Absolute Alcohol (SEILER, *Journ. Roy. Mic. Soc.*, 1882, p. 126).—Absolute alcohol is recommended by Seiler for preparing objects for mounting in balsam, the balsam being in this case dissolved in warm absolute alcohol (*see* No. 398). The method is said by Seiler to give very good results but it is obvious that it is only applicable to cases in which it is not desired to make a preliminary examination of the cleared objects (for the sake of selecting the best or the like) before mounting them.

I have several times tried this process, with results in no wise distinguishable from those obtained by ordinary methods, and cannot recommend it for ordinary purposes.

CHAPTER XIX.

INDIFFERENT LIQUIDS, EXAMINATION AND PRESERVATION MEDIA.

342. Introductory.—I comprehend under this heading all the media in which an object may be examined. The old distinction of "indifferent" liquids, and those which have some action on tissues, appears to be misleading more than helpful; inasmuch as it is now well understood that *no* medium is without action on tissues except the plasma with which they are surrounded during the life of the organism; and this plasma itself is only "indifferent" whilst all is *in situ;* as soon as a portion of tissue is dissected out and transferred to a slide in a portion of plasma the conditions become evidently artificial.

It does not appear necessary to create a separate group for mounting media, as all preservative media may be used for mounting.

343. Water.—To preserve it from mould, a lump of thymol or camphor should be kept in the supply. Water may be employed without inconvenience, and sometimes (on account of its low index of refraction, with great advantage) for the examination of all structures that have been fixed with osmic or chromic acid, or some salt of the heavy metals; but it is by no means applicable to the examination of fresh tissues, that is, tissues that have not been so fixed. It is important that the beginner should bear in mind that water is very far from being an "indifferent" liquid; many tissue elements are greatly changed by it (nerve-end structures for instance), and some are totally destroyed by its action if prolonged (for instance, red blood-corpuscles).

346. Theory of Indifferent Liquids.—In order to render water inoffensive to such tissues as these it must, firstly, have dissolved in it some substance that will give it a density equal to that of the liquids of the tissue, so as to prevent the occurrence of osmosis, to which process the destructive action of pure water is mainly due. Salt solution is a medium suggested by this necessity. But salt solution by no means fulfils all the conditions implied in the notion of an "indifferent" liquid. In so far as it possesses a density approaching to that of the liquids of the tissues, one cause of osmosis is eliminated; but there remains another, due to the difference of composition of the liquids within the tissues and that without. Cell contents are a mixture of colloids and crystalloids, salt solution contains only a crystalloid, whose high diffusibility causes it to diffuse over into the colloids of the tissues. In order to reduce the consequent osmotic processes to a minimum, it is necessary that the examination medium contain in addition to a due proportion of salt or other crystalloid, also a due proportion of colloids. By adding, for instance, white of egg to salt solution, this end may be attained, and, as a matter of fact, the liquids recommended as indifferent are found invariably to contain both crystalloids and colloids. Thus (as stated by Frey) vitreous humour contains 987 parts of water to about 4·6 of colloid matters and 7·8 of crystalloids (common salt). In 1000 parts of the juice of fruits are contained about 3·8 parts of colloid matter (albumen), 5·8 of salt, and 3·4 of urea. In blood serum, 8·5 of colloids and 1 of crystalloid substance are found.

345. Salt Solution.—("Normal salt solution," "physiological salt solution") 0·75 per cent. sodium chloride in water. Carnoy recommends the addition of a trace of osmic acid.

346. Iodised Serum.—Iodised serum was first recommended by Max Schultze (*Virchow's Archiv*, xxx, 1864, p. 263). I take the following instructions concerning it from Ranvier (*Traité*, p. 76).

The only serum that gives really good results is the amniotic liquid of mammals. A gravid uterus of a sheep or cow having been obtained (in large slaughter-houses such can be obtained without difficulty), an incision is made through the wall of the uterus and the fœtal membranes. A jet of serum issues from the incision, and is caught in a flask prepared for the purpose. Flakes of iodine are then added, and the flask is frequently agitated for some days. Two points should be noted. A perfectly fresh amnios must be taken; for the merest incipience of putrefaction will spoil the preparation. The flask should have a wide bottom, so that the serum may form only a shallow layer in it; otherwise the upper layers will not be sufficiently exposed to the action of the iodine.

Another method is as follows:—Serum is mixed with a large proportion of tincture of iodine; the precipitate of iodine that forms is removed by filtration, and there remains a strong solution of iodine in serum. This should be kept in stock, and a little of it added every two or three days to the serum that is intended for use.

Ranvier explains that at the outset serum dissolves very little iodine; but if an excess of iodine be kept constantly present in the solution, it will be found that after two or three weeks iodides are formed, and allow fresh quantities of iodine to dissolve; so that after one or two months a very strongly iodised serum is obtained. It should be dark brown. Such a solution is the most fitting for the purpose of iodising fresh serum in the manner directed above, and for making the different strengths of iodised serum that are required for different purposes. In general, for maceration purposes, a serum of a pale brown colour should be employed.

347. Artificial Iodised Serum (FREY, *Le Microscope*, p. 131).

Distilled water	135 grammes.
White of egg	15 ,,
Sodium Chloride	0·20 ,,

Mix, filter, and add—

Tincture of iodine	3 ,,

There is formed a precipitate, which is removed by filtering through flannel; and a little iodine is added to the filtrate.

348. Kronecker's Artificial Serum (from VOGT et YUNG, *Traité d'Anat. comp. prat.*, p. 473. I have been unable to discover the original source).

Common salt	6 grammes.
Caustic soda	0·06 ,,
Distilled water	1000 ,,

349. Aqueous Humour, Fruit Juice, Simple White of Egg.—Require no preparation beyond filtering. They may be iodised if desired.

350. Syrup.—An excellent medium for examining many

structures in the fresh state. To preserve it from mould, chloral hydrate may conveniently be dissolved in it (1 to 5 per cent.). I have used as much as 7 per cent., and found no disadvantage.

Carbolised Syrup.—Carbolic acid may be employed instead of chloral; 1 per cent. is sufficient.

Either of these syrups may be used as a mounting medium, but they are not to be recommended for that purpose, as there is always risk of the sugar crystallising out.

A good strength for syrup is equal parts of loaf sugar and water. Dissolve by boiling.

351. Saliva.—Saliva has been recommended with the idea of its being innocuous to delicate structures; it is of course a macerating agent (see MACERATING AGENTS, **Artificial Saliva,** § 504).

352. Carbolic Acid.—1 per cent. in water. Is a mounting medium.

353. Kreasote.—5 per cent. in water.

354. Thwaites' Kreasote Fluid (see BEALE, *How to Work*, &c., p. 55).

355. Beale's Naphtha and Kreasote (*ibid*, p. 56).

356. Quekett's Wood-Naphtha Fluid (*ibid.*).

357. Alum Sea-Water.—A saturated solution of alum in sea-water is useful for the examination and preservation of the tissues of many marine organisms (Medusæ, Siphonophora, Ctenophora, Pelagic Tunicata). The animals may be killed in the fluid, which is a fair fixing agent.

358. Acetate of Alumina (GANNAL'S SOLUTION, BEALE, *ibid.*).
    Acetate of alumina . . . . . . 1 part.
    Water . . . . 10 „

359. Acetate of Potash (MAX SCHULTZE, *Arch. mik. Anat.*, vii, 1872, p. 180).—A nearly saturated solution in water. It is used by letting a drop run in under the cover-glass to the object, which is in water. After twenty-four hours the mount may be closed. The index of refraction is lower than that of glycerin.

This medium has been frequently recommended as having the property of preventing the blackening of objects that have been treated with osmium; but it seems extremely doubtful whether this is really the case.

360. Calcium Chloride (*Micro. Dict.*, Art. " *Calcium*, chloride ").—Either about 1 part of the salt to 2 of water, or a saturated solution may be used. A lump of camphor should be added to the solution to preserve it. As this salt is very hygroscopic, its solution presents the advantage of not drying up, so that it is not necessary to close the mounts until it is desired to put them away.

361. Chloral Hydrate.—5 per cent. in water (LAVDOWSKY, *Arch. f. mik. Anat.*, 1876, p. 359).

Or, 2·5 per cent. in water (BRADY, *British Copepods*).

Or, 1 per cent. in water (MUNSON, *Journ. Roy. Mic. Soc.*, 1881, p. 847).

362. Alcohol.—Not very recommendable for mounting, as if taken weak it is not a very efficient preservative, and if taken strong it attacks the cement of mounts.

CARPENTER (*The Microscope*) recommends a strength of 1 part to 5 of water.

The chief use of alcohol for preservation purposes is of course for preserving specimens in till wanted for further preparation and study. KULTSCHITZKY has lately pointed out (*Zeit. f. wiss. Mik.*, iv, 3, 1887, p. 349) that alcohol is not without some defects for this purpose. It alters the structure of tissues by continuously dehydrating their albuminoids, and Kultschitzky therefore proposes for preservation some substance that has not this action, such as ether, toluol, or xylol. After fixation and washing out with alcohol, objects may be put up in one of these till wanted.

Mercurial Liquids.

363. Corrosive Sublimate Solution (HARTING'S FLUID, *Micro. Dict.*, Art. "Preservation," p. 640).—One part of sublimate to from 200 to 500 of water. (For blood-corpuscles of frog 1—400, of birds 1—300, of mammals 1—200.) "Harting recommends this as the best preservative for the corpuscles of the blood, nerve, muscular fibre, &c."

364. Pacini's Fluids (*Journ. de Mic.*, iv, 1880; *Journ. Roy. Mic. Soc.*, (N.S.), ii, 1882, p. 702). Pacini remarks that "bichloride of mercury coagulates and precipitates the albuminous matter that exists in the interstitial fluids of the tissues," and therefore in order to prevent this coagulation it is well to associate with it salt for certain preparations, or acetic acid for others. On this principle are prepared the following classical fluids of Goadby and Pacini.

FLUID No. 1 is identical with that of Harting given above, viz. 1·200 sublimate in water. Pacini uses it for removing, when desired, the salt or acid from preparations that have been placed in one of the other solutions.

FLUID No. 2—

Bichloride of mercury	1 part.
Common salt	2 "
Water	200 "

Of general employment, but especially useful for blood-corpuscles of cold-blooded animals, as it has a less density than the following fluid. It preserves spermatic fluid, epithelia, nerves, and muscle-fibres. It is also used for fixing Infusoria, a small quantity being added to the water containing them.

MODIFICATIONS OF THE FOREGOING SOLUTIONS. 197

FLUID No. 3—
 Bichloride of mercury . . . . . 1 part.
 Common salt . . . . 4 ,,
 Water . . . . 200 ,,
For blood-corpuscles of warm-blooded animals.

FLUID No. 4—
 Bichloride of mercury . . . . . 1 part.
 Acetic acid . . . . 2 ,,
 Water . . . . 300 ,,
"Serves best for the nuclei of animal tissues, but it swells up the fibres and distorts the forms of the cells."

FLUID No. 5 (FREY, *Le Microscope*, 1867, p. 233).—In the place here quoted, Frey speaks of the liquids of Pacini as differing from those of Goadby through their containing glycerin in lieu of alum. He gives the following directions. Take—
 Sublimate . . . . 1 part.
 Sodium chloride . . . . . . 2 ,,
 Glycerin (25° Beaumé) . . . . . 13 ,,
 Water . . . . 113 ,,
Allow the mixture to remain undisturbed for at least two months. At the end of that time, take for use 1 part, mix with 3 parts of water, and filter. This mixture is said to be a good preservative of all delicate tissues.

FLUID No. 6 (*Ibid*)—
 Sublimate . . . . 1 part.
 Acetic acid . . . . 2 ,,
 Glycerin (25° Beaumé) . . . . . 43 ,,
 Water . . . . 115 ,,
This mixture is to be employed in the same way as the last. It is said to destroy red blood-corpuscles, but to preserve white blood-corpuscles.

365. **Modifications of the foregoing Sublimate Solutions.**—The following formulæ are quoted by Frey from Cornil as being in use at the Pathological Institute of Berlin.

 1. Sublimate . . . . 1 part.
  Sodium chloride . . . . . . 2 ,,
  Water . . . . 100 ,,
For the more vascular tissues of warm-blooded animals.

 2. Sublimate . . . . 1 part.
  Sodium chloride . . . . . . 2 ,,
  Water . . . . 200 ,,
For similar tissues of cold-blooded animals.

 3. Sublimate . . . . 1 part.
  Sodium chloride . . . . . . 1 ,,
  Water . . . . 300 ,,
For pus-corpuscles and analogous elements.

4. Sublimate . . . . 1 part.
   Water . . . . . . 300 „
For blood-corpuscles.
5. Sublimate. . . . . . . 1 part.
   Acetic acid . . . . . . 1 „
   Water . . . . . . 300 „
For epithelia, connective tissue, and pus-corpuscles, when it is desired to demonstrate the nuclei.
6. Sublimate . . . . . . 1 part.
   Acetic acid . . . . . . 3 „
   Water . . . . . . 300 „
For ligaments, muscles, and nerves.
7. Sublimate . . . . . . 1 part.
   Acetic acid . . . . . . 5 „
   Water . . . . . . 300 „
For glandular tissues.
8. Sublimate . . . . . . 1 part.
   Phosphoric acid . . . . . 1 „
   Water . . . . . . 30 „ (sic.)
For cartilaginous tissues.

366. Goadby's Fluids (*Micro. Dict.*, Art. " Preservation ").
1ST FLUID—
   Bay salt (coarse sea salt) . . 4 ounces.
   Alum . . . . . . 2 „
   Corrosive sublimate . . . 2 grains.
   Boiling water . . . . 1 quart.

This is found to be "too strong" for most purposes, and therefore the following is recommended for general purposes.

2ND FLUID—
   Bay salt . . . . . 4 ounces.
   Alum . . . . . . 2 „
   Corrosive sublimate . . . 4 grains.
   Water . . . . . . 2 quarts.

"Schultze recommends it for preserving *Medusæ, Echinodermata,* Annelid larvæ, *Entomostraca, Polythalamia,* and *Polycystina,* and advises the use of glycerin afterwards to produce transparence."

3RD FLUID.—When carbonate of lime exists in the preparations, the alum must be omitted. The following formula is recommended:
   Bay salt . . . . . 8 ounces.
   Corrosive sublimate . . . 2 grains.
   Water . . . . . . 1 quart.

4TH FLUID.—" Marine animals require a stronger fluid of this kind, made by adding about 2 ounces more salt to the last."

367. Owen's Fluid (quoted from VOGT et YUNG, *Traité d'Anat. comp. pratique*, p. 19).—

    Corrosive sublimate . . 0·014 grammes.
    Alum . . . . . 79 ,,
    Salt . . . . . 137 ,,
    Water . . . . 1680 ,,

Said to be very useful for the preservation of soft-bodied animals.

368. Gilson's Fluid (CARNOY's *Biologie Cellulaire*, p. 94).—

    Alcohol of 60 per cent. . . . 60 c.c.
    Water . . . . . . 30 ,,
    Glycerin . . . . . . 30 ,,
    Acetic acid (15 parts of the glacial to
        85 of water) . . . . . 2 ,,
    Bichloride . . . . . 0·15 grammes.

A really excellent medium for the study of fine cellular detail with well-fixed objects.

369. Gage's Albumen Fluid (*Zeit. f. wiss. Mik.*, 1886, p. 223).—

    White of egg . . . . . 15 c.c.
    Water . . . . . . 200 ,,
    Corrosive sublimate . . . . 0·5 grammes.
    Salt . . . . . . 4 ,,

Mix, agitate, filter, and preserve in a cool place. Recommended for the study of red blood-corpuscles and ciliated cells.

Other Fluids.

370. Chloride and Acetate of Copper (RIPART et PETIT's fluid, *Brebissonia*, 1880, p. 92 ; CARNOY's *Biol. Cell.*, p. 95).—

    Camphor water (not saturated) . 75 grammes.
    Distilled water . . . . 75 ,,
    Crystallised acetic acid . . . 1 ,,
    Acetate of copper . . . . 0·30 ,,
    Chloride of copper . . . . 0·30 ,,

This is certainly a most valuable medium for work with delicate fresh tissues. It may be used in combination with methyl green, which it does not precipitate. The most delicate elements are perfectly preserved in it; the addition of a drop of osmic acid or corrosive sublimate does not cause the least turbidity and enhances its *fixing* action.

371. Tannin (CARNOY, *l. c.*)—
Water . . . . 100 grammes.
Powdered tannin . . . . . 0·50 „

372. Picro-Carmine.—Picro-carmine has been recommended by Ranvier as a medium for teasing fresh tissues in, in the belief that it possesses sufficient fixing action to preserve the forms of cells. Carnoy finds that cells live in it for a considerable time, and become gorged with water and deteriorated to a considerable degree. Unfortunately, too, picro-carmine cannot be combined with a good fixing agent, as it is precipitated by alcohol and by acids, and especially by osmic acid.

373. Methyl Green.—See under STAINING AGENTS. The aqueous solution is sometimes very useful as an examination medium for fresh tissues. It should be taken fairly concentrated, in which state it has sufficient fixing power, which is enhanced by the addition of a trace of osmic acid.

374. Wickersheimer's Fluid (*Zool. Anz.*, 1879, p. 670; *cf. Journ. Roy. Mic. Soc.*, 1882, p. 427, *id.*, 1880, p. 355; and *Entomol. Nachr.*, 1880, p. 129). This once famous fluid appears to be quite unsuccessful for histological purposes.

375. Meyer's Salicylic Vinegar Preservative Solutions (*Arch. mik. Anat.*, xiii, 1876, p. 868).—"Salicylic vinegar" is a solution of 1 part of salicylic acid in 100 parts of pyroligneous acid. The pyroligneous acid should be of 1·04 specific gravity, and should be of a pale yellow colour. This product is found in commerce and may be obtained from Herrn J. M. Andreæ, Droguerie-Handlung, Frankfurt-a.-M.

1ST FLUID—
One vol. salicylic vinegar to 10 vols. of the following dilute glycerin: viz. glycerin 1 vol., water 2 vols.
For various Larvæ, Hydræ, Nematodes, &c.

2ND FLUID—
One vol. salicylic vinegar to 10 vols. of the following dilute glycerin: viz. glycerin 1 vol., water 4 vols.
For Infusoria.

376. Noll's Salicylic Vinegar and Gum Medium (*Zool. Anz.*, 1883, p. 472).—A mixture of equal vols. of Meyer's second fluid (*ante*, last formula) and Farrant's medium (*post*, 379).

This mixture never becomes turbid and does not dry up. The covers may be luted with asphalt or any other cement. The fluid answers admirably for delicate Crustacea and their larvæ, the preparations do not shrink, and are not too much cleared. It also answers well for hardened and stained preparations of Hydroids, small Medusæ, and other Cœlenterates.

377. Dean's Medium, see *Micro. Dict.*, Art. "Preservation." Appears to be now superfluous.

378. Hoyer's Gum with Chloral Hydrate or Acetate of Potash (*Biol. Centralb.*, ii, 1882, pp. 23-4; *Journ. Roy. Mic. Soc.* (N.S.), iii, 1883, pp. 144-5).—A high 60 c.c. glass with a wide neck is filled two thirds full with gum arabic (in pieces), and then *either* a solution of chloral (of several per cent.) containing 5—10 per cent. of glycerin, is added, *or* acetate of potash or ammonia. The gum with frequent shaking dissolves in a few days, and forms a syrupy fluid, which is slowly filtered for twenty-four hours. The clear filtered fluid will keep a long time, but if spores of fungi begin to develop a little chloral can be added and the fluid refiltered. The solution with chloral is for carmine or hæmatoxylin objects, that with acetate for anilin objects.

379. Gum and Glycerin Medium (FARRANT's medium; BEALE, *How to Work, &c.*, p. 58):

Picked gum arabic . . . 4 ounces.
Water . . . . . . 4 „
Glycerin . . . . . 2 „

To be kept in a stoppered bottle with a lump of camphor.

This medium is quoted by Frey as consisting of equal parts of gum, glycerin, and saturated aqueous solution of arsenious acid.

The *Micrographic Dictionary* gives the following directions:

Gum arabic 1 ounce, glycerin 1 ounce, water 1 ounce, arsenious acid 1½ grains; dissolve the arsenious acid in the water, then the gum (without heat), add the glycerin, and incorporate with great care to avoid forming bubbles.

380. Gum and Glycerin Medium (LANGERHANS' formula, modification of FARRANT's medium, *Zool. Anzeig.*, ii, 1879, p. 575).

Gummi arab. . . . . . 5·0
Aquæ . . . . „

To which after twelve hours are added—
Glycerini . . . . . . 5·0
Sol. aquosa acid. carbol. (5·100) . . 10·0

Marine animals may be preserved in this by simply running in a drop under the cover, and next day or later adding what is necessary to make up for evaporation, and closing the mount. Shrinkage is very slight, and most colours keep well.

381. Cole's Gum and Syrup Medium (*see* above, 309).

382. Fabre-Domergue's Glucose Medium (*La Nature*, No. 823, 9 Mars, 1889, supp.).

Glucose syrup diluted to twenty-five degrees of the areometer (sp. gr. 1·1968) . . . 1000 parts.
Methyl alcohol . . . 200 ,,
Glycerin . . . . . 100 ,,
Camphor, to saturation.

The glucose is to be dissolved in warm water, and the other ingredients added. The mixture, which is always acid, must be neutralised by the addition of a little potash or soda.

This medium is said to preserve without change almost all animal pigments. If it really performs this, its great value is evident.

Glycerin Media.

383. Glycerin.—Glycerin diluted with water is frequently employed as an examination and mounting medium. Dilution with water is sometimes advisable from an optical point of view, on account of the increased visibility that it gives to many structures by lowering the index of refraction of the glycerin. But from the point of view of efficacious preservation, it is always advisable to use undiluted glycerin, the strongest that can be procured.

Long soaking of tissues in glycerin of gradually increased strength is a necessary preliminary to mounting in all cases in which it is desired to obtain the best possible preparations and to ensure that they shall keep well. If this soaking is done on the slide (the cover being removed and the object treated with fresh glycerin every one or two days), it is well to take the precaution recommended by Beale, of luting the edges of the cover so as to make the preparation air-tight, as glycerin is so highly hygroscopic that a drop of it exposed to the air rapidly diminishes in strength to a very considerable degree. In order to facilitate the removal of the cover in this process, the slide may be gently warmed by passing it two or three times through the flame of a spirit-lamp. No preparation can be considered to be made *secundum artem* until every part of the object has been thoroughly impregnated with strong pure glycerin.

The shrinking that frequently occurs when delicate structures are brought into glycerin may generally be cured by this treatment; cells which at first appear hopelessly collapsed gradually swell out to their normal forms and dimensions.

For closing glycerin mounts, the edges of the cover should first (after having been cleansed as far as possible from superfluous glycerin) be painted with a layer of *glycerin jelly*; as soon as this is set a coat of any of the usual cements may be applied. This has of course been for the last twenty years one of the common places of histological technic; but that has not prevented somebody from recently describing the process at great length as new.

Glycerin dissolves carbonate of lime, and is therefore to be rejected in the preparation of calcareous structures that it is wished to preserve.

The already high index of refraction of glycerin (Price's glycerin, $n = 1\cdot46$) may be raised to about that of crown glass by dissolving suitable substances in the glycerin. Thus the refractive index of a solution of chloride of cadmium ($CdCl_2$)* in glycerin may be $1\cdot504$; that of a saturated solution of sulpho-carbonate of zinc in glycerin may be $1\cdot501$; that of a saturated solution of Schering's† chloral hydrate (in crusts) in glycerin is $1\cdot510$; that of iodate of zinc in glycerin may be brought up to $1\cdot56$.‡ The clearing action of glycerin may thus be greatly increased, and the full aperture of homogeneous objectives brought to bear on objects mounted in one of the above-named solutions.

The sulpho-carbolate of zinc solution§ may be prepared by taking equal parts by weight of Price's glycerin and sulpho-carbolate of zinc crystals, mingling the two, and applying sufficient heat to boil the glycerin. The solution can be made in about an hour, but no fear need be had about boiling too long, as the longer this is done the less liability will there be for the solution to deposit crystals on the bottom of the bottle when cooled, which

---

* *Journ. Roy. Mic. Soc*, ii, 1879, p. 346.
† *Ibid.* (N.S.), i, 1881, p. 943.
‡ *Ibid.*, p. 366.
§ *Ibid.*, iii, 1880, p. 1051.

it will do if the temperature is only kept up long enough to dissolve the crystals. Filter while hot. The index may be brought up to 1·525 if desired, by evaporating the solution somewhat, or by adding more carbolate.

384. Barff's Boroglyceride (see *Journ. Roy. Mic. Soc.*, 1882, p. 124).

385. Glycerin and Alcohol Mixtures.—These most useful fluids afford one of the best means of bringing delicate objects gradually from weak into strong glycerin. The object is mounted in a drop of the liquid, and left for a few hours or days, the mount not being closed. By the evaporation of the alcohol the liquid gradually increases in density, and after some time the mount may be closed, or the object brought into pure glycerin or glycerin jelly.

1. CALBERLA'S LIQUID—
   Glycerin . . . . . . 1 part
   Alcohol . . . . . . 1 ,,
   Water . . . . . . 1 ,,

   A most valuable examination fluid.

2. I strongly recommend the following for very delicate objects.
   Glycerin . . . . . . 1 part
   Alcohol . . . . . . 1 ,,
   Water . . . . . . 2 ,,

3. HÆNTSCH'S LIQUID—
   Glycerin . . . . . . 1 part
   Alcohol . . . . . . 3 ,,
   Water . . . . . . 2 ,,

4. JÄGER'S LIQUID (quoted from VOGT and YUNG'S *Traité d'Anat. comp. prat.*, p. 16).
   Glycerin . . . . . . 1 part
   Alcohol . . . . . . 1 ,,
   Sea-water . . . . . . 10 ,,

386. Deane's Glycerin Jelly (from FREY'S *Le Microscope*, p. 231).—120 grammes glycerin, 60 grammes water, 30 grammes gelatin. Dissolve the gelatin in the water, and add the glycerin. This, and the following glycerin-jellies, must of course be used warm.

387. Lawrence's Glycerin Jelly (DAVIES, *Preparation and Mounting of Microscopic Objects*, p. 84).—"He takes a quantity of Nelson's gelatin, soaks it for two or three hours in cold water, pours off the superfluous water, and heats the soaked gelatin until melted. To each fluid ounce of the gelatin, *whilst it is fluid but cool*, he adds a fluid drachm of the white of an egg. He then boils this until the albumen coagulates and the gelatin is quite clear, when it is to be filtered through fine flannel, and to each ounce of the clarified solution adds 6 drachms of a mixture composed of 1 part of glycerin to two parts of camphor water."

388. Beale's Glycerin Jelly (*How to Work, &c.*, p. 57).—Gelatin or isinglass, soaked, melted, and clarified if desired, as in the last formula. To the clear solution add an equal bulk of strong glycerin.

389. Brandt's Glycerin Jelly (*Zeit. f. Mik.*, ii, 1880, p. 69; *Journ. Roy. Mic. Soc.*, iii, 1880, p. 502).—Melted gelatin 1 part, glycerin $1\frac{1}{2}$ parts.

The gelatin to be soaked in water and melted in the usual way. After incorporating the glycerin, the mixture is to be filtered. This is a point of vital importance, as the gelatin of commerce is always mixed with particles of dust and minute threads. Swedish filtering paper does not allow the fluid to pass through sufficiently, and flannel produces more threads than before. The following simple apparatus is found effective. A wide-necked bottle is broken in two, and the upper part taken. The neck is stopped with a cork having two holes bored in it. In the first hole a glass tube, about 20 cc. long, is inserted so as to project a little into the inside of the bottle, and on the outside it is bent sharply to one side and drawn out into a point of about $1\frac{1}{2}$ to 2 mm. diameter. In the second hole a funnel-shaped filter is inserted so that the conical part is inside the bottle and the tube projects a few centimetres beyond the cork and the neck of the bottle. The apparatus is then placed so that the wide opening of the bottle and of the funnel is uppermost, and some spun glass is pressed into the lower conical part of the filter. In using the apparatus the funnel is filled with glycerin gelatin, and the bottle with hot water, which runs off slowly through the tube in the first hole and is constantly replenished.

Some drops of carbolic acid should be added to the fluid product of the filtering. For mounting, use warm, by melting a small portion on the slide, the object having been previously soaked for some time in a small bottle of the medium warmed with a suitable apparatus.

390. Kaiser's Glycerin Jelly (*Bot. Cent.*, i, 1880, p. 25; *Journ. Roy. Mic. Soc.*, iii, 1880, p. 504).—One part by weight finest French gelatin is left for two hours in 6 parts by weight distilled water, 7 parts of glycerin are added, and for every 100 grammes of the mixture 1 gramme of concentrated carbolic acid. Warm for ten to fifteen minutes, stirring all the while, until the whole of the flakes produced by the carbolic acid have disappeared. Filter whilst warm through the finest spun glass laid wet in the filter. Use for mounting as above.

I prepared some of this jelly many years ago, and find it is still perfectly clear.

391. Seaman's Glycerin Jelly (*Amer. Mon. Mic. Journ.*, ii, 1881, pp. 45; *Journ. Roy. Mic. Soc.* (N.S.), i, 1881, p. 534). —Dissolve isinglass in water, so that it makes a stiff jelly when at the ordinary temperature of the room, add one tenth as much glycerin, and a little solution of borax, carbolic acid, or camphor water. Filter whilst warm through muslin, and add a little alcohol.

392. Fol's Glycerin Jellies (*Lehrb.*, p. 138).

1. Melt together 1 volume of Beale's jelly (§ 388) and one half to 1 volume of water, and add 2 to 5 per cent. of salicylic acid solution, or carbolic acid or camphor.

2. Gelatin . . . . . . 30 parts.
    Water . . . . . . 70 ,,
    Glycerin . . . . . . 100 ,,
    Alcoholic solution of camphor . . 5 ,,

Prepare as before, adding the camphor last.

3. Gelatin . . . . . . 20 ,,
    Water . . . . . . 150 ,,
    Glycerin . . . . . . 100 ,,
    Alcoholic solution of camphor . . 15 ,,

393. Castor Oil.—This has been lately recommended as a mounting medium for certain delicate tissues (sections of eyes of Cephalopods) by GRENACHER (*Abhandl. naturf. Ges. Halle-a.-S.*, Bd. xvi; *Zeit. f. wiss. Mik.*, 1885, p. 244). This was with the idea that its low refractive index ($n = 1\cdot49$, whilst Canada balsam $n = 1\cdot54$) would give a useful augmentation of visibility for the more refractive elements of the tissues.

With the objects with which I have experimented I have not found this to be the case.

394. Stephenson's Biniodide of Mercury and Iodide of Potassium (*Journ. Roy. Mic. Soc.*, 1882, p. 167).—Interesting, as giving a solution which when saturated has an index of 1·680, the highest index of any known aqueous fluid. I have experimented both with strong and weak solutions, and doubt whether much practical advantage can be derived from them. Tissues are well preserved, but the preparations are ruined by a precipitate which forms in the fluid.

395. Monobromide of Naphthalin (see *Journ. Roy. Mic. Soc.*, 1880, p. 1043 (ABBÉ and VAN HEURCK), and *Zool. Anz.*, 1882, p. 555 (MAX FLESCH).

Resinous Media.

396. Resins and Balsams.—Resins and balsams consist of a vitreous or amorphous substance held in solution by an essential oil. By distillation or drying in the air they lose the essential oil and pass into the solid state. It is these solidified resins that should in my opinion (and that I believe of the best microscopists) be employed for microscopical purposes. For the raw resins always contain a certain proportion of water, which makes it difficult to obtain a clear solution with the usual menstrua, is injurious to the optical properties of the medium, and to its preservative qualities, and further, especially hurtful to the preservation of stains. I therefore recommend that all solutions be made by heating gently the balsam or resin in a stove until it becomes brittle when cold, and then dissolving in an appropriate menstruum.

FOL (*Lehrb.*, pp. 138-9) is of a different opinion.

Solutions made with volatile menstrua, such as xylol and chloroform, set rapidly, but become rapidly brittle. Solutions made with non-volatile media, such as turpentine, set much less rapidly and pass much less rapidly into the brittle state.

As to the old dispute about the respective merits of damar and balsam, the case appears after all to lie in a nutshell. Damar gives the better definition of delicate detail; balsam has greater clearing action, and affords perhaps more solid mounts.

It may be remarked here that for some of the purposes for which these media are employed, **Oil of Cedar** may be found preferable. The mounts need not be closed (except for immersion work) as the oil soon sets hard enough to keep the cover in place.

397. Canada Balsam.—Prepare with the solid balsam as above described. The usual menstrua are xylol, benzol, chloroform, and turpentine. Dissolve the solid balsam in one of these to the required consistence. The turpentine solution is to be preferred only in cases where it is desired to have a medium that sets very slowly, or in view of the better preservation of certain stains. For all other purposes the xylol solution is the best.

HEYS states that if the chloroform solution be poured into long, thin, half-ounce phials, corked up, and set aside for at least a month, the medium will be clearer and set much quicker than if the balsam is mixed with the chloroform at the time it is required for use (*Trans. Mic. Soc.*, Jan., 1865, p. 19. BEALE, p. 51).

SAHLI (*Zeit. f. wiss. Mik.*, 1885, p. 5) recommends oil of cedar as a menstruum.

398. Seiler's Alcohol Balsam (*Proc. Amer. Soc. Mic.*, 1881, pp. 60-2; *Journ. Roy. Mic. Soc.* (N.S.), ii, 1882, pp. 126-7).—"Take a clear sample of Canada balsam and evaporate it in a water or sand-bath to dryness; *i.e.* until it becomes brittle and resinous when cold. Dissolve this while warm in warm absolute alcohol and filter through absorbent cotton."

The advantage of this medium is stated to be that objects may be mounted in it direct from absolute alcohol, without previous treatment with an essential oil or other clearing agent; Seiler considers that by this means "shrivelling is avoided, as well as *the solution of fat in the cells.*"

The process is not very easy to carry out, and I cannot recommend it for general work.

399. Damar (Gum Damar, or Dammar, or d'Ammar).—The menstrua are the same as for balsam, and the solution should be prepared in the same way. The most beautiful of all these mounting media is the solution of damar in xylol. Heat is not necessary to make the solution.

Minute directions (which I think unnecessary) for preparing a working solution are given by MARTINOTTI in *Zeit. f. wiss. Mik.*, iv, 2, 1887, p. 156, and in *Malpighia*, ii, 1888, p. 270; cf. also *Journ. Roy. Mic. Soc.*, 1889, p. 163.

FLEMMING, PFITZNER, and a writer signing C. J. M., all employ a mixture of benzol and turpentine (see *Arch. mik. Anat.*, xix, 1881, p. 322; *Sci. Gossip*, 1882, p. 257; *Journ.*

Roy. Mic. Soc. (N.S.), iii, 1883, p. 145; *Morphol. Jahrb.*, vi, 1880, p. 469; *Journ. Roy. Mic. Soc.* (N.S.), ii, 1882, p. 583).

MAX FLESCH notes hereon (*Zool. Jahresber. f.* 1880, p. 51) that at Würzburg the ordinary dammar varnish of arists is employed.

JAMES (*Engl. Mech.*, 1887, p. 184; *Journ. Roy. Mic. Soc.*, 1887, p. 1061) also gives some I think superfluous formulæ for damar solutions.

400. Balsam-Damar.—It has been recommended to mix equal volumes of benzol balsam and turpentine damar; but this medium seems to me superfluous.

401. Colophonium.—A solution of colophonium in turpentine has been recommended by Kleinenberg. I find it works very pleasantly. The palest kinds of colophonium should of course be taken.

This medium sets very slowly, so that ample time is afforded for arranging objects in it. Kleinenberg warns against the employment of absolute alcohol as a solvent; the preparations are beautiful at first, but soon become spoiled by the precipitation of crystals or of an amorphous substance.

The turpentine solution keeps perfectly limpid, gives very good definition, and is altogether so excellent a medium that I am surprised that it is not more used. It should be recommended to beginners.

402. Venice Turpentine (*see* the quotation in *Zeit. f. wiss. Mik.*, iii, 3, 1886, p. 400).

403. Copal Varnish.—I have seen tissues very instructively mounted in this medium, which is probably worthy of further study. "Berry's Hard Finish," which is an easily obtainable copal varnish, has been highly praised for mounting purposes (see *Journ. Roy. Mic. Soc.*, 1887, p. 1064).

404. Photographic Negative Varnish (for mounting large sections without cover-glasses) (*see* WEIGERT, *Zeit. f. wiss. Mik.*, iv, 2, 1887, p. 209).

405. Styrax and Liquidambar (see *Journ. Roy. Mic. Soc.*, 1883, p. 741; ib. 1884, pp. 318, 475, 655, and 827; and the places there quoted. Also *Bull. Soc. Belge de Mic.*, 1884, p. 178; and FOL, *Lehrb.* p. 141).

406. Tolu Balsam (see *Zeit. f. wiss. Mik.*, iv, 4, 1887, p. 471).

CHAPTER XX.

CEMENTS AND VARNISHES.

407. Thanks to the efforts of the dilettanti to outshine one another with neatly gaudy " rings," microscopical literature contains a goodly show of receipts for cements and varnishes. I have collected such as appear likely to be useful, rejecting all that relates merely to ornament.

Two, or at most three, of the media given below, will certainly be found sufficient for all useful purposes. For many years I have used only one cement (Bell's). I recommend this as a cement and varnish; gold size may be found useful for turning cells; and Ziegler's white cement or zinc white may be kept for occasions on which the utmost solidity is required.

Marine glue is necessary for making glass cells.

Carpenter lays great stress on the principle that the cements or varnishes used for fluid mounts should always be such as contain *no mixture of solid particles;* he has always found that those that do, although they might stand well for a few weeks or months, yet always became porous after a greater lapse of time, allowing the evaporation of the liquid and the admission of air. All fluid mounts *should be ringed with glycerin jelly before applying a cement; by this means all danger of running-in is done away with.*

The above passage stands as it stood, italicised as here, in the 1st edition. It was translated and amplified, in a special paragraph, in the *Traité des Méth. Techniques*. I may therefore be excused from hunting up the name of the anatomist who recently published as new this old, old method, or the pages of the journals which reproduced his paper without protest.

408. Gelatin Cement (MARSH's *Section-cutting*, 2nd ed., p. 104).—Take half an ounce of Nelson's opaque gelatin, soak well in water, melt in the usual way, stir in 3 drops of kreasote, and put away in a small bottle. It is used warm.

When the ring of gelatin has become quite set and dry, which will not take long, it may be painted over with a solution of bichromate of potash made by dissolving 10 grains of the salt in an ounce of water. This should be done in the daytime, as the action of daylight is necessary to enable the bichromate to render the gelatin insoluble in water. The cover may then be finished with Bell's cement.

This process is particularly adapted for glycerin mounts.

409. The Paper Cell Method.—According to my experience, the best way to make a fluid mount safe is the following. By means of two punches I cut out rings of paper of about a millimetre in breadth, and of about a millimetre smaller in diameter than the cover-glass. *Moisten* the paper ring with mounting fluid, and centre it on the slide. Fill the cell thus formed with mounting fluid; arrange the object in it; put the cover on; fill the annular space between the paper and the margin of the cover with glycerin jelly (a turn-table may be useful for this operation); and as soon as the gelatin has set turn a ring of Bell's or other cement on it.

For greater safety, the gelatin may of course be treated with bichromate according to Marsh's plan.

410. Comparative Tenacity of Cements (*see* BEHRENS, *Zeit. f. wiss. Mik.*, ii, 1885, p. 54, and AUBERT *Amer. Mon. Mic. Journ.*, 1885, p. 227; *Journ. Roy. Mic. Soc.*, 1886, p. 173).—Behrens gives the palm to amber varnish; Aubert places Miller's cauotchouc cement at the head of the list, Lovett's cement coming half way down, and zinc white cement at the bottom, with less than one quarter the tenacity of the caoutchouc cement.

411. Bell's Cement.—Composition unknown. May be obtained from the opticians or from J. Bell and Co., chemists, 338, Oxford Street, London.

This varnish flows easily from the brush, and sets quickly. For glycerin or other fluid mounts, the cover should be ringed as above described with glycerin jelly before applying the varnish. This precaution is especially necessary with glycerin. This is the best varnish for fluid mounts known to me. It is soluble in ether or chloroform. It is not attacked by oil of cedar.

412. Asphalt Varnish (*Bitume de Judée*).—Unquestionably one of the best of these media, either as a cement or a varnish, *provided it be procured of good quality*. It can be procured from the opticians or from the oilshops. KITTON (*Month. Mic. Journ.*, 1874, p. 34) recommends asphalt dissolved in benzol with the addition of a small quantity of gold size.

413. Brunswick Black.—Best obtained from the opticians. A receipt for preparing it is given in BEALE, *How to Work, &c.*, p. 49.

"If a little solution of india rubber in mineral naphtha be added to it, there is no danger of the cement cracking when dry." Carpenter states that without this addition it is brittle when dry. Brunswick black is soluble in oil of turpentine. A most useful cement, works easily and dries quickly. It can be recommended for turning cells.

414. Brunswick Black and Gold Size (EULENSTEIN, BEALE, *How to Work, &c.*, p. 49).—Equal parts of Brunswick black and gold size with a very little Canada balsam.

415. Gold Size.—Receipts for preparing it may be found in the *Micrographic Dict.* or in COOLEY's *Cyclopædia*; but it is certainly best to obtain it from the opticians or oilshops. It is soluble in oil of turpentine. A good cement, *when of good quality*, and very useful for turning cells.

416. Marine Glue.—Found in commerce. Carpenter says the best is that known as G K 4.

It is soluble in ether, naphtha, or solution of potash. Its use is for attaching glass cells to slides, and for all cases in which it is desired to cement glass to glass.

Receipts for preparing it may be found in BEALE, p. 49, or in COOLEY's *Cyclopædia*.

417. Harting's Gutta-percha Cement (*see* BEALE's *How to Work, &c.*, p. 49).—Marine glue serves the same purpose, viz. that of attaching cells to slides.

418. India Rubber and Lime French Cement (*see* BEALE, p. 58).

419. Knotting (*Journ. Roy. Mic. Soc.*, 1882, p. 745).— "Patent knotting" from oil and colour stores, exposed to the air until it has become of the proper consistency;—for mending cells and for preventing running-in of the finishing varnish (*Northern Microscopist*, ii, 1882).

420. Turpentine, Venice Turpentine (CSOKOR, *Arch. mik. Anat.*, xxi, 1882, p. 353; PARKER, *Amer. Mon. Mic. Journ.*, ii, 1881, pp. 229-30; *Journ. Roy. Mic. Soc.* (N.S.), ii, 1882, p. 724).—Venice turpentine (Terebinthina veneta) is the liquid resinous exudation of *Abies larix*. It is seldom met with in a pure state. The following are the directions for preparing and using it given by Parker:

Dissolve true Venice turpentine in enough alcohol, so that after solution it will pass readily through a filter, and, after filtering, place in an evaporating dish, and by means of a sand-bath, evaporate down to about three quarters of the quantity originally used. (The best way to tell when the evaporation has gone far enough is to drop some of the melted turpentine, after it is evaporated down to about three quarters its original volume, into cold water; if on being taken out of the water it is hard and breaks with a vitreous fracture on being struck with the point of a knife, cease evaporation and allow to cool.)

Or (CSOKOR), common resinous turpentine of commerce is put in small pieces to melt over a water-bath, then poured into a suitable vessel and allowed to cool. It should yield a brittle, dark brown mass, not yielding to the pressure of a finger. It is sometimes useful, in order to attain the right degree of hardness in the cold mass, to add a little resinous oil of turpentine to the melted mass, and then to evaporate for several hours over the water-bath.

This cement is used for closing glycerin-mounts; it is applied in the following manner:—Square covers are used and superfluous glycerin is cleaned away from the edges in the usual way.

The cement is then put on with a piece of wire bent at right angles (No. 10—12 wire is taken, and copper is the best, as it gives to the turpentine a greenish tinge); the short arm of the wire should be just the length of the side of the cover-glass. The wire is heated in a spirit-lamp, plunged into the cement, some of which adheres to it, and then brought down flat upon the slide at the margin of the cover.

The turpentine distributes itself evenly along the side of the cover, and hardens immediately, so that the slide may be cleaned as soon as the four sides are finished. It is claimed for this cement that it is perfectly secure, very handy, and never runs in. Parker saw this cement, or a similar one known as *Venedischer Damarlack*, exclusively used for glycerin mounts in the Pathological Laboratory at Vienna.

This is an extremely valuable method. It is very rapid and very safe. The cement sets hard in a few seconds.

421. Colophonium and Wax (Krönig, *Arch. f. mik. Anat.*, 1886, p. 657; *Journ. Roy. Mic. Soc.*, 1887, p. 344).—Seven to nine parts of colophonium are added piecemeal to two parts of melted wax, the whole filtered and left to cool. For use, the mass is melted by placing the containing vessel in hot water. The cement is not attacked by water, glycerin, or caustic potash.

422. Amber Varnish.—As above mentioned, Behrens finds this cement to possess an extreme tenacity. He does not give the composition of his varnish, which was procured from E. Pfannenschmidt at Dantzic. The following is from Cooley's *Cyclopædia*, Art. "Varnish":

"Take of amber (clear and pale) 6 lbs., fuse it; add of hot clarified linseed oil 2 gallons, boil it until it "strings" well, then let it cool a little and add of oil of turpentine 4 gallons or q. s."

Other receipts, l. c.

423. Amber and Copal Varnish (Heydenreich, *Zeit. f. wiss. Mik.*, 1885, p. 338).—An extremely complicated mode of preparation. The varnish may be obtained from Ludwig Marx, at 110, Moskowskaja Sastawa, St. Petersburg, or 79, Gaden, Vienna, or 1, Römerthal, Mayence.

424. Shellac Varnish (Beale, p. 28).—Shellac should be broken into small pieces, placed in a bottle with spirit of wine, and frequently shaken until a thick solution is obtained. The *Micro. Dictionary* says that the addition of 20 drops of castor oil to the ounce is an improvement.

Untrustworthy, but useful for protecting balsam mounts from the action of oil of cedar.

For a method of preparing chemically pure shellac (a somewhat important matter), *see* WITT, *Zeit. f. wiss. Mik.*, 1886, p. 199.

For SEAMAN's shellac cement for attaching metal to glass, see *Journ. Roy. Mic. Soc.*, 1888, p. 520.

425. Sealing-wax Varnish (*Micro. Dict.*, "Cements").—Add enough spirit of wine to cover coarsely powdered sealing-wax, and digest at a gentle heat. This should only be used as a varnish, never as a cement, as it is apt to become brittle and to lose its hold upon glass after a time.

426. Tolu Balsam Cement (CARNOY's *Biol. Cell.*, p. 129).

Tolu balsam	2 parts.
Canada balsam . . .	1 ,,
Saturated solution of shellac in chloroform	2 ,,

Add enough chloroform to bring the mixture to a syrupy consistence. Carnoy finds this cement superior to all others.

427. Stieda's White Zinc Cement (*Arch. f. mik. Anat.*, 1866, p. 435).—Rub up oxide of zinc with turpentine, and add, stirring continually for every drachm of the zinc oxide, 1 ounce of a solution of damar in turpentine (of the consistency of thick syrup). This gives a *white* cement like Ziegler's. For a red cement, take, instead of zinc, *cinnabar*, and take 2 drachms of the metal for each ounce of the damar solution. If the cement has become too thick with age, dilute with turpentine, ether, or chloroform.

428. Ziegler's White Cement.—Composition unknown. Is very much used on the Continent.

429. Kitton's White Lead Cement (*Month. Mic. Journ.*, 1876, p. 221).—Equal parts of white lead, red lead, and litharge (all in powder), ground together with a little turpentine until thoroughly incorporated, then mixed with gold size. The mixture should be thin enough to work with a brush. No more of the cement should be made than is required for present use, as it soon sets and becomes unworkable; but a stock of the materials may be kept ready ground in a bottle.

430. Lovett's Cement (*Journ. Roy. Mic. Soc.*, 1883, p. 786).— Two parts white lead, 2 parts red oxide of lead (minium), 3 parts litharge. To be ground very fine, mixed dry, and kept so in a bottle. When required for use mix a little of the powder with gold-size to the consistency of paint, taking care that no grit gets into it.

430 a. Apáthy's Cement for Glycerin Mounts (*Zeit. f. wiss. Mik.*, vi, 2, 1889, p. 171).—Equal parts of hard (60° C. melting point) paraffin and Canada balsam. Heat together in a porcelain capsule until the mass takes on a golden tint and no longer emits vapours of turpentine. On cooling this forms a hard mass, which is used by warming and applying with a glass rod or brass spatula. One application is enough. The cement does not run in and never cracks.

CHAPTER XXI.

INJECTIONS; GELATIN MASSES.

431. Introduction.—Injection masses are composed of a coloured substance, technically termed the *colouring mass*, and of a substance with which that is combined, technically termed the *vehicle*.

The following formulæ are grouped according to the nature of the vehicle. A note on the employment of nitrite of amyl for provoking the dilatation of vessels will be found at § 473.

432. Robin's Gelatin Vehicle. (*Traité*, p. 30).—Take some gelatin, of the sort known as "colle de Paris." (This gelatin is found in commerce in the form of thin sheets, marked with lozenge-shaped impressions of the cords which supported them whilst drying.) Soak it in cold water, then heat in water over a water-bath. One part of gelatin should be taken for every 7, 8, 9, or even 10 parts of water; it is a common error to employ solutions containing too much gelatin. The solution is now to be combined with one of the colouring-masses given below.

This vehicle, like all gelatin masses, is liable to be attacked by mould if kept long; camphor and carbolic acid do not suffice to preserve it.

Chloral hydrate added to the mass will preserve it (HOYER). A sufficient dose, at least 2 per cent., should be employed (*see* below, No. 445).

433. Robin's Glycerin-Gelatin Vehicle (*Traité*, p. 32).—Dissolve in a water-bath 50 grammes of French gelatin ("colle de Paris") in 300 grammes of water in which has been dissolved some arsenious acid; add of glycerin 150 grammes, and of carbolic acid a few drops. Unlike the pure gelatin vehicles, this mass does keep indefinitely.

The colouring masses recommended for combination with the vehicles above described are made as follows:

434. Carmine Colouring Mass (*Traité*, p. 33).—Rub up in a mortar 3 grammes of carmine with a little water and enough ammonia to dissolve the carmine. Add 50 grammes of glycerin, and filter.

Prepare 50 grammes of acid glycerin (containing 5 grammes of acetic acid for every 50 grammes of glycerin), and add it by degrees to the carmine glycerin, until a slightly acid reaction is obtained (as tested by very sensitive blue test-paper, moistened and held over the mixture).

One part of this mixture is to be added to 3 or 4 parts of the gelatin injection vehicle (*ante*, Formula 432), or of the glycerin-gelatin (No. 433), or glycerin-alcohol vehicle described below, No. 475.

435. Ferrocyanide of Copper Colouring Mass (*Ibid.*, p. 34).
Take—

(1) Ferrocyanide of potassium (concentrated solution) . . . . . 20 c.c.
    Glycerin . . . . . 50 ,,
(2) Sulphate of copper (concentrated solution) . . . . . 35 ,,
    Glycerin . . . . . 50 ,,

Mix (1) and (2) slowly, with agitation; at the moment of injecting combine with 3 volumes of vehicle.

436. Blue Colouring Mass (Prussian Blue) (*Robin's modification of Beale's formula, Ibid.*, p. 35):
Take—

(A) Sulphocyanide of potassium (sol. sat.) . 90 c.c.
    Glycerin . . . . . 50 ,,
(B) Liquid perchloride of iron at 30° . 3 ,,
    Glycerin . . . . . 50 ,,

Mix slowly and combine the mixture with three parts of vehicle. It is well to add a few drops of HCl.

437. Cadmium Colouring Mass (*Ibid.*, p. 36).
Take—

Sulphate of cadmium (sol. sat.) . . 40 c.c.
Glycerin . . . . . 50 ,,

and

Sulphide of sodium (sol. sat.) . . 30 c.c.
Glycerin . . . . . . 50 ,,
Mix with agitation and combine with 3 vols. of vehicle.

438. Scheele's Green Colouring Mass (*Ibid.*, p. 37).
Take—
Arseniate of potash (saturated solution) 80 c.c.
Glycerin . . . . . . 50 ,,
and
Sulphate of copper (saturated solution). 40 ,,
Glycerin . . . . . . 50 ,,
Mix and combine with 3 vols. of vehicle.

438 a. Anilin Colouring Masses (*Ibid.*, p. 37).—The anilin colours have, for injections, the great fault of being soluble in alcohol; fuchsin is soluble in water, in alcohol, and in glycerin; it therefore cannot be employed with a gelatin or glycerin vehicle. Anilin blue, violet, yellow may be combined with these vehicles after dissolving in a small quantity of alcohol; and (alcohol being avoided for hardening purposes) the injected organs may be preserved in glycerin.

Carmine Gelatin Masses.

439. Ranvier's Carmine Gelatin Mass (*Traité technique*, p. 116).—Take 5 grammes Paris gelatin, soak it in water for half an hour, or until quite swollen and soft; wash it; drain it; put it into a test-tube and melt it, in the water it has absorbed, over a water-bath. When melted add slowly, and with continual agitation, a solution of carmine in ammonia, prepared as follows :—$2\frac{1}{2}$ grammes of carmine are rubbed-up with a little water, and just enough ammonia, added drop by drop, to dissolve the carmine into a *transparent* solution.

When the carmine has been added to the gelatin you will have about 15 c.c. of ammoniacal solution of carmine in gelatin, if the operations have been properly performed. This solution is to be kept warm on the water-bath, whilst you proceed to neutralise it by adding cautiously, drop by drop, with continual agitation, a solution of 1 part of glacial acetic acid in 2 parts of water. (When the mass is near neutrality, dilute the acetic acid still furthur.) The instant of saturation is determined by the smell of the solution, which gradually changes from ammoniacal to sour. As soon as the sour smell is perceived, the addition of acetic acid must cease, and the liquid be examined under the microscope. If it contains a granular precipitate of carmine, too much acid has been added, and the mass must be thrown away.

Ranvier states that by practice the operator learns to attain to perfect neutralisation almost infallibly in this way; and that this is the only way to attain to it. Trust must not be put in certain formulæ that profess to indicate the proportions of ammonia and acetic acid necessary for neutralisation, on account of the variation in strength of the solutions of ammonia kept in laboratories. The method proposed by Frey of determining beforehand the quantity of a known acetic solution that is necessary for neutralisation of a given quantity of the ammonia employed, is not infallible because it often happens that commercial gelatin is acid; in which case the proposed method would cause the operator to overpass the point of saturation.

The mass having been perfectly neutralised is strained through new flannel.

440. How to Neutralise a Carmine Mass (VILLE, *Gaz. hebd. d. Sci. méd. de Montpellier*, Fév., 1882; may be had separately from Delahaye et Lecrosnier, Paris).—Ville is of Ranvier's opinion that the method of titration recommended by Frey is defective, but for a different reason. When carmine is treated with ammonia a certain proportion of the ammonia combines with the carmine to form a transparent purple compound, and the rest of the ammonia remains in excess. It is this *excess* that it is required to neutralise precisely. In Frey's method a quantity of acid sufficient for the neutralisation of the *whole* of the ammonia employed is taken; hence, naturally, the point of neutralisation is over-stepped, and a granular mass is the result.

As to the acidity accidentally found in commercial gelatin, that source of error is easily eliminated. Instead of soaking the gelatin in water, it should be placed in a large funnel with a narrow neck, or better, in a stop-cock funnel, and the whole should be placed under a tap, and a stream of water arranged in such a manner that the gelatin be constantly completely immersed. Washing for an hour or so in this way will remove all traces of acids mechanically retained in the gelatin.

As to the neutralisation of the colouring mass, Ville is of opinion that the criterion of neutrality given by Ranvier—the sour smell that takes the place of the ammoniacal odour—cannot be safely relied on in practice. He considers it greatly preferable to employ dichroic litmus paper (litmus paper sensibilised so as to be capable of being used equally for the demonstration of acids and bases).

To prepare such a paper, the tincture obtained by decoction of cake litmus is slightly acidified by an excess of sulphuric acid. By this means the excess of alkali, or of alkaline carbonates, that is always present in litmus decoction, and which diminishes its sensibility as a reagent, is neutralised. The decoction is then heated and agitated with an excess of precipitated carbonate of baryta, and filtered.

The solution of litmus thus obtained is exposed to the air in wide vessels until its intense blue colour has given place to a reddish tint. Strips of white unsized paper are then dipped in it, and dried in the shade on stretched threads, in a place free from vapour of ammonia.

A shorter method consists in adding very dilute sulphuric acid, drop by drop, to the ordinary laboratory tincture of litmus, until the colour changes to red. Then, by adding successively traces of alkali and very dilute sulphuric acid, the reddish, dichroic tint may be obtained, and the paper prepared with the solution as before.

The paper is used in the same way as ordinary litmus paper. A strip is moistened with distilled water and held as close as possible to the injection mass kept melted on a water-bath. It becomes blue at first, very rapidly and decidedly; but as fast as fresh quantities of acid are added, this reaction becomes less evident, and at a certain moment the change of colour becomes very slow in making its appearance. It is then that the addition of acid should cease, and the operation is ended.

Very delicate sensitised paper may also be prepared with other reagents than litmus, for instance, with Nessler's reagent*

---

* Nessler's reagent may be prepared as follows :—Mercuric chloride, in powder, 35 grammes ; iodide of potassium, 90 grammes ; water, 1750 c.c. Heat gently till dissolved in a large basin ; then add of stick caustic potash

or with alcoholic solution of hæmatoxylin, or with a solution made by adding a trace of dilute sulphuric acid to "liqueur orange No. 3" (a liquid found in commerce, and used for detecting acids); the solution takes on a gooseberry red colour.

The preparation of the injection mass is *facilitated* by employing acetic acid and ammonia of known strength. For the acetic acid it is sufficient to keep the glacial acid in a well-stoppered bottle. But this will not suffice for the ammonia, which is notably lowered in strength through the mere pouring from one bottle into another. Ville has imagined an apparatus which allows of withdrawing a known quantity without permitting any access of air to the stock solution. Description and figures, l. c.

With the exception of the processes above described, Ville prepares the injection mass exactly as Ranvier.

441. Gerlach's Carmine Gelatin Mass (see *Arch. f. mik. Anat.*, 1865, p. 148; and Ranvier's *Traité*, p. 113).

442. Thiersch's Carmine Gelatin Mass (see *Ibid*).

443. Carter's Carmine Gelatin Mass (*see* BEALE, p. 113).

444. Davies' Carmine Gelatin Mass (*see* his *Prep. and Mounting of Mic. Objects*, p. 138).

445. Hoyer's Carmine Gelatin Mass (*Biol. Centralb.*, 1882, p. 21).—Take a concentrated gelatin solution and add to it a corresponding quantity of the neutral carmine (staining solution) (Formula No. 138). Digest in a water-bath until the dark violet-red colour begins to pass into a bright red tint. Then add 5—10 per cent. by volumes of glycerin, and at least 2 per cent. by weight of chloral, in a concentrated solution. After passing through flannel it can be kept in an open vessel under a bell-glass.

446. Fol's Carmine Gelatin Mass (*Zeit. f. wiss Zool.*, xxxviii, 1883, p. 492).

The following method of preparation has the advantage of producing masses that can be kept in the *dry state* for an indefinite length of time. (Fol finds that the addition of chloral hydrate to wet masses is not an efficient preservative.)

320 grammes, and 50 c.c. of saturated solution of mercuric chloride (WANKLYN). From COOLEY's *Cyclopædia*, s. v. "Nessler's Test."

One kilog. of Simeon's photographic gelatin* is soaked for a couple of hours, until thoroughly soft, in a small quantity of water. The water is then poured off and the gelatin melted over a water-bath, and one litre of concentrated solution of carmine in ammonia is poured in with continual stirring. (The carmine solution is prepared by diluting strong solution of ammonia with three or four parts of water and adding carmine to saturation; the undissolved excess of carmine is removed by filtration just before the solution is added to the gelatin.)

To the mixture of gelatin and carmine, which should have a strong smell of ammonia, sufficient acetic acid is added to turn the dark purple colour of the mixture into the well-known blood-red hue. Exact neutralisation is not necessary. The mass is set aside until it has become firm, and is then cut up into pieces, which are tied up in a piece of tulle or fine netting. By means of energetic compression with the hand under water (it must be *acidulated* water, 0·1 per cent. acetic acid, otherwise the carmine will wash out; cf. 'Journ. Roy. Mic. Soc.,' iv, part 3, 1884, p. 474) the mass is driven out through the meshes of the stuff in the shape of fine strings, which are washed for several hours in a sieve placed in running water in order to free them from any excess of acid or ammonia. The strings are then again melted, and the molten mass is poured on to large sheets of parchment paper soaked with paraffin, and the sheets are hung up to dry in an airy place. When dry the gelatin can easily be separated from the sheets, and may be cut into long strips with scissors and put away, protected from dust and damp, until wanted for use. In order to get the mass ready for use, all that is necessary is to soak the strips for a few minutes in water and melt them over a water-bath.

The process may be simplified, without giving very greatly inferior results, as follows (*Lehrb.*, p. 13). Gelatin in sheets is macerated for two days in the above-described carmine

---

* This gelatin may be obtained either from the ordinary furnishers of articles used in photography, or direct from Simeon's Gelatin-fabrik, Winterthur, Switzerland. Two sorts, a hard and a soft, are sold; the softer is to be preferred on account of its lower point of fusion. Probably the photographic gelatins of Hinrichs, of Frankfurt, and of Coignet, of Paris, would answer equally well; as also the best English preparations.

solution, then rinsed and put for a few hours into water acidulated with acetic acid. It is then washed on a sieve for several hours in running water, dried on parchment paper, and preserved as above.

This injection mass is very well spoken of.

Blue Gelatin Masses.

447. Robin's Prussian Blue Gelatin Mass (*see above*, No. 436).

448. Ranvier's Prussian Blue Gelatin Mass. (*Traité*, p. 119). —Twenty-five parts of a concentrated aqueous solution of soluble Prussian blue (prepared as directed below), mixed with 1 part of solid gelatin.

The mixture of the Prussian blue with the vehicle is effected in the following manner:

Weigh the gelatin, soak it in water for half an hour or an hour, wash it, and melt it in a test-tube, in the water it has absorbed, by heating over a water-bath. Put the solution of Prussian blue into another test-tube, and heat it on the same water-bath as the gelatin, so as to have the two at the same temperature. Pour the gelatin gradually into the Prussian blue solution, stirring continually with a glass rod. Continue stirring until the disappearance of the curdy precipitate that forms at first. (Some gelatins produce a *persistent* precipitate; these must be rejected; but it must be borne in mind that the precipitate that invariably forms in even the best gelatins disappears if the heating be continued. It is essential to remember this when preparing Prussian blue and gelatin mass.) As soon as the glass rod has ceased to show blue granulations on its surface on being withdrawn from the liquid, it may be concluded that the Prussian blue is completely dissolved. Filter through new flannel, and keep the filtrate at 40° over a water-bath until injected.

The soluble Prussian blue for the above mass is prepared as follows:

449. Soluble Prussian Blue for Injection Masses (RANVIER, *Ibid*).—Make a concentrated solution of sulphate of peroxide of iron in distilled water, and pour it gradually into a concentrated solution of yellow prussiate of potash. There is produced a precipitate of insoluble Prussian blue. (An excess of prussiate of potash ought to remain in the liquid; in order to ascertain whether this is the case take a small quantity of the liquid and observe whether a drop of sulphate of iron still precipitates it.) Filter the liquid through a felt strainer, underneath which is arranged a paper filter in a glass funnel. The liquid at first runs clear and yellowish into the lower funnel; distilled water is then poured little by little on to the strainer; gradually the liquid issuing from the strainer acquires a blue tinge, which, however, is not visible in that which issues from the lower filter. Distilled water is continually added to the strainer for some days until the liquid begins to run off blue from the second filter. The Prussian blue has now become soluble. The strainer is turned inside out and agitated in distilled water; the Prussian blue will dissolve if the quantity of water be sufficient.

The solution may now be injected just as it is, or it may be kept in bottles till wanted, or the solution may be evaporated in a stove, and the solid residuum put away in bottles.

For injections, if a simple aqueous solution be taken, it should be *saturated*. Such a mass never transudes through the walls of vessels. Or, it may be combined with one fourth of glycerin, or with the gelatin vehicle above described.

450. Soluble Prussian Blue (GUIGNET, *Journ. de Microgr.*, 1889, p. 94; *Journ. Roy. Mic. Soc.*, 1889, p. 468).—Guignet gives two methods:

1. To a boiling solution of 110 grammes of ferridcyanide of potassium, are added gradually 70 grammes of crystallised sulphate of iron. After boiling two hours it is filtered, and the filtrate washed with fresh water until the washings are strongly blue. It is then dried at 100° C.

2. A saturated solution of oxalic acid is mixed to a pasty consistence with an excess of pure Prussian blue. The liquid is filtered and allowed to stand for two months until all the blue is precipitated. It is then filtered and washed with weak spirit in order to remove any oxalic acid, then dried.

A similar result may be at once obtained by precipitating the oxalic solution with 95 per cent. alcohol, or with a concentrated solution of sodium sulphate, and then washing the precipitate with weak spirit.

451. Brücke's soluble Berlin Blue (*Arch. f. mik. Anat.*, 1865,

p. 87).—Brücke first prepared it by taking a 10 per cent. solution of ferrocyanide of potassium, and precipitating by means of a dilute solution of sesquichloride of iron (taken in such a quantity as to contain just half as much chlorine as is necessary for the decomposition), and washing the precipitate on the filter until solubility is attained.

Later on, he employed a greater excess of ferrocyanide, and took just so much dilute solution of chloride of iron that the weight of the dry chloride employed came to $\frac{1}{10}$th or $\frac{1}{8}$th of that of the ferrocyanide. The precipitate was washed on a filter (using the filtrate to wash with), until nothing but a clear yellow liquid filtered off, then washed with water, until the water began to run off blue, then dried, pressed between blotting-paper in a press, the resulting mass broken in pieces and dried by exposure to the air.

A cheaper method is the following:

Make a solution of ferrocyanide of potassium containing 217 grammes of the salt to 1 litre of water.

Make a solution of 1 part commercial chloride of iron in 10 parts water.

Take equal volumes of each, and add to each of them twice its volume of a cold saturated solution of sulphate of soda. Pour the chloride solution into the ferrocyanide solution, stirring continually. Wash the precipitate on a filter until soluble, and treat as above described.

The concentrated solution of the colouring matter is to be gelatinised with just so much gelatin that the mass forms a jelly when cold.

452. Thiersch's Prussian Blue Gelatin Mass (*Arch. f. Mik. Anat.*, i, 1865, p. 148).

Take—

(1) A solution of 1 part gelatin in 2 parts water.

(2) A saturated aqueous solution of sulphate of iron.

(3) A saturated aqueous solution of red prussiate of potash.

(4) A saturated aqueous solution of oxalic acid.

Now (A) mix 12 c.c. of the iron solution with one ounce of the gelatin solution at the temperature of 25° R.

Then (B) mix, at the same temperature, 24 c.c. of the prussiate solution with two ounces of the gelatin solution.

(c.) To the latter mixture add first 24 c.c. of the oxalic-acid solution, stir well, and then add the gelatin and iron mixture (A). Stir continually, keeping the temperature at from 20° to 25° R. until the whole of the Prussian-blue is precipitated. Finally, heat over a water-bath to about 70° R. and filter through flannel.

453. Fol's Berlin Blue Gelatin Mass (*Zeit. f. wiss. Zool.*, xxxviii, 1883, p. 494).—A modification of Thiersch's formula, No. 452. 120 c.c. of a cold saturated solution of sulphate of iron are mixed with 300 c.c. of the warm gelatin solution. In a separate vessel 600 c.c. of the gelatin solution are mixed with 240 c.c. of a saturated solution of oxalic acid, and 240 c.c. of a cold saturated solution of red prussiate of potash are added to the mixture. The first mixture is now gradually poured into the second, with vigorous shaking, the whole is warmed for a quarter of an hour over a boiling water-bath, the mass is allowed to set, is pressed out into strings through tulle or netting, as described for the carmine mass, *supra*, § 446, and the strings are washed and spread out to dry on the prepared paper. (It is necessary to dry the strings without remelting in this case, because the mass does not readily melt without the addition of oxalic acid.) In order to prepare the mass for injection, the strings are put to swell up in cold water, and then warmed with the addition of enough oxalic acid to allow of complete solution.

454. Hoyer's Soluble Berlin Blue Gelatin Mass (*Arch. f. mik. Anat.*, 1876, p. 649).—The filtered and not too much washed precipitate of soluble Berlin blue is brought in a little water on to a Graham's dialyser, and the external water changed until the solution begins to pass through the parchment. Dilute the solution and filter through filter-paper, an operation which becomes easy *after* dialysis. The solution may be injected pure (for lymphatics, for instance) or may be combined with gelatin. To do this, warm the solution almost to boiling-point, and add *gradually* a warm, thin solution of gelatin until coagulation begins to set in. Strain through wetted flannel.

Gelatin Masses of other Colours.

455. Robin's Cadmium Gelatin Mass (see § 437).

456. Thiersch's Lead Chromate Gelatin Mass (*Arch. f. mik. Anat.*, 1865, p. 149).

Make—

(A) A solution of 1 part gelatin in 2 parts water.

(B) A solution of 1 part neutral chromate of potash in 11 parts water.

(C) A solution of 1 part nitrate of lead in 11 parts water.

Mix 4 parts of the gelatin solution with 2 parts of the lead solution, and in another vessel mix 4 parts gelatin solution with 1 part of the chromate solution. Heat both the mixtures to 25° R., mix them together with continual stirring until all the chromate of lead is precipitated; heat over a water-bath to 70° R. and filter through flannel.

457. Hoyer's Lead Chromate Gelatin Mass (*Ibid.*, 1867, p. 136).

Take—

One volume of a solution of gelatin containing 1 part of gelatin to 4 of water.

One volume of cold saturated solution of bichromate of potash.

And one volume cold saturated solution of sugar of lead (neutral plumbic acetate).

Filter the gelatin solution through flannel and mix in the bichromate solution. Then *warm almost to boiling point*, and add gradually the (warmed) sugar of lead solution. Allow the mass to cool down to body temperature and inject at once. Another mode of preparation is as follows: Mix the sugar of lead solution with part of the gelatin solution, mix the bichromate solution with the remaining gelatin solution, heat the latter mixture, and pour into it the former mixture (gradually), stirring continually.

If the solutions are mixed at a low temperature a lumpy granular precipitate is formed. Further, when solution of sugar of lead is added to a *hot* solution of bichromate of potash, a rich orange-red precipitate is obtained; whilst if the solutions be mixed *cold*, the precipitate is bright yellow

If the solutions of the two salts be kept ready prepared,

the injection mass may be mixed in less than a quarter of an hour. Its advantages are that, on account of the extremely fine state of division of the precipitate, the mass is almost transparent, and runs so freely that even lymphatics may be perfectly injected with it, whilst its intensity of colour makes the vessels much more distinct than the very pale mass of Thiersch (No. 456). It is also easier to manage than Thiersch's mass, as it does not solidify so quickly. It shows well in the vessels by reflected, as well as by transmitted, light.

458. **Fol's Lead Chromate Gelatin Mass** (*Lehrb.*, p. 15).

459. **Hoyer's Silver Nitrate Yellow Gelatin Mass** (*Biol. Centralbl.*, ii, 1882, pp. 19, 22; *Journ. Roy. Mic. Soc.* (N.S.), iii, 1883, p. 142).—" A concentrated solution of gelatin is mixed with an equal volume of a 4 per cent. solution of nitrate of silver and warmed. To this is added a very small quantity of an aqueous solution of pyrogallic acid, which reduces the silver in a few seconds; chloral and glycerin are added as before " (see *ante*, HOYER's formula for carmine gelatin, No. 445).

This mass is yellow in the capillaries and brown in the larger vessels. It does not change either in alcohol, chromic or acetic acid, or bichromate of potash, &c.

460. **Hoyer's Green Gelatin Masses** (*Ibid.*).—Made by mixing a blue mass and a yellow mass.

461. **Thiersch's Green Gelatin Mass** (*Arch. f. mik. Anat.*, 1865, p. 149).—Made by mixing the blue mass, § 452, and the yellow mass, § 456.

462. **Robin's Scheele's Green Gelatin Mass** (see § 438).

463. **Hartig's White Gelatin Mass** (FREY, *Le Microcope*, p. 190).—Dissolve 125 grammes of acetate of lead in so much water that the whole shall weigh 500 grammes.

Dissolve 95 grammes of carbonate of soda in so much water that the whole shall weigh 500 grammes.

Take equal volumes of the two solutions, and add two volumes of gelatin solution.

464. **Frey's White Gelatin Mass** (*Ibid*).—Put into a tall glass cylinder 125 to 185 grammes of cold saturated solution of

chlorate of baryta. Add drop by drop, very carefully, sulphuric acid. Allow the precipitate that forms to settle for twelve hours, then decant almost all the clear supernatant liquid. The remaining mucilaginous mass containing the precipitate is to be mixed with an equal part of concentrated gelatin solution.

Frey states that this is a very finely-grained mass. Injected organs may be preserved in chromic acid.

465. Teichmann's White Gelatin Mass (*Ibid.*, p. 191).—"Take 3 parts of nitrate of silver dissolved in the gelatin solution, and add 1 part of common salt."

The mass is very fine-grained, and is not decomposed by chromic acid; the disadvantage of it is that it blackens under the influence of light and of sulphurous solutions.

466. Fol's Brown Gelatin Mass (*Zeit. f. wiss. Zool.*, xxxviii, 1883, p. 494).—Five hundred grammes of gelatin are soaked, and allowed to swell up, in two litres of water in which 140 grammes of common salt have previously been dissolved; the mass is melted over a water-bath, and a solution of 300 grammes of nitrate of silver in a litre of water is gradually added, with vigorous shaking. (If it be desired to have an extremely fine-grained mass, both the solutions should be diluted with three or four volumes of water.) The mass is pressed out into strings as before (§ 446), and the strings are stirred up, in clear daylight, with the following mixture: 1½ litres of cold saturated solution of potassic oxalate to 500 c.c. of cold saturated solution of sulphate of iron. As soon as the whole mass is thoroughly black the operation is at an end. The strings are then washed for several hours, re-melted, and poured on to the prepared paper.

467. Miller's Purple Silver Nitrate Gelatin Mass. See *Amer. Mon. Mic. Journ.*, 1888, p. 50; *Journ. Roy. Mic. Soc.*, 1888, p. 518; *Zeit. f. wiss. Mik.*, v, 3, 1888, p. 361.

468. Robin's Mahogany Gelatin Mass (*see* § 435).

469. Ranvier's Gelatin Mass for Impregnation (*Traité*, p. 123). —Concentrated solution of gelatin, 2, 3, or 4 parts; 1 per cent. nitrate of silver solution, 1 part.

470. Fol's Metagelatin Vehicle (*Lehrb.*, p. 17).—The operation of injecting with the ordinary gelatin masses is greatly complicated by the necessity of injecting them warm. Fol proposes to employ metagelatin instead of gelatin.

If a slight proportion of ammonia be added to a solution of gelatin, and the solution be heated for several hours, the solution passes into the state of metagelatin, that is, a state in which it no longer coagulates on cooling. Colouring masses may be added to this vehicle, which may also be thinned by the addition of weak alcohol. After injection, the preparations are thrown into strong alcohol or chromic acid, which sets the mass.

CHAPTER XXII.

INJECTIONS—OTHER MASSES.

471. Joseph's White-of-Egg Injection Mass (Carmine) (*Ber naturw. sect. Schles. Ges.*, 1879, pp. 36—40; *Journ. Roy. Mic. Soc.* (N.S.), ii, 1882, p. 274).—" Filtered white of egg, diluted with 1 to 5 per cent. of carmine solution. . . . . This mass remains liquid when cold; it coagulates when immersed in dilute nitric acid, chromic or osmic acids, remains transparent, and is sufficiently indifferent to reagents."

For Invertebrates.

472. Bjeloussow's Gum Arabic Mass (*Arch. f. Anat. u. Phys.*, 1885, p. 379).—Make a syrupy solution of gum arabic and a saturated solution of borax in water. Mix the solutions in such proportions as to have in the mixture 1 part of borax to 2 of gum arabic. Rub up the transparent, almost insoluble mass with distilled water, added little by little, then force it through a fine-grained cloth. Repeat these operations until there is obtained a mass that is free from suspended gelatinous clots. (If the operation has been successful, the mass should coagulate in the presence of alcohol, undergoing at the same time a dilatation to twice its original volume.)

The vehicle thus prepared may be combined with any colouring mass, except cadmium and cobalt.

After injection the preparation is thrown into alcohol, and the mass sets immediately, swelling up as above described, and consequently showing vessels largely distended.

Cold-blooded animals may be injected whilst alive with this mass. It does not flow out of cut vessels. Injections keep well in alcohol. Glycerin may be used for making them transparent.

If it be desired to remove the mass from any part of a preparation, this is easily done with dilute acetic acid, which dissolves it.

Glycerin Masses (cold).

473. As to Glycerin Masses.—Glycerin masses are certainly very convenient, and give very good results from the scientific —not from the æsthetic—point of view. They have a great defect for the injection of fresh specimens, that is, those in which rigor mortis has not set in; they stimulate the contraction of arteries. In these cases it may be advisable to use nitrite of amyl as a vaso-dilatator. The animal may be anæsthetised with a mixture of ether and nitrite of amyl, and finally killed with pure nitrite. Or, after killing in any way, a little nitrite of amyl in salt solution may be injected before the injection mass is thrown in. In any case it is advisable to add a little nitrite to the mass just before using. The relaxing power is very great (*see* OVIATT and SARGENT, in *St. Louis Med. Journ.*, 1886, p. 207; and *Journ. Roy. Mic. Soc.*, 1887, p. 341).

474. Beale's Carmine Glycerin Mass (*How to Work, &c.*, p. 95).—Five grains of carmine are dissolved in a little water with the aid of about five drops of ammonia, and added to half an ounce of glycerin. Then add half an ounce of glycerin with eight or ten drops of acetic or hydrochloric acid, gradually, with agitation. Test with blue litmus paper, and if necessary add more acid till the reaction is decidedly acid. Then add half an ounce of glycerin, two drachms of alcohol, and six drachms of water.

475. Robin's Carmine Glycerin Mass (*Traité*, p. 33).—Consists of the following vehicle:

Glycerin	2 parts.
Alcohol	1 ,,
Water	1 ,,

Combined with one third or one fourth its volume of the carmine colouring mass, *ante*, formula No. 434.

476. Beale's Prussian Blue Glycerin Mass (*How to Work, &c.*, p. 93).

Common glycerin	1 ounce.
Spirits of wine	1 ,,
Ferrocyanide of potassium	12 grains.
Tincture of perchloride of iron	1 drachm.
Water	4 ounces.

Dissolve the ferrocyanide in 1 ounce of the water and glycerin, and add the tincture of iron to another ounce. "These solutions should be mixed together *very gradually* and well shaken in a bottle, *the iron being added to the solution of the ferrocyanide of potassium.* Next, the spirit and the water are to be added very gradually, the mixture being constantly shaken."

"*The water*" spoken of in the last sentence appears to mean the remaining 3 ounces of water that were not mixed with the glycerin at first.

Injected specimens should be preserved in acidulated glycerin, otherwise the colour may fade.

477. Beale's Acid Prussian Blue Glycerin Mass (*Ibid.*, p. 296).

Price's glycerin	2 fluid ounces.
Tinct. of sesquichloride of iron	10 drops.
Ferrocyanide of potassium	3 grains.
Strong hydrochloric acid	3 drops.
Water	1 ounce.

Proceed as directed above, dissolving the ferrocyanide in one half of the glycerin, the iron in the other, and adding the latter drop by drop to the former. Finally add the water and HCl. Two drachms of alcohol may be added to the whole if desired.

I consider this a most admirable formula. I possess some of this mass prepared many years ago, in which not the smallest flocculus has made its appearance. The Prussian blue appears to be in a state of true solution. The mass runs well, and has not so much tendency to exude from cut capillaries as might be supposed.

478. Ranvier's Prussian Blue Glycerin Mass (*Traité*, p. 120).—Consists of the Prussian blue fluid, § 449, mixed with one fourth of glycerin.

479. Other Colours.—Any of the colouring masses, §§ 434 to 438*a*, or other suitable colouring masses, combined with the vehicle § 475.

Aqueous Masses.

480. Ranvier's Prussian Blue. Aqueous Mass (*Traité*, p. 120).—The soluble Prussian blue, § 449, injected without any vehicle. It does not extravasate.

481. Müller's Berlin Blue (*Arch. f. mik. Anat.*, 1865, p. 150).—Precipitate a concentrated solution of Berlin blue by means of 90 per cent. alcohol.

The precipitate is very finely divided; the fluid is *perfectly neutral*, and much easier to prepare than the formula of Beale.

482. Mayer's Berlin Blue (*Mitth. zool. Stat. Neapel*, 1888, p. 307).—A solution of 10 c.c. of tincture of perchloride of iron in 500 c.c. of water is added to a solution of 20 gr. of yellow prussiate of potash in 500 c.c. of water, allowed to stand for twelve hours, decanted, the deposit washed with distilled water on a filter until the washings come through dark blue (1 to 2 days), and the blue dissolved in about a litre of water.

483. Emery's Aqueous Carmine (*Ibid.*, 1881, p. 21).—To a 10 per cent. ammoniacal solution of carmine is added acetic acid, with continual stirring, until the colour of the solution changes to blood-red through incipient precipitation of the carmine. The supernatant clear solution is poured off, and injected cold without further preparation. The injected organs are thrown at once into strong alcohol to fix the carmine. For injection of Fishes.

484. Letellier's Vanadate of Ammonia and Tannin (*Journ. Roy. Mic. Soc.*, 1889, p. 151).—Vanadate of ammonia is soluble in warm, and tannin in hot water. The two solutions are kept apart until required for use, when they are mixed according to the tint required. A black mass, very fine. The walls of vessels are stained black by it.

485. Taguchi's Indian Ink (*Arch. f. mik. Anat.*, 1888, p. 565; *Zeit. f. wiss. Mik.*, 1888, p. 503).—Chinese or (better) Japanese ink well rubbed up on a hone until a fluid is obtained that does not run when dropped on thin blotting-paper nor form a grey ring round the drop. Inject until the preparation appears quite black, and throw it into some hardening liquid (not pure water).

I believe this will be found useful for many purposes, especially for work amongst invertebrates, as well as for lymphatics, juice-canals, and the like.

Celloidin Masses.

486. Schiefferdecker's Celloidin Masses (*Arch. Anat. u. Phys.*, 1882 (*Anat. Abth.*), p. 201). (For Corrosion preparations).—
1. **Asphalt-celloidin** is the best of these injections. To prepare it—
Pulverise asphalt in a mortar, and put it for twenty-four hours into a well-closed vessel with some ether, shaking occasionally. After the twenty-four hours pour off the ether into another vessel, and dissolve in it small pieces of celloidin until the solution is of the consistency of one of the thicker fatty oils. (The undissolved asphalt may be employed for colouring a fresh quantity of ether, in which substance it is not very soluble.)

2. **Vesuvianin Celloidin Brown Injection.**—Make a concentrated solution of Vesuvianin in absolute alcohol and dissolve celloidin in it. (This colour is not fast.)

3. **Opaque Blue Celloidin Injection.**—Dissolve celloidin in equal parts of absolute alcohol and ether, and add pulverised Berlin blue.

4. **Opaque Red Celloidin Injection.**—Proceed as above (3), taking pulverised cinnabar instead of Berlin blue. The two last pigments should be rubbed up in a mortar with a little absolute alcohol, and the paste added to the celloidin-mass. Be careful not to take more pigment than is absolutely necessary, or the injection will become brittle. To filter (if this be thought necessary), strain the mass through flannel wetted with ether. Syringes must be free from grease, which would render the mass brittle. The nozzles to be filled with ether. Inject quickly, as the mass soon sets on contact with watery tissues. Clean syringes and nozzles with ether.

Corrosion of the Preparations.—The injected organs are thrown into unrectified hydrochloric acid, where they remain (the acid being changed from time to time if necessary) until all the tissues are destroyed. Wash under a slow stream of water from a tap furnished with an india-rubber tube. Leave for some weeks in water, rinse, and put up in glycerin, or a mixture of glycerin, alcohol, and water in equal volumes.

487. Hochstetter's Modification of Schiefferdecker's Mass (*Anat. Anz.*, 1886, p. 51; *Journ. Roy. Mic. Soc.*, 1888, p. 159). —Kaolin is rubbed up with ether, to which cobalt blue, chrome yellow, or cinnabar is added. To this, celloidin solution of the consistence of honey is added.

Other Masses.

488. Budge's Asphaltum Mass (*Arch. f. mik. Anat.*, xiv, 1877, p. 70).—A large quantity of asphaltum has benzol poured on it, and is allowed to stand for several days, and then preserved for use. Before injecting add ⅛ to ½ benzol, and filter. Chloroform and turpentine may also be used as solvents. Used for injecting the juice-canals of cartilage by the method described l. c., or by puncture.

489. Hoyer's Shellac Mass (*Arch. f. mik. Anat.*, 1876, p. 645). —Place a quantity of good shellac in a wide-necked flask and add just enough alcohol (of about 80 per cent. strength) to cover the shellac. Leave it for twenty-four hours, and then warm it in a water-bath to complete the solution. When cool, dilute, if necessary, with alcohol to the consistency of a thin syrup and strain through moderately thick muslin. The solution thus obtained may be coloured by the addition of anilin colours in (filtered) concentrated alcoholic solution, or of granular pigments suspended in alcohol. Of these, cinnabar gives the finest colouration, and may be employed for corrosion preparations (anilin colours may also be used for this purpose, but then they are not permanent). Berlin-blue and yellow sulphide of arsenic are useful. A mixture of the two gives green. Freshly-precipitated sulphide of cadmium gives a fine permanent yellow. The pigments should be rubbed up to fine powder with water, and alcohol added; let the mixture settle, pour off the dilute alcohol and add strong alcohol. Shake in a flask, by which means the coarser particles are brought to the bottom of the liquid, and at this moment pour off the supernatant fluid which contains the finer particles only. Add this to the shellac solution and strain through muslin. For very minute injections dilute the mixture with alcohol, filter through filter-paper on a covered funnel, and evaporate down to the desired consistency. Common moist water-colours, such as are sold in tin tubes, may be employed; they are to be well washed through several changes of water to get rid of the medium with which the pigments are mixed, and then suspended in alcohol as above directed. (These are to be recommended for injections into the blood of living animals.)

The shellac-solution is not attacked by hydrochloric acid; hence its applicability to corrosion preparations. To correct the brittleness of the corroded mass it is well to add to the injection-fluid some 5 per cent. of a filtered alcoholic solution of Venetian turpentine. This may also be of use for preparations that are not to be corroded. For corrosion, concentrated (fuming) hydrochloric acid may be used, and small objects left in it for one day, large ones many days or even weeks.

For hardening injections, of which it is desired to cut sections, chromic acid may be used, or a mixture of chromic and hydrochloric acid (1 part of each to 250—500 parts water). Sections are best mounted in glycerin.

This method, with some slight modifications of detail, has lately been recommended by BELLARMINOW (*Anat. Anz.*, 1888, p. 650; see also *Zeit. f. wiss. Mik.*, v, 4, 1888, p. 523, and *Journ. Roy. Mic. Soc.*, 1889, p. 150).

490. Hoyer's Oil-Colour Masses (*Internat. Monatschr. f. Anat.*, 1887, p. 341; see also *Zeit. f. wiss. Mik.*, 1888, p. 80, and *Journ. Roy. Mic. Soc.*, 1888, p. 848).—5 g. artist's Berlin blue oil-colour are rubbed up with 5 g. thickened linseed oil, and mixed with about 30 g. of lavender oil, fennel oil, thyme oil, or rosemary oil, allowed to stand for twenty-four hours in a well-stoppered vessel, and decanted. Shake before using.

For injection of the vessels of the spleen and other difficult objects.

Good results were also obtained with chrome-yellow oil-colour.

491. Pansch's Starch Mass (see *Arch. f. Anat. u. Entw.*, 1877, p. 480; 1881, p. 76; 1880, pp. 232, 371; 1882, p. 60; 1883, p.

265; and a modification of the same by GAGE, *Amer. Mon. Mic. Journ.*, 1888, p. 195, and *Journ. Roy. Mic. Soc.*, 1888, p. 1056).

492. Teichmann's Linseed Oil Masses (see *S. B. Math. Kl. Krakau Akad.*, vii, pp. 108, 158; *Journ. Roy. Mic. Soc.*, 1882, pp. 125 and 716).

493. Olive Oil for Corrosion Preparations (*see* below, § 530).

494. Natural Injections (ROBIN, *Traité*, p. 6).—To preserve these, throw the organs into a liquid composed of 10 parts of tincture of perchloride of iron and 100 parts of water.

CHAPTER XXIII.

MACERATION AND DIGESTION.

Maceration.

495. Methods of Dissociation.—It is sometimes necessary, in order to obtain a complete knowledge of the forms of the elements of a tissue, that the elements be artificially separated from their place in the tissue and separately studied after they have been isolated both from neighbouring elements and from any interstitial cement-substances that may be present in the tissue. Simple teasing with needles is often insufficient to effect the desired isolation, as the cement-substances are often tougher than the elements themselves, so that the latter are torn and destroyed in the process. In this case recourse must be had to maceration processes, by which is here meant treatment with media which have the property of dissolving or at least softening the cement substances or the elements of the tissue that it is not wished to study, whilst preserving the forms of those it is desired to isolate. When this softening has been effected the isolation is completed by teasing, or by agitation with liquid in a test-tube, or by the method of tapping, which last gives in many cases (many epithelia for instance) admirable results which could not be attained in any other way. The macerated tissue is placed on a slide and covered with a thin glass cover supported at the corners on four little feet made of pellets of soft wax. By tapping the cover with a needle it is now gradually pressed down, whilst at the same time the cells of the tissue are segregated by the repeated shocks. When the segregation has proceeded far enough, mounting medium may be added, and the mount closed.

The student will do well not to neglect this simple method, which is one that it is most important to be acquainted with.

496. Iodised Serum.—The preparation of this reagent has been given above, § 346.—The manner of employing it for maceration is as follows:—A piece of tissue smaller than a pea must be taken, and placed in 4 or 5 c.c. of weakly iodised serum in a well-closed vessel. After one day's soaking the maceration is generally sufficient, and the preparation may be completed by teasing or pressing out, as indicated above; if not, the soaking must be continued, fresh iodine being added as often as the serum becomes pale by the absorption of the iodine by the tissues. By taking this precaution, the maceration may be prolonged for several weeks.

It is obvious that these methods are intended to be applied to the preparation of *fresh* tissues, the iodine playing the part of a fixing agent with regard to protoplasm, which it slightly hardens.

497. Artificial Iodised Serum (FREY, *Le Microscope*, p. 131; RANVIER, *Traité*, p. 77).

The formula has been given above, § 347. Ranvier states that he has been unable to obtain good results, for purposes of maceration, by this method.

498. Alcohol.—Ranvier employs one-third alcohol (1 part of 36° alcohol to 2 parts of water). Epithelia will macerate well in this in twenty-four hours. Ranvier finds that this mixture macerates more rapidly than iodised serum.

Other strengths of alcohol may be used, either stronger (equal parts of alcohol and water) or weaker ($\frac{1}{4}$ alcohol, for isolation of the nerve-fibres of the retina, for instance,—*Thin*).

All observers are agreed that one-third alcohol is a macerating medium of the highest order; LIST (*Zeit. f. wiss. Mik.*, 1885, p. 511) states that for glandular structures it should be used with precaution, on account of swellings that it produces in the cells, and that Müller's solution, or osmic acid, should be preferred for such objects.

499. Salt Solution.—Ten per cent. solution of sodium chloride is a well known and valuable macerating medium.

500. MOLESCHOTT and PISO BORME's Sodium Chloride and Alcohol (MOLESCHOTT's *Untersuchungen zur Naturlehre*, xi, pp. 99—107; RANVIER, *Traité*, p. 242).—Ten per cent. solution of sodium chloride, 5 volumes; absolute alcohol, 1 volume.

For vibratile epithelium, Ranvier finds the mixture inferior to one third alcohol.

501. Chloral Hydrate.—In not too strong solution, from 2 to 5 per cent., for instance, chloral hydrate is a mild macerating agent that admirably preserves delicate elements. LAVDOWSKY (*Arch. f. mik. Anat.*, 1876, p. 359) recommends it greatly for salivary glands. HICKSON (*Quart. Journ. Mic. Sci.*, 1885, p. 244) recommends it for the study of the retina of Arthropods.

502. Caustic Potash, Caustic Soda.—These solutions must be employed *strong*, 35 to 50 per cent. (Moleschott) : so employed they do not greatly alter the forms of cells, whilst weak solutions destroy all the elements. (Weak solutions may, however, be employed for dissociating the cells of epidermis, hairs, and nails.) The strong solutions may be employed by simply treating the tissues with them on the slide. It should be remembered that preparations obtained by means of these alkalies cannot be permanently preserved.

503. Sulpho-cyanides of Ammonium and Potassium (STIRLING, *Journ. Anat. and Phys.*, xvii, 1883, p. 208).—Ten per cent. solution of either of these salts is an admirable dissociating medium for epithelium. Macerate small pieces for twenty-four to forty-eight hours, stain with fuchsin, eosin, or picro-carmine.

If a crystalline be macerated as above its fibres become beaded or moniliform.

504. Saliva, Artificial (for embryology of nerve and muscle) (CALBERLA's formulæ, *Arch. f. mik. Anat.*, xvi, 1879, p. 471, *et seq*).—After having made trial of various different macerating agents, with the object of obtaining isolation of the developing muscle and nerve of embryos of *Amphibia* and *Ophidia*, Calberla found that the best results were obtained by means of Czerny's mixture of saliva and solutio Mülleri. This led him to imagine an artificial saliva, which on trial gave results as good as those obtained by natural saliva, or even better.

Second formula (the first formula is suppressed, as being more complicated, and not giving better results) :

Potassium chloride	0·4
Sodium chloride	0·3
Phosphate of soda	0·2
Calcium chloride	0·2
	1·1

This is dissolved in 100 parts of water, saturated with carbonic acid, and the solution combined with water and solutio Mülleri, one volume of the solution being combined with half a volume of Müller's solution and a volume of water.

In either case the Müller's solution may be replaced by a 2½ per cent. solution of chromate of ammonia. The best results were obtained when the solutions were saturated with the CO_2 just before using.

The tissues are isolated by teasing and shaking, and specimens mounted in concentrated acetate of potash.

505. Landois's Solution (*Arch. f. mik. Anat.*, 1885, p. 445).

Saturated sol. of neutral chromate of ammonia	5 parts.
Saturated sol. of phosphate of potash	5 ,,
Saturated sol. of sulphate of soda	5 ,,
Distilled water	100 ,,

To be used in the same way as chromic acid:—small pieces of tissue are macerated for one to three, or even four or five days, in the liquid, then brought for twenty-four hours into ammonia carmine diluted with one volume of the macerating liquid.

Gierke particularly recommends this liquid for all sorts of macerations, but especially for the central nervous system, for which he finds it superior to all other agents. It is also recommended for the same purpose by Nansen (v. *Zeit. f. wiss. Mik.*, v, 2, 1888, p. 242).

506. Permanganate of Potash.—Has an action similar to that of osmic acid, but more energetic. Is recommended, either alone or combined with alum, as the best dissociating agent for the fibres of the cornea (Rollett, *Stricker's Handbuch*, p. 1108).

507. Chromic Acid.—Generally employed of a strength of about 0·02 per cent. Specially useful for nerve tissues and smooth muscle. Twenty-four hours' maceration will suffice for nerve tissue. About 10 c.c. of the solution should be taken for a cube of 5 mm. of the tissue (Ranvier).

508. Bichromate of Potash.—0·2 per cent.

509. Müller's Solution.—Same strength.

510. Müller's Solution and Saliva (*see* above, § 504).

511. BROCK'S **Medium** (for nervous system of Mollusca; *Intern. Monatsch. f. Anat.*, i, 1884, p. 349).—Equal parts of 10 per cent. solution of bichromate of potash and visceral fluid of the animal.

512. MÖBIUS' **Media** (quoted from *Zeit. f. wiss. Mik.*, iii, 3, 1886, p. 402).

1. 1 part of sea-water with 4 to 6 parts of 0·5 per cent. solution of bichromate of potash.

2. 0·25 per cent. chromic acid, 0·1 per cent. osmic acid, 0·1 per cent. acetic acid, dissolved in sea-water. For Lamellibranchiata. Macerate for several days.

513. Osmic Acid.—0·1 per cent., for from a few minutes to a fortnight (cortex of cerebrum, Rindfleisch). May be followed by maceration in glycerin.

514. Osmic and Acetic Acid (*the* HERTWIGS' *Liquid, Das Nervensystem u. die Sinnesorgane der Medusen,* Leipzig, 1878, and *Jen. Zeitschr.*, xiii, 1879, p. 457; *Journ. Roy. Mic. Soc.*, iii, 1880, p. 441, and (N.S), iii, 1883, p. 732).

0·05 per cent. osmic acid . . . . 1 part.
0·2 ,, acetic acid . . . . . ,,

Medusæ are to be treated with this mixture for two or three minutes, according to size, and then washed in repeated changes of 0·1 per cent. acetic acid, until all traces of free osmic acid are removed. They then remain for a day in 0·1 per cent. acetic acid, are washed in water, stained in Beale's carmine (in order to prevent the osmium from over-blackening and to assist the maceration), and are preserved in glycerin.

For *Actiniæ*, the osmic acid is taken weaker, 0·04 per cent.; both the solutions are made with sea-water; and the washing-out is done with 0·2 per cent. acetic acid. If the maceration is complete, stain with picro-carmine; if not, with Beale's carmine.

515. Béla Haller's Mixture (*Morphol. Jahrb.*, xi, p. 321).—One part glacial acetic acid, 1 part glycerin, 2 parts water. Specially recommended for the central nervous system of mollusca (Rhipidoglossa). A sufficient degree of maceration is obtained in thirty to forty minutes, the cells showing less shrinkage than with other liquids.

516. Nitric Acid.—Most useful for the maceration of muscle. The strength used is 20 per cent. After twenty-four hours' maceration in this, isolated muscle-fibres may generally be obtained by shaking the tissue with water in a test-tube.

517. Nitric Acid and Chlorate of Potash (Kühne's method, *Ueber die peripherischen Endorgane, &c.*, 1862; Ranvier, *Traité*, p. 79).—Chlorate of potash is mixed, in a watch-glass, with four times its volume of nitric acid. A piece of muscle is buried in the mixture for half an hour, and then agitated with water in a test-tube, by which means it entirely breaks up into isolated fibres.

518. Sulphuric Acid (Ranvier, *Traité*, p. 78).—Sulphuric acid has been employed by Max Schultze for isolating the fibres of the crystalline.

Macerate for twenty-four hours in 30 grammes of water, to which are added 4 to 5 drops of concentrated sulphuric acid. Agitate.

Odenius found very dilute sulphuric acid to be the best reagent for the study of nerve-endings in tactile hairs. He macerated hair-follicles for from eight to fourteen days in a solution of from 3 to 4 grains of "English sulphuric acid" to the ounce of water.

Hot concentrated sulphuric acid serves to dissociate horny epidermic structures (horn, hair, nails).

519. Oxalic Acid.—Maceration for many days in concentrated solution of oxalic acid has been found useful in the study of nerve-endings.

520. Schiefferdecker's Methyl Mixture (for the retina) (*Arch. f. mik. Anat.*, xxviii, 1886, p. 305).—Ten parts of glycerin, 1 part of methyl alcohol, and 20 parts of distilled water. Macerate for several days (perfectly fresh tissue).

Digestion.

521. BEALE's Digestion Fluid (*Archives of Medicine*, i, 1858, pp. 296—316).—The mucus expressed from the stomach glands of the pig is rapidly dried on glass plates, powdered, and kept in stoppered bottles. It retains its properties for years. Eight-tenths of a grain will dissolve 100 grains of coagulated white of egg.

To prepare the digestion fluid, the powder is dissolved in distilled water, and the solution filtered. It filters readily. Or the powder may be dissolved in glycerin. The tissues to be digested may be kept for some hours in the liquid at a temperature of 100° F. (37° C.).

522. BRÜCKE's Digestion Fluid (from CARNOY's *Biologie Cellulaire*, p. 94).

Glycerinated extract of pig's stomach . . 1 vol.
0·2 per cent. solution of HCl . . . 3 vols.
Thymol, a few crystals.

523. BICKFALVI's Digestion Fluid (*Centralbl. f. d. med. Wiss.*, 1883, p. 833).—One gramme of dried stomachal mucosa is mixed with 20 c.c. of 0·5 per cent. hydrochloric acid, and put into an incubator for three or four hours, then filtered. Macerate the tissue in the solution for not more than half an hour to an hour.

524. KUSKOW's Digestion Fluid (*Arch. f. mik. Anat.*, xxx, p. 32, cf. *Zeit. f. wiss. Mik.*, iv, 3, 1887, p. 384).—One part of pepsin dissolved in 200 parts of 3 per cent. solution of oxalic acid. The solution should be freshly prepared, and the objects (sections of hardened Ligamentum Nuchæ) remain in it at the ordinary temperature for ten to forty minutes.

525. SCHIEFFERDECKER's Pancreatin Digestion Fluid (*Zeit. f. wiss. Mik.*, iii, 4, 1886, p. 483).—Solution of pancreatin in water. Schiefferdecker employs the " Pankreatinum siccum " prepared by Dr. Witte, Rostock. A saturated solution is made in distilled water, cold, and filtered. Pieces of tissue (epidermis) are macerated in it for three to four hours at about body temperature. Nuclei are preserved, and the forms of prickle-cells well shown.

526. Kühne's Methods (see *Unters. a. d. Phys. Inst. Univ. Heidelberg*, i, 2, 1877, p. 219).—Trypsin is prepared by extracting ox-pancreas with alcohol and ether, and evaporating to dryness. One part of the product is heated for three to four hours at a temperature of 40° C. with five to ten parts of 0·1 per cent. solution of salicylic acid, and the solution forced through linen and filtered cold.

CHAPTER XXIV.

CORROSION, DECALCIFICATION, AND BLEACHING.

Corrosion.

527. Caustic Potash, Caustic Soda, Nitric Acid.—Boiling, or long soaking in a strong solution of either of these, is an efficient means of removing soft parts from skeletal structures (appendages of arthropods, spicula of sponges, &c.).

528. Eau de Javelle (Hypochlorite of Potash) (NOLL'S METHOD, *Zool. Anzeig.*, 122, 1882, p. 528).—Noll remarks that the usual method of preparing the skeleton of siliceous sponges and similar structures by corroding away the soft parts by means of caustic potash has many disadvantages, of which a principal one is that the spicula are not preserved in their normal positions. He therefore proceeds as follows : A piece of sponge is brought on to a slide and treated with a few drops of eau de Javelle, in which it remains until all soft parts are dissolved. (With thin pieces this happens in twenty to thirty minutes.) The preparation is then cautiously treated with acetic acid, which removes all precipitates that may have formed, and treated with successive alcohols and oil of cloves, and finally mounted in balsam.

The same process is stated to be applicable to calcareous structures. I feel convinced, however, that if the structures are *delicate*, they will suffer, or be entirely destroyed.

529. Eau de Labarraque (Hypochlorite of Soda) may be used in the same way as eau de Javelle. Looss (*Zool. Anz.*, 1885, p. 333) finds that either of these solutions will completely dissolve chitin in a short time with the aid of heat. For this purpose the commercial solution should be taken concentrated and boiling. A formula for making it is given in § 548.

If solutions diluted with four to six volumes of water be taken, and chitinous structures be macerated in them for twenty-four hours or more, according to size, the chitin is not dissolved, but becomes transparent, soft, and permeable to staining fluids, aqueous as well as alcoholic. The most delicate structures, such as nerve-endings, are stated not to be injured by the treatment. The method is applicable to Nematodes, and their ova, an object well known for the resistance they oppose to ordinary reagents.

This is undoubtedly a valuable method.

530. Altmann's Corrosion Method (*Arch. f. mik. Anat.*, 1879, p. 471).—Whilst almost all animal tissues are very quickly destroyed by eau de Javelle, yet fats, and particularly fats hardened by osmic acid, withstand its action for a long time. If, then, you introduce some fat or other into a tissue, harden it with osmic acid and corrode the tissue with eau de Javelle, you will obtain a mould, in osmium-blackened and hardened fat, of the spaces you had filled with the fat introduced.

The method may be of much use in certain special researches, such as those on the choroid, iris, and pigmented organs. I recommend the reader to carefully study the article, which does not well bear abstracting. A good abstract will be found in *Journ. Roy. Mic. Soc.*, 1879, p. 610, with plate.

Decalcification and Desilicification.

531. Decalcification of Bone (*Arch. f. mik. Anat.*, xiv, 1877, p. 481).—I take the following historical sketch from Busch's article "On the Technique of the Histology of Bone."

The most widely-used agent for decalcification is hydrochloric acid. Its action is rapid, even when very dilute, but it has the disadvantage of causing serious swelling of the tissues. To remedy this chromic acid may be combined with it, or alcohol may be added to it. Or a 3 per cent. solution of the acid may be taken and have dissolved in it 10 to 15 per cent. of common salt. Or (Waldeyer) to a $\frac{1}{1000}$ per cent. solution of chloride of palladium may be added $\frac{1}{10}$th of its volume of HCl.

Chromic acid is also much used, but has a very weak decalcifying action and a strong shrinking action on tissues. For this latter reason it can never be used in solutions of more than 1 per cent. strength, and for delicate structures much lower strengths must be taken.

Phosphoric acid has been recommended for young bones.

Acetic, lactic, and pyroligneous acid have considerable decalcifying power, but cause great swelling. Picric acid has a very slow action, and is only suitable for very small structures.

532. Nitric Acid (BUSCH, *l. c.*).—To all other agents Busch prefers nitric acid, which causes no swelling and acts most efficaciously, whilst at the same time it does not injuriously attack tissue-elements.

One volume of chemically pure nitric acid of sp. gr. 1·25 is diluted with 10 vols. water. It may be used of this strength for very large and tough bones; for young bones it may be diluted down to 1 per cent.

Fresh bones are first laid for three days in 95 per cent. alcohol; they are then placed in the nitric acid, *which is changed daily*, for eight or ten days. They must be *removed as soon as* the decalcification is complete, or else they will become stained yellow. When removed they are washed for one or two hours in running water and placed in 95 per cent. alcohol. This is changed after a few days for fresh alcohol.

Young and fœtal bones may be placed in the first instance in a mixture containing 1 per cent. bichromate of potash and $\frac{1}{10}$ per cent. chromic acid, and decalcified with nitric acid of 1 to 2 per cent., to which may be added a small quantity of chromic acid ($\frac{1}{10}$ per cent.) or chromate of potash (1 per cent.). By putting them afterwards into alcohol the well-known green stain is obtained.

Staining agents.—Sections of bone treated in the last-described manner are stained five or ten minutes in a weak aqueous solution of eosin. The ground-substance and small cells of cartilage remain colourless, the nuclei of the large cells are stained red, and so is periosteum, bone-tissue, and the cellular contents of the medullary spaces. Hæmatoxylin may be used in conjunction with eosin (before or after it), to obtain double-stains, which, however, are seldom successful.

Sections are dehydrated in absolute alcohol, and mounted (*without* clearing by oil of cloves or the like) in a benzol-solution of Canada balsam.

533. Nitric Acid and Alcohol.—3 per cent. of nitric acid in 70 per cent. alcohol. Soak specimens for several days or weeks. I do not know who first recommended this admirable medium.

534. Chromic Acid is employed in strengths of from 0·1 per cent. to 1 per cent., the maceration lasting two or three weeks (in the case of bone). It is better to take the acid weak at first, and increase the strength gradually.

535. Chromic and Nitric Acid.—Dissolve 15 gr. pure chromic acid in 7 oz. of distilled water, to which 30 minims of nitric acid are afterwards to be added. Macerate for three or four weeks, changing the fluid frequently (Marsh).

FOL takes 70 volumes of 1 per cent. chromic acid, 3 of nitric acid, and 200 of water (*Lehrb.*, p. 112).

536. Hydrochloric Acid may be taken of 50 per cent. strength, and then has a very rapid action (Ranvier).

537. Hydrochloric Acid and Chromic Acid (BAYERL, *Arch. f. mik. Anat.*, 1885, p. 35).—Equal parts of 3 per cent. chromic acid and 1 per cent. hydrochloric acid. For ossifying cartilage.

538. Picric Acid should be taken saturated.

Picro-sulphuric acid should of course be avoided on account of the formation of gypsum.

Picro-nitric or Picro-hydrochloric acid.—The reader will perhaps reflect that the two last fluids appear likely to be very useful for decalcifications. Mayer points out that the action is very rapid, and that the copiously evolved CO_2 often produces, mechanically, lesions in tissues; so that in many cases in which calcareous structures are concerned chromic acid is to be preferred, the more so as it more effectually hinders any *collapsing* of the structures that might result from the withdrawal of their supporting calcareous elements.

539. Glycerin. Alum-Carmine.—It should be remembered that these commonly used reagents dissolve carbonate of lime; they must therefore be avoided in the preparation of structures containing calcareous elements that it is wished to preserve (calcareous sponges, echinodermata, &c.).

540. Phoroglucin (ANDEER, *Centralbl. f. d. med. Wiss.*, xii, xxxiii, pp. 193, 579; *Intern. Monatschr.*, i, p. 350; *Zeit. f. wiss. Mik.*, 1885, pp. 375, 539; *Journ. Roy. Mic. Soc.*, 1887, p. 504).
—Andeer recommends a mixture of phoroglucin with hydrochloric acid. He takes, for bones of Batrachia, 5 to 10 per cent. of acid; for Chelonia and Birds, 10 to 20 per cent.; for Mammalia, 20 to 40 per cent. In these mixtures the bone is decalcified in a few hours, becoming so soft that it has to be subsequently hardened.

Desilicification.

541. Hydrofluoric Acid (MAYER's method, *Zool. Anz.*, 1881, No. 97, p. 593).—The objects from which it is desired to remove siliceous parts are brought in alcohol into a glass vessel coated internally with paraffin (otherwise the glass would be corroded by the acid). Hydro-fluoric acid is then added drop by drop (the operator taking great care to avoid the fumes, which attack mucous membranes with great energy). A *Wagnerella borealis* may thus be completely desilicified in a few minutes. Small pieces of siliceous sponges will require a few hours or at most a day. The tissues do not suffer; and if they have been previously stained with acetic acid carmine, the stain does not suffer; at least, this was so in the case of *Wagnerella.*

This dangerous method is best avoided. As regards sponges, I would point out that if well-imbedded, good sections may be made from them without previous removal of the spicula. The spicula appear to be cut; probably they break very sharply when touched by the knife. Knives are of course not improved by cutting such sections.

Bleaching.

542. MAYER's **Chlorine Method** (*Mitth. Zool. Stat. Neapel.*, ii, 1881, p. 8).—This is a process imagined for the purpose of getting rid of the blackening that often occurs as a consequence of treatment by osmic acid.

The specimens are put into alcohol (either of 70 or 90 per cent). Crystals of chlorate of potash are added until the bottom of the vessel is covered with them. A few drops of concentrated hydrochloric acid are then added by means of a pipette, and mixed-in by shaking the vessel as soon as the green colour of the evolving chlorine has begun to show itself. Warm if necessary; but most objects, even large ones, may be bleached in half a day without the employment of heat. The tissues do not suffer.

Instead of hydrochloric acid, nitric acid may be used; in which case the bleaching agent is the freed oxygen, instead of chlorine.

The first method may be used for the purpose of removing pigment from the eyes of insects.

543. MARSH's **Chlorine Method** (*Section Cutting*, p. 89).—Marsh generates chlorine in a small bottle by treating crystals of chlorate of potash with strong HCl, and leads the gas (by means of a piece of glass tubing bent twice at right angles) to the bottom of a bottle containing the sections in water. (See a fig. of the apparatus in *Journ. Roy. Mic. Soc.*, iii, 1880, p. 854.)

544. **Chlorine Solution** (SARGENT's method).—Hydrochloric acid, 10 drops; chlorate of potash, ½ dr.; water 1 ounce. Soak for a day or two. Wash well.

This method is intended for "bleaching insects;" it will be seen that it is only applicable to the preparation of hard parts, as soft tissues would be destroyed by the solution.

545. **Kreasote** (POUCHET's method, *Journ. de l'Anat.*, 1876, p. 8, *et seq.*).—I gather from the paper here quoted that most of the granular animal pigments are soluble in kreasote. Other solvents are mentioned in this paper ("On the Change of Colouration through Nervous Influence"), but this appears to be the only one capable of general histological application.

546. **Nitric Acid.**—Nitric acid has a similar action.

547. **Oxygenated Water** (POUCHET's method, M. DUVAL, *Précis, &c.,*' p. 234).—Macerate in glycerin to which has been added a little oxygenated water (5 to 6 drops to a watch-glass of glycerin). (Oxygenated water may be procured from perfumers or hair-dressers, by whom it is sold as a hair dye under the name of "Auréoline," "Golden hair-wash," or the like.)

The brownish-green colour communicated to tissues by

chromic solutions may be changed to yellow by means of oxygenated water (see *Arch. f. mik. Anat.*, 1887, p. 47 and *Journ. Roy. Mic. Soc.*, 1887, p. 1060).

548. Eau de Labarraque. Eau de Javelle (*see* §§ 528, 529).—These are bleaching agents. For the manner of preparing a similar solution see *Journ. de Microgr.*, 1887, p. 154, or *Journ. Roy. Mic. Soc.*, 1887, p. 518. It is, shortly, as follows:—8 parts of caustic soda are dissolved in 100 parts of distilled water, and chlorine is passed through to saturation. During the passage of the chlorine the solution must be surrounded with a mixture of salt and ice, otherwise the temperature rises, and chloride and chlorate of soda are produced. The resulting solution contains 7·45 per cent. of hypochlorite of soda. It is green; and the more effectual the cold, the greener is the colour. The energy of the decolourising action is proportional to the greenness of the solution.

549. Chloroform helps to clear strongly pigmented chitin, and combined with nitric acid will decolourise it entirely (*see* below, in the chapter on Arthropods, Part II).

550. GRENACHER'S **Mixture for Eyes of Arthropods and Other Animals** (*Abh. nat. Ges. Halle-a.-S.*, xvi; *Zeit. f. wiss. Mik.*, 1885, p. 244).

   Glycerin . . . . . . 1 part
   80 per cent. alcohol . . . . 2 „

Mix and add 2 to 3 per cent. of hydrochloric acid.

Pigments dissolve in this fluid, and so doing form a stain which suffices in twelve to twenty-four hours for staining the nuclei of the preparation. You may, if you like, first stain the objects with borax-carmine, and then put them into the liquid—the pigment being washed out more rapidly than the carmine. But the progress of the decolouration must be carefully watched.

PART II.

SPECIAL METHODS AND EXAMPLES.

CHAPTER XXV.

EMBRYOLOGICAL METHODS.

551. Artificial Fecundation.—This practice, which affords the readiest means of obtaining the early stages of development of many animals, may be very easily carried out in the case of the Amphibia anura, Teleostea, Cyclostomata, Echinodermata, and many Vermes and Coelenterata.

In the case of the Amphibia, both the female and the male should be laid open, and the ova should be extracted from the uterus and placed in a watch-glass or dissecting dish, and treated with water in which the testes, or better, the vasa deferentia, of the male have been teased.

Females of Teleostea are easily spawned by manipulating the belly with a gentle pressure; and the milt may be obtained from the males in the same way. (It may occasionally be necessary, as in the case of the Stickleback, to kill the male and dissect out the testes and tease them). The spermatozoa of fish, especially those of the Salmonidae, lose their vitality very rapidly in water; it is therefore advisable to add the milt immediately to the spawned ova, then add a little water, and after a few minutes put the whole into a suitable hatching apparatus with running water.

Artificial fecundation of invertebrates is easily performed in a similar way. It is sometimes possible to perform the operation under the microscope and so observe the penetration of the spermatozoon and some of the subsequent phenomena;—as has been done by Fol, the Hertwigs, Selenka, and others, for the Echinodermata and other forms.

552. Superficial Examination.—The development of some animals, particularly some invertebrates, may be to a certain extent followed by observation of the living ova under the microscope. This may usefully be done in the case of various

Teleosteans, such as the Stickleback, the Perch, *Macropodus*, and several pelagic forms, and with *Chironomus*, *Asellus aquaticus*, Ascidians, *Planorbis*, many Cœlenterata, &c. I advise the student to carefully draw the different stages so observed, for such drawings are most important aids to the study of the same stages by the section method.

Some ova of Insecta and Arachnida which are completely opaque under normal conditions become transparent if they are placed in a drop of oil; if care be taken to let their surface be simply impregnated with the oil, the normal course of development is not interfered with (BALBIANI).

553. Preparation of Sections.—Osmic acid, employed either alone or in combination with other reagents, is an excellent fixing agent for small embryos, but not at all a good one for large ones. It causes cellular elements to shrink somewhat, and therefore brings out very clearly the slits that separate germinal layers, and any channels or other cavities that may be in course of formation.

In virtue of its property of blackening fatty matters, myelin amongst them, it is of service in the study of the development of the nervous system.

Chromic acid is indispensable for the study of the external forms of embryos; it brings out elevations and depressions clearly, and preserves admirably the mutual relations of the parts; but it does not always preserve the forms of cells faithfully, and is a hindrance to staining in the mass.

Picric liquids have an action which is the opposite of that of osmic acid; they cause cellular elements to swell somewhat, and thus have a tendency to obliterate spaces that may exist in the tissues. But notwithstanding this defect, the picric compounds, and especially Kleinenberg's picro-sulphuric acid, are amongst the best of embryological fixing agents.

For imbedding, the celloidin-chloroform method of Viallanes gives excellent results, and so does paraffin. This latter is preferable in so far as it lends itself better to the rapid production of series of sections, and allows of the use of the Cambridge Rocking Microtome, which is perhaps the microtome *par excellence* of the embryologist.

As to staining, my eminent fellow-worker, Dr. Henneguy,

writing the chapter on embryological methods for the French edition of this work, advised staining in the mass with borax-carmine or alum-carmine (Henneguy's acetic-acid formula, § 151); or, as an alternative, the staining of sections by Flemming's method. The improvements that have in recent times been worked out in this method give still greater weight to the latter recommendation.

554. Reconstruction of Embryos from Sections.—The study of a series of sections of any highly differentiated organism of unknown structure is so complicated that it is often necessary to have recourse to elaborate methods of geometrical or of plastic reconstruction in order to obtain an idea or a model of the whole. These methods have now been brought to so high a degree of complexity that a volume rather than a paragraph would be necessary to describe them. See BORN, Die Plattenmodellirmethode, in *Arch. f. mik. Anat.*, 1883, p. 591, and *Zeit. f. wiss. Mik.*, v, 4, 1888, p. 433; STRASSER, in *Zeit. f. wiss. Mik.*, iii, 2, 1886, p. 179, and iv, 2 and 3, pp. 168 and 330; KASTSCHENKO, in *Zeit. f. wiss. Mik.*, iv, 2 and 3, 1887, pp. 235-6 and 353, and v, 2, 1888, p. 173 (abstracts of all these papers may be found in *Journ. Roy. Mic. Soc.* of the years quoted).

A simple, but in many cases quite efficient plan, has been described by FOL (*Lehrb.*, p. 35), as follows. Before cutting your sections, you make an outline drawing of your object, under the magnification that you intend to employ for the reconstructed drawing, and in a plane perpendicular to that of the intended sections. For instance, if you intend to make transverse sections of an embryo, begin by making a profile drawing of it, that is, a drawing of the outline of an ideal sagittal section of it. Then make your series of sections, and make drawings of them all under the same magnification as the sagittal drawing. Then trace over your sagittal drawing a series of equidistant parallel lines in positions corresponding to the sections that have been made. If your sections are one hundredth of a millimetre thick, and your drawing be magnified one hundred times, the lines should be one millimetre apart (if you intend to reconstruct the whole of your sections, but the operation may frequently be abridged by only reconstructing say every fifth or every tenth section).

You have now to fill in your outline drawing with details borrowed from the drawings of the sections. You may help yourself greatly in the following way. A plate of glass, of a size suitable to the intended drawings, is covered with a layer of gelatin, and dried. On this is ruled a series of parallel lines, very close together, and ruled with differently coloured inks, the colours recurring in regular order. The plate is then cut into two unequal parts by a diamond, on a line perpendicular to the coloured lines.

Lay one of the parts of the plate on the outline drawing so that the cut edge covers the line that corresponds to the first section you are going to fill in; then lay the other part of the plate on the drawing of the section in such a position that the limits of the drawing correspond to the same coloured lines that cover the limits of the outline drawing on the other part of the plate already placed.

Trace on the plate that covers the drawing of the section the outline of the internal organs. Lay it against its fellow plate on the outline drawing, making the coloured lines correspond, and you will easily be able to mark off accurately on the outline drawing a series of dots that correspond in position to the outlines of the internal organs. This operation having been repeated for each of the sections that you desire to bring into your reconstruction, nothing remains but to join your dots by lines, and you will have filled up your outline drawing with a representation of the internal organs in the same plane.

If any reader think this process complicated, he needs but to spend five minutes in trying it with a piece of tracing paper, and will find it to be in reality extremely simple.

Another simple plan is to gum the drawings of the section on cardboard of a thickness proportional to the thickness of the section and the magnification, cut out all the cavities of the drawing with a knife or fretsaw, and gum all the fretwork thus obtained together. This gives, of course, a model of the object.

Mammalia.

555. Rabbit.—The rabbit may conveniently be taken as a type for this kind of work.

Dissection.—For the study of the early stages the ova must be sought for in the *tubæ* a certain number of hours after copulation. The dehiscence of the follicles takes place about ten hours after the first coitus. The *tubæ* and *cornua* having been dissected out should be allowed to cool and remain until the muscular contractions have ceased. Then, with the aid of fine scissors or a good scalpel, all the folds of the genital duct are carefully freed from their peritoneal investment.

The tubæ are then (if the ova are still within them, which is the case up to the end of the third day after coition) laid out on a long slip of glass, and slit up longitudinally by means of a pair of fine sharp scissors. By means of needles and forceps the tubal mucosa is spread out so as to smooth out its folds as much as possible, and is carefully looked over with a strong lens or with a low power of the microscope. When the ova are found a drop of some "indifferent" liquid is dropped on each, and it is carefully taken up with the point of a scalpel, a cataract needle, or a small pipette. They may be examined in the peritoneal fluid of the mother, if the animal has been killed, or in its aqueous humour, or in amniotic liquid, or in blood serum, or in Kronecker's or other artificial serum.

If you have not been able to find the ova with the lens or the microscope, scrape off the epithelium of the tubal mucosa with a small scalpel, mix it with a little indifferent liquid, and look for the ova under the microscope by transmitted light.

Another method, employed by Kölliker, consists in injecting solution of Müller or weak osmic acid into the oviduct by means of a small syringe, and collecting the liquid that runs out in a series of watch-glasses, in which the ova can very easily be found by the microscope.

The same doe may be made to serve for two observations, at some hours' or days' interval. A longitudinal incision of 8 to 10 centimetres length is made on the median or a lateral line of the abdomen; an assistant keeps the intestines in their place; a ligature is placed at the base of one of the uterine cornua, beneath the neck, and a second ligature around the mesometrium and mesovarium. The ovary, the tuba, and the cornu of that side are then detached with scissors. The abdomen is then closed by means of a few sutures passing through the muscle layers and the skin. The animals support the operation perfectly well, and the development of the ova of the opposite side is not in the least interfered with. When it is desired to study these, the animal may be killed, or may be subjected to a second laparotomy if it be desired to preserve it for ulterior observations.

During the *fourth, fifth,* and *sixth* days after copulation the ova of the rabbit are free in the uterine cornua; they are easily visible to the naked eye, and may be extracted by the same manipulations as those of the tubes. After the sixth day they are at rest in the uterus, but have not yet contracted adhesions with the mucosa, so that they can still be extracted whole. At this stage the parts of the cornua where the ova are lodged are easily distinguishable by their peculiar aspect, the ova forming eminences of the size of a pea. The cornua should be cut up transversely into as many segments as there are eminences, care being taken to have the ova in the centre of the segments. You then fix each segment by means of two pins on the bottom of a dissecting dish, with the mesometrial surface downwards and the ovular eminence upwards. The dissecting dish is then filled up with serum or liquid of

Müller, or 0·1 per cent. solution of osmic acid, or Kleinenberg's picro-sulphuric acid, or nitric acid, or acetate of uranium solution. With a small scalpel a longitudinal incision is made on the surface of the ovular eminence, not passing deeper than the muscular layer; the underlying uterine mucosa is then gently dilacerated with two pairs of small forceps, and the ovum set free in the liquid.

From the moment the ova have become adherent to the uterine mucosa they can no longer be extracted whole. The embryo being always situated on the mesometrial surface, the ovular eminence is opened by a *crucial* incision, and the strip of mucosa to which the embryo remains adherent is fixed with pins on the bottom of the dish. ED. V. BENEDEN (see *Arch. de Biol.*, v., fasc. iii, 1885, p. 378) has been able, by operating in this way in serum of Kronecker and keeping the whole at blood temperature, to observe the circulation of the embryo for hours together. (If this be desired to be done, the crucial incision should not be too extended, so as to leave the terminal sinus intact.)

Preparation.—In order to make permanent preparations of the different stages of fecundation and segmentation, v. BENEDEN (*Arch. de Biol.*, i, 1, 1880, p. 149) recommends the following process: The living ovum is brought into a drop of 1 per cent. osmic acid on a slide, and thence into solution of Müller (or bichromate of ammonia or solution of Kleinenberg). After an hour the liquid is changed, and the whole is put into a moist chamber, where it remains for two or three days. It is then treated with glycerin of gradually increasing strength, and at last mounted in pure glycerin acidified with formic acid. Ova may be stained with Beale's carmine or picro-carmine after removal from the osmic acid and careful washing.

In order to bring out the outlines of blastoderm cells, the living ovum may be brought into one third per cent. solution of nitrate of silver. After remaining there for half a minute to two minutes, according to the age of the vesicle, it is brought into pure water and exposed to the light. The preparations thus obtained are instructive, but blacken rapidly and cannot be permanently preserved.

After the end of the third day the blastodermic vesicle can be opened with fine needles, and the blastoderm washed, stained, and mounted in glycerin or balsam. v. Beneden has also obtained good preparations by means of chloride of gold.

For embryonic areas and more advanced embryos Kölliker recommends putting the ovum into 0·5 per cent. solution of osmic acid until it has taken on a somewhat dark tint, which happens in about an hour, and then treating it with successive alcohols for several hours. If the ovum be adherent to the uterine mucosa the portion of the membrane to which it is fixed should be left, stretched out with pins, in 0·1 per cent. solution of osmic acid for from four to six hours. The blastodermic vesicle can then easily be removed, and immersed for a few hours more in 0·5 per cent. solution of osmic acid, and finally be brought into alcohol. For sections Kölliker fixes with osmic acid. v. Beneden treats the ova for twenty-four hours with 1 per cent. solution of chromic acid, then washes well, and brings them through successive alcohols. Chromic acid has the advantage of hardening thoroughly the vesicle, and maintaining at the same time the epiblast cells perfectly adherent to the zona pellucida. v. Beneden also recommends the liquid of Kleinenberg. Henneguy writes that he frequently employs it for embryonic areas and embryos of various ages, always with excellent results. Fol's modification of the liquid of Flemming, and Ranvier and Vignal's osmic acid and alcohol mixture (§ 29), also give excellent results. For staining, Henneguy recommends borax-carmine or Delafield's hæmatoxylin for small embryos; for large ones, Henneguy's acetic acid alum-carmine is the only reagent that will give a good stain in the mass.

For sections, pure paraffin. Cut in series and mount in balsam.

PIERSOL (*Zeit. f. wiss. Zool.*, xlvii, 2, 1888, p. 155) has been lately using for fixation either Kleinenberg's solution, or, for young stages, Altmann's 3 per cent. nitric acid. Staining and cutting as above.

Aves.

556. Superficial Examination.—Excellent instructions on this head are given in FOSTER and BALFOUR's *Elements of Embryology*, to which, as it is certain to be in the student's hands, he may be referred. What follows here is given merely as being of more recent publication.

If it be desired to observe a living embryo by transmitted light, the egg should be opened under salt solution, as described in Foster and Balfour. A little of the white is then removed through the window, the egg is lifted out of the liquid, and a ring of gummed paper is placed on the yolk so as to surround the embryonic area. As soon as the paper adheres to the vitelline membrane, which will be in a few minutes, a circular incision is made in the blastoderm outside the paper ring. The egg is put back into the salt solution, and the paper ring removed, carrying with it the vitelline membrane and the blastoderm, which may then be brought into a watch-glass or on to a slide and examined under the microscope (DUVAL).

556a. Gerlach's Window Method (*Nature*, 1886, p. 497; *Journ. Roy. Mic. Soc.*, 1886, p. 359).—Remove with scissors the shell from the small end of the egg: take out a little white by means of a pipette; the blastoderm will become placed underneath the window thus made; and the white that has been taken out may be replaced on it. Paint the margins of the window with gum mucilage, and build up on the gum a little circular wall of cotton wool; place on it a small watch-glass (or circular cover-glass), and ring it with gum. When the gum is dry, the cover is further fixed in its place by means of collodion and amber varnish, and the egg is put back in its normal position in the incubator. The progress of the development may be followed up to the fifth day through the window.

A description of further developments of this method, with figures of special apparatus, will be found in *Anat. Anz.*, ii, 1887, pp. 583, 609; see also *Zeit. f. wiss. Mik.*, iv, 3, 1887, p. 369.

557. Preparation.—During the first twenty-four hours of incubation, it is extremely difficult to separate the blastoderm from the yolk, and they should be fixed and hardened together. In later stages, when the embryo is conspicuous, the blastoderm can easily be separated from the yolk, which is very advantageous. The egg should be opened in salt solution, then lifted up a little, so as to have the blastoderm above the surface of the liquid; the blastoderm is then treated with some fixing solution dropped on it from a pipette (1 per cent. solution of osmic acid, or Ranvier and Vignal's osmic acid and alcohol mixture, iodised serum, solution of Kleinenberg, 10 per cent. nitric acid, &c.). By keeping the upper end of the pipette closed, and the lower end in contact with the liquid on the blastoderm, the blastoderm may be kept well immersed for a few minutes, and should then be found to be sufficiently fixed to be excised.

The egg is put back into the salt solution, and a circular incision made round the embryonic area. The blastoderm may then be floated out and got into a watch-glass, in which it may be examined, or may be brought into a hardening liquid.

Before putting it into the hardening fluid, the portion of vitelline membrane that covers the blastoderm should be removed with forceps and shaking.

In order to counteract the turning up of the edges of the blastoderm that generally happens during the process of hardening, it is well to get the blastoderm spread out on the *convex* surface of a watch-glass, and leave it so during the hardening.

For hardening, FOSTER and BALFOUR recommend solution of Kleinenberg for five hours, followed by alcohol. Or chromic acid, a solution of 0·1 per cent. for twenty-four hours, followed by a solution of 0·3 per cent. for twenty-four hours more, then by 70 per cent. alcohol for a day, 90 per cent. alcohol for two days, and lastly absolute alcohol. They also recommend a 0·5 per cent. solution of osmic acid, in which the embryo remains for two hours and a half in the dark, and after washing is brought into absolute alcohol.

HENNEGUY prefers the osmic acid and alcohol mixture of Ranvier and Vignal, or Flemming's mixture followed by successive alcohols.

Staining and imbedding may be performed by the usual methods.

Up to about the fiftieth hour embryos may be mounted entire, in glycerin or balsam.

559. M. Duval's Orientation Method (*Ann. d. Sc. nat. Zool.*, 1885).—In the early stages of the development of the ova of Aves, before the appearance of the primitive streak, it is difficult to obtain a correct orientation of the hardened cicatricula, so as to be able to make sections in any desired direction. Duval, starting from the fact that during incubation the embryo is almost always found to be lying on the yolk in such a position that the big end of the egg is to the left, and the little end to the right of it, marks the position of the blastoderm in the following way.

With a strip of paper 5 millimetres wide and 50 millimetres long you construct a sort of triangular bottomless box. You lay this on the yolk, enclosing the cicatricula in such a position that the base of the triangle corresponds to what will be the anterior region of the embryo, and its apex to the posterior region; that is to say, if the big end of the egg is to your left, the apex of the triangle will point towards you. You now, by means of a pipette, fill the paper triangle with 0.3 solution of osmic acid. As soon as the preparation begins to darken you put the whole egg into weak chromic acid, remove the white, and put the rest into clean chromic acid solution for several days. After hardening you will find on the surface of the yolk a black triangular area, which encloses the cicatricula, and marks its position; you cut out this area with scissors and a scalpel, and complete the hardening with chromic acid and alcohol.

Another way of hardening is to place the egg, after the action of the osmic acid, in a solution of chromic acid which is then raised to boiling-point on a water-bath; after cooling, the blackened region is cut out, and the hardening completed in the usual way with chromic acid and alcohol.

559a. Köller's Method (*Arch. f. mik. Anat.*, xx, 1881, p. 182).—Chromic acid, 0.1 per cent., twenty-four hours; *ibid.*, 0.2 per cent., twenty-four hours; and so forth, with daily increments of 0.1 per cent. up to 0.5 per cent. When hard, remove the blastoderm together with a segment of the yolk. Water, twenty-four hours. Stain, and imbed.

Reptilia.

560. General Directions.—The methods described above for the embryology of birds are applicable to the embryology of reptiles. During the early stages the blastoderm should be hardened *in situ* on the yolk; later the embryo can be isolated, and treated separately with Kleinenberg's solution and alcohol (STRAHL, *Arch. f. Anat. u. Phys.*, 1881, p. 123).

561. Kupffer's Method (*Ibid.* 1882, p. 4).—The ova are opened and the albumen removed under osmic acid of $\frac{1}{10}$ per cent. The yolk is put for twenty-four hours into an ample quantity of ¼ per cent. chromic-acid solution; the blastoderm is removed, washed out in water, and put for three hours into Calberla's liquid (āā glycerin, water, and alcohol), and finally hardened in 90 per cent. alcohol.

The preparations are then stained with Böhn's neutral carmine (§ 142) for twenty-four hours (or more if of a greater thickness than 1 mm.), and afterwards may, if desired, be washed out with a mixture of equal parts of glycerin and water containing ½ per cent. of hydrochloric acid, which will ensure a perfectly nuclear stain. Karyokinetic figures are brought out with great distinctness.

562. Sarasin's Method (SEMPER's *Arbeiten*, 1883, p. 159).—Fix with chromic acid or hot water, and harden with alcohol. Stain with Bismarck brown, alum-carmine, or hæmatoxylin, or picro-carmine. Imbed in collodion, and collodionise the sections as cut.

Amphibia.

563. Preliminary.—In order to prepare the ova of Amphibia for section cutting, it is essential to begin by removing their thick coats of albumen. This may be done by putting them for two or three days into 1 per cent. solution of chromic acid, and shaking well; but ova thus treated are very brittle and do not afford good sections. A better method is that described by WHITMAN (*Amer. Natural.*, xxii, 1888, p. 857), and by BLOCHMANN (*Zool. Anz.*, 1889, p. 269). Whitman puts the fixed eggs into a 10 per cent. solution of sodium hypochlorite diluted with 5 to 6 volumes of water, and leaves them there till they can be shaken free, which happens (for *Necturus*) in a few minutes. Blochmann takes *eau de Javelle* (potassium hypochlorite), and dilutes it with 3 to 4 volumes of water, and agitates the eggs, previously fixed with solution of Flemming, for fifteen to thirty minutes in it. The ova are afterwards preserved in alcohol in the usual way.

564. Axolotl.—The ova are easier to prepare than those of the Anura, because the yolk is separated from the albuminous layer by a wide space filled with a liquid that is not coagulated by reagents. Put the eggs for a few hours into picrosulphuric acid, then pierce the inner chorion with fine scissors or needles, and gently press out the ovum. Harden in alcohol.

Another method that gives good results is that of O. HERTWIG for *Rana* (below, § 568). Stain in the mass with borax-carmine, or Henneguy's acetic acid alum-carmine, and imbed in paraffin or celloidin, collodionising the sections (HENNEGUY).

565. Triton (SCOTT and OSBORN, *Quart. Journ. Mic. Sci.*, 1879, p. 449).—The albumen is here present in the form of several concentric coats, which are very delicate. Incise each of them separately with fine scissors, turn out the ovum, and fix it. Solution of Kleinenberg is the reagent that gives the best results.

566. Triton (HERTWIG, *Jen. Zeit. f. Naturw.*, 1881-2, p. 291).—Put the eggs into a mixture of equal parts of 2 per cent. acetic acid and 0·5 per cent. chromic acid. After ten hours incise the membranes, opening one end of the inner chorion, and turn out the embryos, and bring them into successive alcohols.

567. Salamandra (RABL, *Morphol. Jahrb.*, xii, 2, 1886, p. 252).—Fix in chloride of platinum of 0·25 to 0·3 per cent., kept warm for from three to twenty-four hours, according to the size of the embryos, wash well with water, treat with successive alcohols, make sections, and stain on the slide.

568. Rana (O. HERTWIG, *Jen. Zeit. f. Naturw.*, xvi, 1883, p. 249).—The ova are thrown into nearly boiling water (90° to 96° C.) for five to ten minutes. The albuminous envelope of the ovum is then cut open and the ovum extracted under water. The ova are then brought into 0·5 per cent. osmic acid, or into alcohol of 70, 80, and 90 per cent. Chromic acid makes ova brittle; they ought not to remain in it for more than twelve hours. Chromic acid destroys or attacks the pigment of the ova, whilst alcohol preserves it, which is frequently important for the study of the germinal layers.

569. Whitman's Method (*Meth. of Research*, 1885, p. 156).—Osmic acid of 0·25 per cent. twenty minutes. Chromo-platinic mixture (§ 575) twenty-four hours. Water two hours. Alcohol. Borax-carmine.

570. Schultze's Method (*Zeit. f. wiss. Zool.*, v, 1887, p. 177).—Flemming's mixture, twenty-four hours. Water. Alcohol 50 per cent. twenty-four hours; 70 per cent. ditto; 85 per cent. ditto; 95 per cent. ditto; the last to be frequently changed. The ova passed through turpentine into paraffin. If the ova are to be stained in the mass, they should be put into borax-carmine for twenty-four hours in the place of the 50 per cent. alcohol. The times of immersion in the various reagents to be strictly observed.

Pisces.

571. Teleostea in General.—The ova of many of the bony fishes can be studied by transmitted light in the living state; but those of the Salmonidæ must be hardened and removed from their envelopes for the study of the external forms of the embryo.

To this end, the ova may be put for a few minutes into water containing 1 to 2 per cent. of acetic acid, and thence into 1 per cent. chromic acid. After three days the capsule of the ovum may be opened at the side opposite to the embryo, and be removed with fine forceps. The ovum is put for twenty-four hours into distilled water, and then into successive alcohols. Embryos thus prepared show no deformation, and their histological elements are fairly well preserved. But the vitellus rapidly becomes excessively hard and brittle, so as greatly to interfere with section cutting.

The following processes give good results as regards section cutting.

Put the ova for a few minutes into 1 per cent. osmic acid; as soon as they have taken on a light-brown colour bring them into Müller's solution. Open them therein with fine scissors—the vitellus, which immediately coagulates on contact with air, dissolves on the contrary in Müller's solution;—and the germ and cortical layer can be extracted from the capsule of the ovum. They should be left in clean Müller's solution for a few days, then washed with water for twenty-four hours, and brought through successive alcohols.

Another method is as follows. The ova are fixed in solution of Kleinenberg containing 10 per cent. of acetic acid. After ten minutes they are opened in water containing 10 per cent. of acetic acid, which dissolves the vitellus. The embryos are put for a few hours into pure solution of Kleinenberg, and are then brought through alcohol of gradually increasing strength.

572. Kowalewsky's Method (*Zeit. f. wiss. Zool.*, xliii, 1886, p. 434).

—Fix for an hour and a quarter in a mixture of eight volumes of picrosulphuric acid with one of 1 per cent. chromic acid. Wash out for twelve hours with 20 per cent. alcohol, and pass the ova very gradually through alcohol of 20, 28, 35, 43, 50, 60, and 70 per cent. strength, the last to be changed frequently until all the picric acid is extracted. Before staining, the capsules of the ova should be opened. Stain with borax-carmine or hæmatoxylin, and imbed in paraffin.

573. Kollmann's Method (see § 40).

574. Perényi's Method (see § 39).

575. Pelagic Fish Ova (WHITMAN's method; *Amer. Natural.*, xvii, 1883, pp. 1204-5; *Journ. Roy. Mic. Soc.* (N.S.), iii, 1883, p. 912, and *Methods of Research*, &c., p. 152).—Fix by treatment first for five to ten minutes with a mixture of equal parts of sea-water and ½ per cent. osmic acid solution, and then for one or two days with a modified Merkel's solution (due to Eisig), consisting of equal parts of 0·25 per cent. platinum chloride and 1 per cent chromic acid. Prick the membrane before transferring to alcohol. Whitman found that the usual Merkel's fluid caused maceration of the embryonic portion of the egg. Picrosulphuric acid causes the embryonic cells to swell, and in many cases to become completely disorganised. The osmic acid treatment is necessary in the case of segmenting ova because the Merkel's fluid does not kill rapidly enough, so that eggs placed in it may even pass through one or two stages of cleavage before dying. This fluid arrests the process of blackening by the osmium, or will even bleach the objects if blackening has set in. (*See* also AGASSIZ and WHITMAN, in *Proc. Amer. Acad. Arts and Sciences*, xx, 1884.) For later stages the authors recommend the method of Perényi.

576. Embryos of Teleostea (RABL-RÜCKHARD, *Arch. f. Anat. u. Entw.*, 1882, p. 67).—Fix in 10 per cent. nitric acid for fifteen minutes. Remove the membranes, to avoid deformation of the embryos, and put the ova back into the acid for an hour. Wash out in 1 to 2 per cent. solution of alum for an hour, and harden in alcohol.

Modification of this method by GORONOWITSCH (see *Morph. Jahrb.*, x, 1884, p. 381).

Tunicata.

577. Distaplia. DAVIDOFF (*Mitth. Zool. Stat. Neapel*, ix, 1, 1889, p. 118) has some important observations on the fixation of the ova of *D. magnilarva*. The best reagent is a mixture of 3 parts of saturated solution of corrosive sublimate and 1 of glacial acetic acid. The ova to remain in it for from half an hour to an hour, and be then washed for a few minutes in water and brought through successive alcohols. Another reagent, almost as good, consists of 3 parts of saturated solution of picric acid and one of glacial acetic acid, the objects

577a. Amarœcium (Maurice and Schulgin, *Ann. Sci. Nat. Zool.*, xvii, 1884).—Stain in borax-carmine, wash out, and stain for fifteen to twenty hours in very weak solution of *bleu de Lyon* in 70 per cent. alcohol with a few drops of acetic acid. In sections the epiblast and hypoblast appear chiefly blue, the mesoblast cells on the contrary appearing almost entirely red.

Mollusca.

578. Cephalopoda (Ussow, *Arch. de Biol.*, ii, 1881, p. 582).—Segmenting ova are placed, without removal of the membranes, in 2 per cent. solution of chromic acid for two minutes, and then in distilled water to which a little acetic acid (one drop to a watch-glassful) has been added, for two minutes. If an incision be now made into the egg-membrane the yolk flows away and the blastoderm remains; if any yolk still cling to it, it may be removed by pouring away the water and adding more. The blastoderms thus prepared show, when appropriately stained, fine karyokinetic figures.

579. Gasteropoda (Henneguy).—Ova of Helix may be fixed for from four to six hours in Mayer's picronitric solution (§ 56). The carbonate of lime that encrusts the external membrane is thus dissolved, and the albuminous coat of the egg is coagulated. The egg is opened with needles; the albumen comes away in bits, and the embryo can be removed. Treat with successive alcohols, and imbed in paraffin.

580. Limax (early stages) (Mark, *Bull. Mus. Comp. Zool., Harvard Coll.*, vi, 1881).—The ova are treated with acetic acid of 1 to 2 per cent. for four or more hours. The two external membranes are incised with fine scissors, and the egg squeezed out in its albumen membrane. This is dissected off on a slide, the egg is separated from the albumen, stained, and mounted in glycerin.

For later stages, or for making sections, osmic acid is used instead of acetic acid, and the egg is hardened within its albuminous coats.

Arthropoda.

581. Fixation of Ova.—In most cases the ova of Arthropods are fixed by heat in a more satisfactory way than by any other means. This may be followed either by alcohol or some watery hardening agent.

582. Removal of Membranes.—This is frequently very difficult, and it may often be advisable not to attempt to remove them, but to soften them with *eau de Javelle* or *eau de Labarraque* (*see* the methods of Looss and List).

MORGAN (*Amer. Natural.*, xxii, 1888, p. 357; *Zeit. f. wiss. Mik.*, vi, 1, 1889, p. 69) recommends (for the ova of *Periplaneta*) eau de Labarraque diluted with five to eight volumes of water, and slightly warmed. Thus used, it will soften the chitin membranes sufficiently in thirty to sixty minutes if employed before fixing. Fixed ova take longer. The fluid must of course not be allowed to penetrate into the interior of the ovum.

583. Lepidoptera (BOBRETZKY, *Zeit. f. wiss. Zool.*, 1879, p. 198).—Ova (of *Pieris cratægi* and *Porthesia chrysorrhœa*) are slightly warmed in water and put for sixteen to twenty hours into 0·5 per cent. chromic acid. The membranes can then be removed, and the ova brought for a few hours into absolute alcohol, stained with carmine, and cut.

584. Phryganida (*Neophalax*) **and Blattida** (PATTEN, *Quart. Journ. Mic. Sci.*, 1884, p. 549).—The ova or larvæ are placed in cold water, which is gradually raised to 80° C. You leave off heating as soon as the ova have become hard and white. Pass very gradually through successive alcohols, beginning with 20 per cent.; stain with Kleinenberg's hæmatoxylin, or Mayer's cochineal (only alcoholic stains will traverse the chorion). The ova may remain in the hæmatoxylin for five or six days, and be washed out in alcohol containing one drop of HCl per 20 grammes, in which they should remain for several days, and then be soaked in pure alcohol until they have regained their violet colour. Penetrate with benzol and imbed in paraffin.

587. Diptera (HENKING, *Zeit. f. wiss. Zool.*, xlvi, 1888, p. 289; *Zeit. f. wiss. Mik.*, 1889, p. 59).—Ova still contained within the fly may be fixed by plunging the animal for some time into boiling water, then dissecting out and bringing them into 70 per cent. alcohol. Laid eggs may have boiling water poured over them, or be put into solution of Flemming in a test-tube which is plunged into boiling water until the eggs begin to darken (about a minute). Cold solution of Flemming easily causes a certain vacuolisation of the contents of the ova. Open the ova at the larger end, stain with borax-carmine for fifteen to thirty hours, and cut in paraffin.

586. **Aphides** (WILL, *Semper's Arbeiten*, 1883, p. 223).—Sections to be made through the entire animals containing the ova and embryos. The animals are killed in water of 70° C., and brought into alcohol. The cuticle may then be pricked with a needle and the animals stained in the mass with borax-carmine or hæmatoxylin. You may imbed in collodion and collodionise the sections as cut.

587. **Araneina** (BALFOUR, *Quart. Journ. Mic. Sci.*, 1880, p. 167).— Balfour hardened the embryos in bichromate of potash, after placing them for a short time in nearly boiling water. · After removal of the membranes they were stained as a whole with hæmatoxylin, and imbedded for cutting in coagulated albumen.

588. **Agelena** (LOCY, *Bull. Mus. Comp. Zool. Harvard*, xii, 3, 1886; *Zeit. f. wiss. Mik.*, iii, 2, 1886, p. 242).—Fix by heating the ova in water to 80° C., and bring them into alcohol. The liquid of Perényi may also be used; it has the advantage of not making the yolk so granular.

589. **Phalangida** (BALBIANI).—The ova of *Phalangium opilio* are enclosed in a chorion covered with yellow corpuscles which renders them quite opaque. They may be cleared by treating them with water containing a little solution of caustic potash and raised to boiling-point. The ova are then laid on blotting-paper, and the chorion is removed by rubbing them gently with a small brush. The vitelline membrane remains intact and transparent, and the embryo may be studied through it.

590. **Phalangida** (HENKING, *Zeit. f. wiss. Mik.*, iii, 4, 1886, pp. 470 et seq.).—Fix with boiling water or "Flemming." Preserve the ova in 90 per cent. alcohol. To open the chorion, bring them back into 70 per cent. alcohol, which causes them to swell up so that the chorion can easily be pierced with needles, and the ovum turned out. Stain with borax-carmine or with "eosin-hæmatoxylin" (*sic*). Penetrate with bergamot oil (rather than chloroform or toluol) and imbed in paraffin.

591. **Astacus** (REICHENBACH, from *Zeit. f. wiss. Mik.*, 1886, p. 400).— Fix in water gradually warmed to 60° or 70° C. (if the chorion should burst, that is no evil), harden for twenty-four hours in 1 to 2 per cent. bichromate of potash or 0·5 per cent. chromic acid, wash out for the same time in running water, and bring into alcohol. Remove the chorion, remove the em-

bryo from the yolk by means of a sharp knife, and stain with picro-carmine and mount in balsam.

592. Amphipoda (*Orchestia*) (ULIANIN, *Zeit. f. wiss. Zool.*, xxxv, 1881, p. 441).—Ova in the earliest stages of development were treated for two hours with picrosulphuric acid (Kleinenberg's formula). This causes the chorion to swell and burst. Wash out with alcohol, stain with Beale's carmine. Make sections. Ova in later stages, in which the embryo is surrounded by a cuticular membrane, which encloses an albuminous liquid, must have this membrane torn with needles and the albuminous liquid allowed to ooze out before placing in the picrosulphuric acid.

593. Maturation of Ova, and other early stages.—These should be studied by the methods given in the chapter on Cytological Methods.

Vermes.

594. Tænia (v. BENEDEN, *Arch. de Biol.*, ii, 1881, p. 187).—Ova in which a chitinous membrane has formed around the embryo are impervious to reagents. They may be put on a slide with a drop of some liquid and covered. Then, by withdrawing the liquid by means of blotting-paper, the cover may be made to gradually press on them so as to burst the membranes, and the embryo may then be treated with the usual reagents.

595. Planaria (IIJIMA, *Zeit. f. wiss. Zool.*, xl, 1884, p. 359).—The capsule containing the ova (of fresh-water Planaria) is opened with needles on a slide, in a drop of 2 per cent. nitric acid. The ova are extracted and covered (the cover being supported by paper, or by wax feet). After half an hour they are treated with successive alcohols under the cover, and finally mounted in glycerin. For sections, the whole of the contents of a capsule is hardened in the mass in 1 per cent. chromic acid and cut together.

596. Lumbricus (KLEINENBERG, *Quart. Journ. Mic. Sci.*, 1879, p. 207).—Fix with Kleinenberg's picrosulphuric acid, or, which is not quite so good, with vapours of osmium, pass through successive alcohols, stain with Kleinenberg's hæmatoxylin, and cut in paraffin.

597. Ascaris.—*See* the chapter on Cytological Methods.

Echinodermata, Cœlenterata, and Porifera.

See the paragraphs treating of these groups in Chapter XXXII.

For the Maturation and Fecundation of the ova of the Echinodermata, see also the chapter on Cytological Methods.

CHAPTER XXVI.

CYTOLOGICAL METHODS.

598. The Methods of Study.—There are three ways of obtaining knowledge of cell-structure—study of living cells, study of fresh unhardened cells, and study of hardened material in sections. Of these the last is the most fruitful; and I advise the beginner to keep as close as possible to the method of Flemming and Rabl, utilising unhardened material chiefly for the purpose of controlling the results obtained by the study of sections, and reserving the study of living cells chiefly for establishing the seriation of already observed phenomena.

599. Observation of Living Cells.—One of the best objects for this purpose is the tail of young larvæ of Amphibia, both Anura and Urodela.

In the living animal the epithelial cells and nuclei (in the state of repose) are so transparent as to be invisible in the natural state. They may, however, be brought out by curarising the larva; or, still better, by placing the curarised larva for half an hour in 1 per cent. chloride of sodium solution. Normal larvæ may be used for the study of the active state of the nucleus, but much time is saved by using curare.

Curare.—Dissolve 1 part of curare in 100 parts water, and add 100 parts of glycerin. Of this mixture add from 5 to 10 drops (according to the size of the larva), or even more for large larvæ, to a watch-glassful of water. From half to one hour of immersion is necessary for curarisation. The larvæ need not be left in the solution until they become quite motionless; as soon as their movements have become slow they may be taken out and placed on a slide with blotting-paper. If they be replaced in water they return to the normal state in eight or ten hours, and may be re-curarised several times.

Etherisation.—Three per cent. alcohol, or 3 per cent. ether, may be used in a similar way. These reagents cause no obstruction to the processes of cell-division, and are useful, but their action as anæsthetics is inconstant.

Indifferent Media.—One per cent. salt solution, iodised serum, syrup, cold water ($+ 1°$ C.), and warm water ($35°$—$40°$ C.). The tail may be excised from the living animal and studied for a long time in these media (PEREMESCHKO, *Arch. f. mik. Anat.*, xvi, 1879, p. 437).

Perhaps (FLEMMING, *ibid.*, pp. 304 *et seq.*) the very best subject for these studies is *Salamandra*. The adult offers for study the thin transparent bladder; in the larva the gills and caudal "fin" may be studied in the living state. The gills are difficult to fix in position for observation, and are obscured by pigment. In the fin there is always a spot, near to the hind limbs, that is free from pigment; and on lightly coloured larvæ other such spots may be found on the ventral half of the fin and on the lateral line. On a flat-finned larva it is possible to study these spots with high-power glasses.

The larva may be fixed in a suitable cell, or wrapped in moist blotting-paper, or may be curarised; or the tail may be excised. (It is preferable to cut through the larva close in *front* of the hind limbs.)

A favorable object for *preparation* is found in the gill-*plates*, delicate laminæ that are to be found attached to the gill-cartilages on the mouth side.

Larvæ may be bred from adults kept in confinement, and supplied with a vessel of water, in which they will place the larvæ of their own accord. In May, gravid females may be killed and the larvæ extracted. The larvæ must be kept in frequently changed water, and fed every day or two. Aquatic worms may be used for feeding them, viz. *Tubifex rivulorum*.

It is extremely important that they should be fed regularly and abundantly, for if not, cell-divisions in the tissues become rare, and may even cease altogether.

600. Stains for Living Cells.—It is sometimes of the very greatest importance to be able to stain a cell in the living state, even though it be but feebly and imperfectly. Methylen blue, dahlia, or gentian violet may be used in solution in pure water, or in an indifferent liquid—the addition of a trace of chloral hydrate will enable you to obtain a clear solution of the last two in saline media. It is sometimes advisable to rub them up with serum, as recommended by v. LA VALETTE ST. GEORGE. These methods are most important for the study of the *Nebenkern*.

The student will remember that no known reagent will stain any part of the nucleus whilst alive. The "Nebenkern" stains sometimes, but feebly. Most frequently the colour is only taken up by certain granules of the cytoplasm, which may or may not be identical with the "granules" or "bioblasts" of ALTMANN.

These matters have already been discussed in the paragraph headed "*Staining 'intra vitam*,'" § 93.

601. Study of Fresh and Lightly fixed Cells.—It has been rightly pointed out by Flemming that so-called "indifferent" liquids must not be believed to be without action on nuclei. Iodised serum, salt-solution, serum, aqueous humour, lymph, better deserve the name of weak hardening agents. Between these, and such energetic hardening agents as Flemming's mixture, come such light fixing agents as picric acid, or very dilute acetic acid. These it is whose employment is indicated for the study of fresh isolated cells.

A typical example of this kind of work is as follows: Tease out a piece of living tissue in a drop of acidulated solution of methyl green (0·75 per cent. of acetic acid). This is a delicate fixing agent, killing cells instantly without change of form. Complete the fixation by exposing the preparation for a quarter of an hour to vapour of osmium; and add a drop of solution of Ripart and Petit, and a cover.

Or you may fix the preparation, after teasing, with vapour of osmium for half a minute to two minutes, then add a drop of methyl green, and after five minutes wash out with 1 per cent. acetic acid, and add solution of Ripart and Petit, and cover.

Or you may kill and fix the cells by teasing in solution of Ripart and Petit (to which you may add a trace of osmium if you like), and afterwards stain with methyl green.

Other fixing agents, such as picric acid, or weak sublimate solution, may of course be used, and in some cases doubtless should be preferred. Other stains, too, such as Bismarck brown, or Delafield's hæmatoxylin, may be used as occasion dictates; and of course other examination media than solution of Ripart may be employed. But, for general purposes, the methyl-green-osmium-and-Ripart's-medium method gives such good results, and is so very convenient, that it may well be called the classical method for the study of fresh cells. I think great credit is due to CARNOY for his frequent insistence on the excellence and handiness of this method.

Other fixing agents and stains that are applicable to this kind of work will be found discussed in the course of the following paragraphs.

602. Some Microchemical Reactions.—Methyl green is a test for nuclein, in so far as it colours nothing but the nuclein *in the nucleus*. It is, however, not a perfect test, for the intensity of the coloration it produces varies greatly in different nuclei, and may in certain nuclei be extremely weak, or (apparently) even altogether wanting. In these cases other tests must be applied in order to establish with certainty the presence or absence of that element. The following suggestions are taken from CARNOY, who is, I believe, the only writer, on the zoological side at all events, who has insisted on the necessity of applying microchemical methods in a systematic manner to the study of cells.

Nuclein is distinguished from the lecithins and from albuminoids by not being soluble, as these are, in water, and in weak mineral acids, such as 0·1 per cent. hydrochloric acid. It is easily soluble in concentrated mineral acids, in alkalies, even when very dilute, and in some alkaline salts, such as carbonate of potash and biphosphate of soda. In the presence of 10 per cent. solution of sodium chloride it swells up into a gelatinous mass, or even, as frequently happens, dissolves entirely (*Biol. Cell.*, pp. 208—9). It is only partially digestible (when *in situ* in the nucleus) in the usual laboratory digestion fluids.

The solvents of nuclein that are the most useful in practice are—1 per cent. caustic potash, fuming hydrochloric acid, or cyanide of potassium, or carbonate of potash. These last generally give better results than dilute alkalies. They may be employed in solutions of 40 to 50 per cent. strength. If it be desired to remove all the nuclein from a nucleus, the reaction must be prolonged—sometimes to as much as two or three days, especially if the operation be conducted on a slide and under a cover-glass, which is the safer plan.

It must be remembered that these operations must be performed on *fresh* cells, for hardening agents bring about very considerable modifications in the nature of nuclein, rendering it almost insoluble in ammonia, potash, or sodic phosphate, &c. Hydrochloric acid, however, still swells and dissolves it, though with difficulty.

Partial digestion may render service in the study of the chromatic elements of nuclei. Nuclein resists the action of digestive fluids much longer than the albumens do; so that a moderate digestion serves to free the chromosomes from any caryoplasmic granulations that may obscure them, whilst at the same time it clears up the cytoplasm.

In the presence of iodine, of hot nitric acid, and of MILLON's test, nuclein gives the reactions of protein matters.

(MILLON's test consists of 1 part mercury, 1 part fuming nitric acid, and 2 parts water.)

Another statement of the micro-chemistry of the cell, by ZACHARIAS, is to be found in *Zeit. f. wiss. Mik.*, iv, 3, 1887, p. 409, and *Journ. Roy. Mic. Soc.*, 1888, p. 505.

603. Cytological Fixing Agents.—The following is in great part taken from the numerous papers of FLEMMING in the *Arch. f. mik. Anat.*, from the year 1879 onwards, and from his *Zellsubstanz, Kern- und Zelltheilung*.

Osmic acid ($\frac{1}{10}$ to 2 per cent.) preserves the form of the entire cell, but swells the nuclei and rounds off nucleoli. It renders the nuclear "reticulum" undiscernible. Picric acid, either concentrated or dilute, and chromic acid, 0·1 to 0·5 per cent., are to be preferred to alcohol and other agents for the study of the cells of *Vertebrates*. Shrinking and distortion of the nuclear figures (and, with picric acid, swellings of them) are to be expected, but other agents have the same defect to a much greater degree; alcohol especially causes *entanglement* of the filaments. Acetic acid does the same, and causes swelling besides. Stronger chromic acid solutions cause shrinking. Neither of these reagents is harmless as regards the nuclei of red blood-corpuscles. The salts of picric acid (potash-, soda-, and baryta-salts) are most harmful. Weak (*i. e.* not more than 1 per cent.) acetic, hydrochloric, or nitric acid combined with clearing in glycerin, and staining, may be useful for bringing out reticula and nucleoli. Chloride of gold preserves the forms well, but generally leaves the nuclear structures unstained. Nitrate of silver is hopelessly uncontrollable in its action. Alcohol has much the effect of chromic acid, but often causes a much greater shrinking of the nuclei. Bichromate of potash and chromate of ammonia bring out very sharply the appearance of a reticulum, but these appearances cannot be accepted as true (l. c., p. 334 *et seq.*).

"*Those who seek to study cell-division by means of bichromate of potash or other chromic salts are hopelessly in the wrong road.*" And this because of the injurious action of the bichromate, *not* on the body of the cell, which it preserves well, but on the chromatin structures. Chromic salts are excellent reagents for general histological work, but not for nuclear structures. They dissolve nucleoli, destroy nuclear "networks," and swell up and distort karyokinetic figures to such a degree that the appearances obtained from them are merely unnatural caricatures of the true structure.

Altmann's nitric acid method is excellent for the purpose of hunting for cell-divisions in tissues; but the minute structure of the figures is not so well preserved as it is by means of chromic or picric acid. The same must be said of Kleinenberg picrosulphuric acid method.

The best fixing agent in general is the **chromo-aceto-osmic** acid mixture (§§ 35, 36). Attempts to omit the chromic acid did not give good results. The omission of acetic acid (as in Max Flesch's formula, § 34) causes the figures to be far less sharply brought out. The presence of acetic or formic acid in all osmium solutions is favorable to the precision of subsequent staining with hæmatoxylin, picro-carmine, or gentian-violet. But mixtures of osmic and acetic acid without chromic acid (Eimer) do not give such good results as the chromo-aceto-osmic acid mixture. Mixtures of picric acid with osmic acid or with osmic and acetic acid (proportions of the latter as in the chromo-acetic osmic mixture (§ 35), but of picric acid about 50 per cent.) fix quite as well as the chromic mixtures, but precise staining is even more difficult than with pure osmic acid preparations. Flemming concludes that the beneficial effects of the osmium in all these mixtures are to be ascribed to the instantaneous rapidity with which it kills, the function of the other acids of the mixture being to render the structures distinctly visible.

Mixtures containing osmic acid should therefore be employed whenever it is desired to fix the chromatic figures as faithfully as possible; whilst pure chromic acid should be taken whenever very sharp staining is the more important point.

For the study of the achromatic figures he recommends the chromo-acetic acid mixture (§ 31), followed by staining in hæmatoxylin (anilins do not give so good results for this purpose).

For the study of polar corpuscles he recommends the osmium mixtures, or pure chromic acid followed by staining with gentian-violet.

The above account stands nearly as it stood in the first edition. The state of things at present is as follows:—It is admitted by all competent observers that the chromo-aceto-osmic mixture is, with at most one or two possible exceptions, by far the best fixing agent for nuclei. But some observers have stated that it does not always preserve the cell-body well. This is a question that has been already discussed in §§ 35 and 36. I will only add here that after considerable experience I see no reason to distrust Flemming's mixture as a preservative of any kind of protoplasm, provided it be used in the proper way. It must be taken of the proper strength, it must be used with very small objects, so that it may act on all parts of them with its full strength, and not be filtered and diluted through thick walls of tissue before coming into contact with the object of study; and it must be allowed to act for the proper time.

This brings us to another point. There are *two* **chromo-aceto-osmic** mixtures—the old *weaker* one, and the new *stronger* one. Flemming recommended the strong one primarily as affording a means of differentiating kinetic chromatin from resting chromatin; he did not recommend it as a reagent for general work. Whether of these two solutions should be used for general work? According to my experience, the strong solution *does* preserve both nuclear structures and caryoplasmic structures quite as faithfully at least as the old formula, and some structures most decidedly much better. Of course the one and the other should be taken according to the nature of the object you are dealing with; but I think it may safely be stated as a general rule that if you take the strong mixture, and fix thoroughly in it, you are not likely to go far wrong. And what is meant by a thorough fixation? Half an hour may be taken to be generally enough; but for very delicate things, such as the Nebenkern and the achromatic figure, at least eighteen hours ought to be given.

It only remains to point out that this doctrine is at variance with that expressed in the first edition and in the *Traité*, and with the earlier recommendations of Flemming; but I feel some confidence that it will not be called in question by the majority of workers at this subject. Of course it goes without saying that further precise evidence on the matter is very much to be desired.

Two or three of the fixing agents proposed by other writers fix quite as faithfully as Flemming's mixture. There is RABL's **chromo-formic** acid (§ 32). Fix in this for twelve to twenty-four hours, wash out well with water, and pass into alcohol. And there is the same observer's **platinum chloride** solution (§ 47). In Rabl's latest communication (*Anat. Anz.*, iv, 1889, p. 21) he recommends that *Salamandra* larvæ be fixed (for twenty-four hours) in a solution of from one tenth to one eighth per cent. strength. In his earlier work he used solutions of 1—300 strength. Platinum chloride has the peculiarity of causing a slight shrinkage of the chromatin, which helps to bring into evidence the granules of Pfitzner and the longitudinal division of the chromosomes.

Acetic Alcohol is a reagent with which some of the most important work in recent cytology has been done—namely, much of that on the maturation and fecundation of the ovum of *Ascaris*.

CARNOY (*La Cellule*, iii, 1, 1886, p. 6) used at first a mixture of three parts of absolute alcohol with 1 of glacial acetic acid; later (*ibid.*, iii, 2, 1887) the chloroform mixture (§ 52). From five to fifteen minutes is enough for even the most resistent ova.

VAN BENEDEN and NEYT (*Nouvelles Rech. sur la Féc. et la Division mitosique*, 1887) employed a mixture of equal parts of absolute alcohol and glacial acetic acid, or even pure acetic acid.

Acetic alcohol may be washed out with either pure alcohol,

or with dilute glycerin (Calberla's formula would be a good one in many cases). For further details see *ante*, § 52.

Sulphurous acid has been used by CARNOY (*La Cellule*, i, 1885, p. 212; ii, 1886, p. 17) and by GILSON (*ibid.*, ii, 1886, p. 84). It is prepared for use by saturating cold absolute alcohol with well-dried SO_2. It may be used in the ordinary way, or by exposing the objects for a few seconds to the vapours of the solution. It kills rapidly, but *only* fixes the chromatin, and is therefore not likely to be generally useful.

Uranium Salts are mild fixing agents, and very penetrating, and may be useful for some special objects (see § 65).

Lemon Juice (fresh, filtered) has lately been warmly recommended as a fixative for nuclei by van GEHUCHTEN (*Anat. Anzeig.*, iv, 1889, p. 52). Fix for five minutes, wash well with water and stain with methyl green, and examine in liquid of Ripart and Petit.

604. Cytological Stains.—For fresh or lightly fixed tissues methyl green is the most generally useful nuclear stain. For the properties of this reagent see *ante*, § 105.

Bismarck brown is another useful stain for such objects. It may be used in aqueous solution with acetic acid or with hydrate of chloral, or dissolved in dilute glycerin. Alum-carmine may occasionally be useful. Delafield's hæmatoxylin will render services for osmium objects. Methyl violet, employed according to the method of GRASER (§ 107), may also be found a very useful stain.

For sections of hardened tissues, the best chromatin stains are those obtained by means of safranin, gentian violet, Victoriablau, and some other anilins, used according to the indirect or Flemming's method. This has been so fully explained in Chapter VIII that it is only necessary to refer the reader back to the paragraphs in question.

BABES's supersaturated safranin stain (*Arch. f. mik. Anat.*, xxii, 1883, p. 361) may also occasionally be useful. It is as follows:—A supersaturated solution of safranin in water is warmed to 60° C., and filtered warm. On cooling, it becomes turbid through the formation of small crystals. Sections are placed in a watch-glass with some of this turbid solution, and the whole is warmed for a few seconds (till the liquid becomes clear) over a spirit-lamp. Allow the whole to remain for one minute, and wash out with water, and treat with alcohol and turpentine in the usual way. Tissues which do not take on the stain at once must be warmed over and over again. Clove oil must be avoided for clearing.

BENDA's copper hæmatoxylin stain may also be employed (*see* below, § 612).

The staining of the *achromatic* figure, including the Nebenkern, is another matter. These structures only stain in a really distinct manner in two reagents, Kernschwarz and hæmatoxylin. I unhesitatingly recommend hæmatoxylin (*see* below, § 607).

For a double-stain I recommend Rabl's combination of hæmatoxylin followed by safranin, if it be wished to demonstrate the achromatic figure at the same time as the chromatic element. This is the only combination known to me that will do so in a really satisfactory way. One would think that safranin and Kernschwarz might be better, but hitherto I have not been able to succeed with this plausible combination.

605. Mounting.—For fresh objects you have so large a choice of mounting media that you may take whatever liquid gives you the best optical results. Sections of hardened tissue should always (if possible) be mounted in damar (or colophonium), not balsam, as the slightly lower index of damar or colophonium solutions gives more powerful images of very delicate details. If you have to deal with objects so delicate that you have cause to fear mechanical injury to them on putting them into damar, they may be mounted in thickened turpentine (or cedar oil, but turpentine is preferable on account of its lower index). Rabl has lately been using methyl alcohol as an examination medium for dehydrated objects; but the preparations do not keep in this. Castor oil may be tried (GRENACHER); I have not had good results with it.

606. Synthetic Review.—The following examples will serve to gather up into one view the directions given in the preceding paragraphs.

FLEMMING's **Method** may be stated as follows. Fix, for twenty-four hours or more, in chromo-aceto-osmic acid or in one of Rabl's liquids. Wash out in running water for an hour or more. Bring the preparation into alcohol, and let it remain in absolute alcohol for at least some hours. Then imbed it by the method of simple imbedding, without penetration, in paraffin, in pith, or in celloidin, and make sections with a knife wetted with alcohol. Bring the sections into water, and stain with safranin or gentian for twenty-four hours or more, or (for the achromatic figure) with Delafield's hæmatoxylin. Mount in damar.

Flemming prefers not to imbed by penetration in paraffin, nor to employ any other of the infiltration methods of imbedding, because he finds that these methods may be somewhat injurious to cells unless great care be taken. I myself have not been able to discover any deformations or other erroneous appearances that can be attributed to the paraffin method, and think that it may legitimately be employed, using cedar oil for clearing, and using a soft paraffin so as to be able to imbed at the lowest temperature possible.

RABL'S **Method** (*Morph. Jahrb.*, x, 1884, p. 215) is nearly the same. He fixes in the chromo-formic mixture or in platinum chloride for twelve or twenty-four hours, washes out, and treats for twenty-four to thirty-six hours with alcohol of 60 to 70 per cent., and then with absolute alcohol, and stains sections with safranin or hæmatoxylin, or with both. This is done by first staining very lightly with dilute Delafield's hæmatoxylin, and then with safranin. This he rightly claims to be perhaps the most beautiful stain that can be produced. He advises that green light (which can be obtained by means of a green glass stage-plate) be used to work with (*see* the paragraph on Green Light in the Appendix).

STRASBURGER's **Methods** (see *Arch. f. mik. Anat.*, xxi, 1881, p. 477).

USKOFF's **Nitric Acid Method** (see *Arch. f. mik. Anat.*, xxi, 1882, p. 292).

PFNITZER's **Sulphate of Soda Method** (see *Morphol. Jahrb.*, xi, 1885, p. 54).

607. The "Nebenkern" (or "Sphère Attractive," or "Archoplasmakugel").—The best objects at present known for the study of this element are the ova (segmenting) of *Ascaris*, and the sperm-cells of *Helix*, or other Pulmonata.

In the ova of *Ascaris* it is best demonstrated, according to VAN BENEDEN and NEYT (*Nouv. Rech. sur la Fécond. et la Div. mitosique*), by fixing with acetic alcohol (*ante*, § 603), and bringing the ova into one-third glycerin in which is dissolved a little malachite green. The "sphères attractives" stain green. See also BOVERI's *Zellen-Studien* (in *Jen. Zeitschr. f. Naturw.*, xxi, 1887, p. 423, or sold separately).

The "Nebenkern" of the sperm-cells of Pulmonata should be studied both living and in sections. According to my experience—and I have studied the most various forms—*Helix* is by far the best subject. The best method for sections is that last recommended by PLATNER (*Arch. f. mik. Anat.*, xxxiii, 1889, p. 125). Fix for an hour in Flemming's *strong* chromo-aceto-osmic mixture, and then for twenty-four hours in a fresh quantity of the mixture diluted with three to four volumes of water. (I can confirm the statement that a thoroughly long fixation is absolutely necessary; I generally used the undiluted fluid for twenty-four to thirty-six hours). Platner then washes out with water, brings the gland into alcohol, stains for twenty-four hours in Apàthy's modification of Heidenhain's hæmatoxylin, and washes out for some hours (twelve to twenty-four) in the alcoholic bichromate solution, then washes out for days in 70 per cent. alcohol in the dark, dehydrates, clears with cedar oil, and imbeds in paraffin. The method followed by me differed from this only in so far as I made sections first, and stained on the slide with dilute Delafield's hæmatoxylin, or double-stained with hæmatoxylin followed by safranin.

In an earlier communication (*Zeit. f. wiss. Mik.*, iv, 3, 1887, p. 350) Platner had recommended Kernschwarz (*see* § 195) for staining. This gives very good results, but I think not so good as hæmatoxylin. I do not think the Nebenkern stains more electively, and the cell body does not seem to me to remain so transparent as with hæmatoxylin.

It is very important to supplement the results thus obtained by study of the living cell. V. LA VALETTE ST. GEORGE (*Arch. f. mik. Anat.*, 1885, p. 584, and 1886, pp. 8 and 9) recommends the use of an indifferent liquid such as serum, with which a little gentian violet or dahlia is rubbed up. Perhaps methylen blue may be found still more useful. If the cells be killed by 1 per cent. acetic acid, the Nebenkern may be stained (lightly) with dahlia or gentian.

608. Nucleus of BALBIANI ("Noyau Vitellin," "Cellule Embryogène") (*Zool. Anz.*, 1883, p. 659).—This may be observed in the fresh state, without the addition of any reagent, in the ova of some animals, amongst others a great number of Arachnida and Myriapoda. It may be brought out more distinctly by treating the ova with a mixture of equal parts of acetic acid and 1 per cent. osmic acid, to which is added a little sodium chloride. This mixture does not render ova so granular as pure dilute acetic acid.

609. Division of Ovum (Echinodermata) (FLEMMING's method,

Arch. f. mik. Anat., xx, 1881, p. 3).—The ova are stained on the slide by adding the stain at the edge of the cover. Safranin or other nitro- or anilin-colours may be used. As soon as the entire ovum is of a dark colour the stain is drawn off with blotting-paper, and acetic acid of 1 per cent. added. Schneider's acetic carmine (*Zool. Anzeig.*, 1880) (*see* § 154) is very convenient, and gives good results. (For the details of the manipulation by which these reagents are added and drawn off on the slide it is well to consult the article quoted, p. 6.)

Another good method is as follows:—Segmenting ova are treated with a mixture of 40 to 50 parts of concentrated nitric acid with 60 to 50 parts water. Wash with water until all the yellow stain of the nitric acid has disappeared; stain with Schneider's acetic carmine, and mount in glycerin. (The preparations cannot be said to be permanent, as after a time the stain darkens in such a way as to render the nuclear figures unrecognisable.)

610. Other General Cytological Methods.

TIZZONI (*Bull. delle Sci. Med. di Bologna*, 1884, p. 259).
BAUMGARTEN (*Zeit. f. wiss. Mik.*, i, 1884, p. 415).
BIZZOZERO (*ibid.*, iii, 1886, p. 24).
BABES (*ibid.*, iv., 4, 1887, p. 470).
ZWAARDEMAKER (*ibid.*, iv, 2, 1887, p. 212).
NISSEN (*Arch. f. mik. Anat.*, 1886, p. 338) stains material fixed in Flemming and sectioned, by the method of GRAM (*see* above, § 102).
SCHOTTLÄNDER (*Arch. f. mik. Anat.*, xxxi, 1888, p. 426; *Zeit. f. wiss. Mik.*, v, 4, 1888, p. 515 (cell-division in the corneal endothelium).

611. Cytology of the Ovum (*Ascaris*).

VAN BENEDEN (*Arch. de Biol.*, iv, 1883, p. 279).
CARNOY (*La Cellule*, ii, 1, 1886, pp. 17, 18; iii, 1, 1886, p. 6; iii, 2, 1887).
VAN BENEDEN et NEYT (*Nouv. Rech. sur la Féc., &c.*, separate, 1887, or *Bull. Acad. roy. d. Sc. de Bruxelles*, 1887, iii sér., xiv; also *Journ. Roy. Mic. Soc.*, 1888, p. 508).
BOVERI (*Zellen-Studien*, in *Jen. Zeit. f. Naturw.*, xxi, 1887, p. 423, or separate; also *Journ. Roy. Mic. Soc.*, 1888, p. 664).
ZACHARIAS (*Anat. Anz.*, iii, 1888, p. 24; also *Journ. Roy. Mic. Soc.*, 1888, p. 663).
VAN GEHUCHTEN (*ibid.*, p. 237).
KULTSCHITZKY (*Arch. f. mik. Anat.*, xxxi, 1888, p. 567) recommends for fixing a mixture of three parts of acetic ether with one part of absolute alcohol. Clears for mounting in concentrated acetic acid, and mounts in a mixture of acetic acid and balsam. An excellent *résumé* of the technical part of these papers is to be found in *Zeit. f. wiss. Mik.*, v, 3, 1888, p. 367.

Other Ova.

HERTWIG, O. (*Morph. Jahrb.*, x, 1884, p 338, *Rana*).
HERTWIG, O. and R. (*Jen. Zeit. f. Naturw.*, xix, 1885, p. 124).—Fecundation of *Strongylocentrotus* (Echinodermata). Careful study of cell-anæsthetics. Abstracts in *Zeit. f. wiss. Mik.*, iii, 4, 1886, p. 505, and *Journ. Roy. Mic. Soc.*, 1887, p. 835.

612. Spermatological Methods.—One of the great difficulties here met with consists in getting a sufficiently rapid fixation of spermatids and spermatozoa, some of which are probably the most rapidly contracting elements that exist. For fresh material, teased on the slide, strong solution of permanganate of potash is very useful, being the most rapidly fixing agent that I know of. Tincture of iodine is also useful. For staining, gentian violet or dahlia should be used as far as possible, as they have a special affinity for spermatic cells. For mounting aqueous preparations, solution of Ripart and Petit, or some such mercurial fluid as those of Pacini, are indicated.

As to the Nebenkern, which is here so important an element, see *ante*, § 607. Besides the papers quoted for cytological methods, *see* the following:

DOWDESWELL (*Quart. Journ. Mic. Sci.*, 1883, p. 336).—Stain in magenta, one part; glycerin, 200; alcohol, 150; water, 150.

CARNOY (*La Cellule*, i, 1885, p. 209, Arthropoda).

GILSON (*ibid.*, i, 1885, pp. 40, 56, 58, 87, 121, 141, and ii, 1886).—Here is a useful fixing medium for fresh teased preparations—*iodised osmic acid*, made by adding a little of a concentrated solution of iodine in iodide of postassium to 2 per cent. osmic acid; and some other formulæ useful for their special objects (most various forms of Arthropoda).

SWAEN ET MASQUELIN (*Arch. de Biol.*, iv, p. 752).—Selacians, *Salamandra, Bos*.

FLEMMING (*Arch. f. mik. Anat.*, xxix, 1887, p. 387; abstracts in *Zeit. f. wiss. Mik.*, v, 2, 1888, p. 236, and *Journ. Roy. Mic. Soc.*, 1888, p. 146).

PRENANT (*Intern. Monatschr. f. Anat.*, iv, 1887, p. 358; abstracts in *Zeit. f. wiss. Mik.*, v, 1, 1888, p. 84; and *Journ. Roy. Mic. Soc.*, 1888, p. 841 (Mammalia)).

RENSON (*Arch. de Biol.*, iii, 1882, p. 302).

BENDA (*Arch. f. Anat. u. Phys.; Phys. Abth.*, 1886, p. 186).—For his modification of Weigert's hæmatoxylin stain, see below.

JENSON (*Arch. de Biol.*, iv, 1883, p. 11, *et passim* (Selacians and Invertebrates)).—Concentrated solution of oxalic acid for hardening.

BALBIANI (*Lec. sur la Génér. des Vertébrés*, 1879, p. 244).—Double-stain sections with picrocarmine and methyl green. The latter stages of differentiating cells stain blue, less advanced ones lilac; and the cells which form the walls of the tubes stain only red.

RYDER, in WHITMAN's *Meth. of Research*, p. 52.—Sections of hermaphrodite gland of Lamellibranchiata stained for two or three hours in a mixture of equal parts of concentrated alcoholic solution of safranin and methyl green, diluted with eight volumes of water, washed out for five to ten minutes in alcohol and mounted in balsam. Nuclei of ova, red; heads of spermatozoa, bluish-green.

If sections of a gland of *Helix* be stained with methyl green and eosin, the ova are stained rose colour, and the spermatozoa bluish green (HENNEGUY in *Traité des Méth. Techn.*, LEE ET HENNEGUY, p. 343).

BENDA (*Arch. f. mik. Anat.*, xxx, p. 49 (Mammalia)). Fix in Flemming. Make paraffin sections. Stain them as follows: Put them for twenty-four hours into concentrated solution of neutral acetate of copper, kept at a temperature of about 40° C. Wash them out well with water, and stain to a dark grey or black tint in aqueous hæmatoxylin solution. Decolourise in dilute hydrochloric acid (0·2 per cent.), until the sections appear of a fairly light yellow. The acid must now be neutralised, which may be done by putting the sections back into the copper solution until they turn of a bluish grey. Then wash, dehydrate, and mount in balsam. *See* also the remarks of PIERSOL, in *Amer. Mon. Mic. Journ.*, 1887, p. 155; or in *Journ. Roy. Mic. Soc.*, 1888, p. 158; or *Zeit. f. wiss. Mik.*, v, 4, 1888, p. 499.

See also, of course, the numerous publications of V. LA VALETTE ST. GEORGE.

CHAPTER XXVII.

TEGUMENTARY ORGANS.

613. Epithelium.—One of the chief methods of obtaining preparations giving instructive surface-views of epithelia is the nitrate of silver method. For this, see *ante* § 198, *et seq.* in the chapter on Impregnation Methods. The reader may also consult with advantage the admirable instructions, given by RANVIER in his *Traité technique*, p. 246 *et seq.*, and the memoir of TOURNEUX and HERMANN in the *Journ. de l'Anat.*, 1876, p. 200.

Sections are easily made by the usual methods. The best hardening agent for *skin* appears to be Müller's solution. This was the conclusion of F. E. SCHULTZE in 1867 (*Arch. f. mik. Anat.*, p. 145), and it is that of TIZZONI, the author of important recent researches on this organ (*Bull. delle Sc. med. di Bologna*, 1884, p. 259).

For *glandular epithelium*, it is frequently better to employ a chromic acid liquid, or osmic acid (*see*, for example, RANVIER, *loc. cit.*, p. 258 *et seq.*), or absolute alcohol (BLAUE, *Arch. f. Anat. u. Phys.*, 1884, p. 231); "Kleinenberg" is not so good.

Prickle-cells and inter-cellular canals.—Besides maceration, which is one of the most important of the methods for the study of these objects, impregnation may be useful. MITROPHANOW (*Zeit. f. wiss. Zool.*, 1884, p. 302, and *Arch. f. Anat. u. Phys.*, 1884, p. 191) recommends the following process :—Wash with distilled water the tail of an Axolotl larva; put it for an hour into 0·25 per cent. solution of gold chloride with one drop of hydrochloric acid to a watch-glassful of the solution; wash; and reduce in a mixture of one part of formic acid with six parts of water.

Macerating Media.—For soft epithelia, mild macerating agents, such as iodised serum, one third alcohol, saliva, or Schultze's mixture of saliva and solution of Müller, or a mixture of saliva with three to four volumes of physiological salt solution (BIZZOZERO, *Intern. Monatschr. f. Anat.*, 1885, p. 278);—for hard epithelia, energetic dissociating agents, such as 40 per cent. solution of caustic potash.

MINOT (*Amer. Natural.*, xx, 1886, p. 575; cf. *Journ. Roy. Mic. Soc.*, 1886, p. 872) recommends maceration for several days in 0·6 per cent. solution of sodium chloride containing 0·1 per cent. of thymol, which allows the isolation of the epidermis of embryos, and is useful for the study of the development of hairs.

Another method, given by MITROPHANOW (see *Zeit. f. wiss. Mik.*, v, 4, 1888, p. 513), is as follows:—An embryo of Axolotl is fixed for a quarter of an hour in 3 per cent. nitric acid, and then brought into one-third alcohol. After an hour the epidermis begins to come away in places; and if the embryo be put for twenty-four hours into stronger spirit, it will come away almost entirely.

The elements of hairs and nails may be isolated by prolonged maceration in 40 per cent. potash solution, or by heating with concentrated sulphuric acid.

Horny tissues stain well in safranin or gentian violet (REINKE, *Arch. f. mik. Anat.*, xxx, 1887, p. 183; *Zeit. f. wiss. Mik.*, iv, 3, 1887, p. 383).

614. Intra-epidermic Nerve-fibres.—Must be studied by the gold-method. RANVIER (*Traité*, p. 900) recommends the boiled-formic-acid and gold-chloride method, § 208.

He also (p. 910) recommends this method for the study of the tactile menisci of the pig's or mole's snout.

Pieces of skin are impregnated as directed, § 208, and after reduction are brought into alcohol, which completes the hardening, and stays the further reduction of the gold. Sections are then made.

For the study of the tactile menisci of the snout, Ranvier also recommends the lemon-juice and gold-chloride method, § 209.

615. Innervation of the Muzzle of the Ox (CYBULSKY, *Zeit. f. wiss. Zool.*, 1883, p. 635; cf. *Journ. Roy. Mic. Soc.*, 1885, p. 555).—Cybulsky employs the method of HENOCQUE given above, § 211, impregnating sections of the fresh organ (made

616. Tactile Corpuscles (FISCHER, *Arch. f. mik. Anat.*, 1875, p. 366).—Fischer employed the gold-method of Löwit, *see* § 207. Ranvier (*Traité*, p. 918) also recommends this method, as well as his two gold-methods, Nos. 208, 209. Pieces of skin are first impregnated whole, then hardened by alcohol, and sectioned. He finds (as do other authors) that osmic acid and picro-carmine are invaluable aids to the study of these structures and to that of the corpuscles of Pacini. Purpurin and hæmatoxylin may also be used for after-staining. *See* RANVIER, *Traité*, p. 919.

617. Tactile Corpuscles and Rete Malpighi (LANGERHANS, *Arch. f. mik. Anat.*, 1873, p. 730).—Pieces of fresh skin are placed for twenty-four hours in a large quantity of ½ per cent. osmic acid, and are then found to be both stained and hardened to the right point for cutting sections.

618. Corpuscles of Herbst and Corpuscles of Grandry (CARRIÈRE, *Arch. f. mik. Anat.*, 1882, p. 146).—Take fresh beaks of ducks, remove the skin and papillæ from the margins, and put pieces for twenty-four hours into 1 per cent. osmic acid, wash in water, and put into 90 per cent. alcohol: or put them at once into alcohol (40 per cent. for a few hours, then 70 per cent., then 90 per cent.). The latter are made into sections and stained with neutral carmine, picro-carmine, fuchsin, or hæmatoxylin. The last gives the best results. Or, the pieces of skin are treated as follows :

Formic acid (50 per cent.) twenty minutes or until transparency is attained; remove the corneous layer of epithelium; rinse in water; gold chloride 1 per cent. (twenty minutes); rinse in water; Pritchard's solution (amyl-alcohol 1 per cent., formic acid 1 per cent., water 98 per cent.) from mid-day till next morning (in the dark); rinse in water; treat with alcohol; imbed in paraffin, and make sections.

It is important to take only *small quantities* of gold chloride, not more than about 10 c.c. of the solution to "quite a number" of pieces of skin and papillæ. On the other hand, *large* quantities of Pritchard's solution should be employed.

619. Corpuscles of Pacini (MICHELSON, *Arch. f. mik. Anat.*, 1869, p. 147).—Michelson found maceration for several days in concentrated solution of oxalic acid useful for isolating the nuclei of Pacinian corpuscles. The preparation may be subsequently stained with carmine.

620. Tactile Corpuscles (KULTSCHIZKY, *Arch. f. mik. Anat.*, 1884, p. 358).—Macerate pieces of duck's tongue for eighteen to twenty-four hours in 0·1 per cent. nitric acid; then treat with 0·1 per cent osmic acid; make sections, and stain with picro-carmine.

621. The Corpuscles of Krause (Conjunctiva) (LONGWORTH, *Arch. f. mik. Anat.*, 1875, p. 655).—A fresh bulbus is carefully extracted *in toto* in such a way as to spare as much conjunctiva as possible; the posterior half is cleaned of its fat, muscle, &c., and the conjunctiva drawn back and stretched over it by means of threads passed through different points of its margin. The whole is then thrown into a ½ per cent. osmic acid solution, or hung up in a cylinder and exposed to the vapour of osmic acid. It is best to let it remain twelve or twenty-four hours. The epithelium is then removed by rubbing with a camel-hair brush or with the finger; and portions of the conjunctiva as large and as thin as possible are removed and examined for corpuscles of Krause either in water or 1 to 2 per cent. acetic acid. They may then be stained and mounted in glycerin if desired. It is advantageous to make a large number of preparations, as the corpuscles are not found equally distributed in all eyes nor in all parts of the conjunctiva. If a conjunctiva be divided into five segments two of them will generally be found quite wanting in corpuscles, whilst the other three will contain thirty to sixty of them.

(*Further information concerning the corpuscles of Pacini will be given, à propos of the corpuscles of Golgi, in the paragraphs dealing with Nerve-endings in Muscle and Tendon,* § 635 *et seq.*)

622. Organs of a Sixth Sense in Amphibia (MITROPHANOW).— See *Zeit. f. wiss. Mik.*, v, 4, 1888, p. 513. This paper contains some details as to staining with "Wasserblau," for which see also *Boil. Centralb.*, vii, 1887, p. 175.

623. Papillæ Foliatæ of the Rabbit (HERMANN).—See *Zeit. f. wiss. Mik.*, v, 4, 1888, p. 524.

623a. Olfactive Organs of Vertebrates (DOGIEL, *Arch. f. mik. Anat.*, 1887, p. 74).

624. Cornea.—Impregnation with gold and with silver is indispensable in the study of the cornea.

Negative images of the corneal cells are easily obtained by the dry silver method (KLEIN). The conjunctival epithelium should be removed by brushing from a living cornea, and the corneal surface well rubbed with a piece of lunar caustic. After half an hour the cornea may be detached and examined in distilled water.

In order to obtain *positive* images of the fixed cells the simplest plan (RANVIER) is to macerate a cornea that has been prepared as above for two or three days in distilled water. There takes place a secondary impregnation, by which the cells are brought out with admirable precision.

The same result may be obtained by cauterising the cornea of a living animal as above, but allowing it to remain on the living animal for two or three days before dissecting it out, or by treating a negatively impregnated cornea with weak salt-solution or weak solution of hydrochloric acid (HIS).

But the best positive images are those furnished by gold chloride. RANVIER prefers his lemon-juice method to all others for this purpose, *see* § 209. It is important that the cornea should *not remain too long in the gold-solution*, or the nerves alone will be well impregnated.

Ranvier also recommends this method as being the best for the study of the nerves.

ROLLETT (Stricker's *Handbuch*, p. 1115) recommends a double impregnation with silver followed by gold for obtaining gold-stained *negative* images. A cornea having been treated *for a short time only* with 0·5 per cent. silver nitrate solution, and the silver reduced, is treated with 0·5 per cent. gold-chloride solution. The brown stain of the silver disappears immediately the preparation is placed in the gold-solution; after a few minutes the preparation is exposed to the light in acidulated water. Reduction of the gold rapidly takes place, and in the place of the former brown stain of the silver the ground-substance shows the well-known blue of reduced gold

· The cells are, however, visible, being recognisable by their granular appearance and pale yellow tint.

RENAUT (*Comptes Rend.*, 1880, 1ʳ sem., p. 137) gives the following process for corneal corpuscles :—Cornea of frog. Formic acid, 20 per cent., ten minutes ; gold chloride, 1 per cent., twenty-four hours ; formic acid, 33·3 per cent., twenty-four hours.

625 Cornea, Other Methods (ROLLETT, Stricker's *Handb.*, p. 1102).—Rollett strongly recommends the following plan :—A fresh cornea is placed (in humour aqueus) in a moist chamber, and exposed to the action of iodine vapour. As soon as it has become brown the epithelium may easily be peeled off. If the reaction is not complete the cornea may be put back into the iodine chamber. When sufficient iodine has been absorbed the preparation may be examined, and it will be found that the network of corneal cells is brought out with an evidence hardly inferior to that of gold preparations. The method never fails, which is not the case with the gold-method. It is admirable as a fixing method.

For dissociation of the fibres Rollett recommends maceration in a solution of permanganate of potash or a mixture of this solution with alum. As soon as the tissue has become brown it is shaken in a test-tube with water, and breaks up into fibres and bundles of fibres.

Cell-division in the membrane of Descemet (*see* SCHOTTLÄNDER's methods, *ante*, § 610).

626. Crystalline (Hardening of) (LÖWE, *Arch. f. mik. Anat.*, 1878, p. 557).—A fresh bulb is placed in a vessel containing *several litres* of 1 per cent. bichromate of potash solution, which is frequently changed for stronger solutions until the strength of a cold-saturated solution is attained. The bulb must remain in this for at least *a year and a half*, in order that the crystalline may attain the right degree of hardness.

627. Crystalline, Maceration.—Use Max Schultze's sulphuric acid solution, *supra*, § 518.

628. Tactile Hairs.—RANVIER (*Traité*, p. 914) recommends for the study of the nerve-endings the boiled formic-acid and gold-chloride method, § 208. A tactile hair having been isolated with its bulb, and its capsule incised, is put for about an hour into the formic-acid and gold-chloride mixture, the gold is reduced in slightly acidulated water, hardening is completed in alcohol, and longitudinal and transverse sections are made. For DRASCH's method see *Zeit. f. wiss. Mik.*, iv, 4, 1887, p. 492.

CHAPTER XXVIII.

MUSCLE AND TENDON—(NERVE-ENDINGS).

Striated Muscle.

629. Sections (ROLLETT, *Denkschr. math. naturw. Kl. k. Acad. Wiss., Wien*, 1885; *Zeit. f. wiss. Mik.*, 1886, p. 92).—Besides the usual section methods, the following methods of Rollett should be noted:—(1) The method mentioned above, § 312, of freezing living tissue in white of egg. (2) The same method applied to recently fixed muscle. (3) For hardened muscle, Rollett prefers celloidin (soak for twenty-four or forty-eight hours in a very thin celloidin solution, then for twenty-four hours in a solution of one part of celloidin in four parts of a mixture of equal parts of alcohol and ether, then gradual evaporation of the mass in a test tube to a gelatinous consistency, followed by hardening for twenty-four hours in a mixture of two parts of 93 per cent. alcohol with one of water). Rollett stains for several hours in Renaut's hæmatoxylic glycerin diluted with water to a very light violet tint. Dehydrate with alcohol, clear with origanum oil, and mount in dammar.

630. Dissociation. *See* Chapter XXIII.

LANGERHANS' methods for *Amphioxus* (*Arch. f. mik. Anat.*, 1875, p. 291). —For isolation of the muscle-plates macerate the fresh animal in 20 per cent. nitric acid.

For isolation of the nervous system macerate an animal for three days in 20 per cent. nitric acid, then place it for twenty-four hours in water, and shake forcibly. The *whole* of the nervous system may thus be separated, almost down to the finest peripheral terminations of nerves.

631. Motor Plates (FISCHER, *Arch. f. mik. Anat.*, 1876, p. 365). —In these researches Fischer used for mammals the gold-method proposed by LÖWIT (*Wien. Sitzgsber.*, Bd. lxxi, Abth.

iii, 1875, p. 1), and employed by himself in his researches on the tactile corpuscles (*Arch. f. mik. Anat.*, xii, p. 366).

For birds, a muscle (viz. the M. complexus) is cut up into strips 1 to 2 mm. thick and 10 mm. long, which are treated with dilute formic acid (1 part of the acid of sp. gr. of 1·06 to 2 parts water) until they become transparent. (During this maceration the strips are teased to facilitate the penetration of the gold). They are then passed direct into 1 per cent. gold chloride, and remain there a quarter of an hour. They are then washed with water and placed, according to Löwit's method, in a solution of formic acid 1 part, water 3 parts, where they remain twenty-four hours. They are *not* treated with the concentrated acid.

For reptilia and for pisces the same method was adopted. For amphibia the same method also, except that dilute acetic acid was used in the first instance in the place of formic acid to produce the necessary swelling of the tissues.

632. Motor Plates (RANVIER, *Traité*, p. 813).—Ranvier finds that for the study of the motor terminations of batrachia the best method is his lemon-juice and gold-chloride process (§ 209). The delicate elements of the arborescence of Kühne are better preserved by this method than by the simple method of Löwit.

For the study of the motor plates of reptiles, fishes, birds, and mammals, he finds (*ibid*, p. 826) that his formic-acid and gold-chloride method, § 208, gives preparations infinitely superior to those obtainable by the method of Löwit, but the lemon-juice method is still better, especially for lizards and mammals. The branches of the terminal arborescence are more regular than in preparations obtained by the formic-acid process.

He finds that the silver-nitrate method of Cohuheim is also useful. He employs it as follows : (*ibid*, p. 810).—Portions of muscle (*gastrocnemius* of frog) having been very carefully teased out in fresh serum are treated for ten to twenty seconds with nitrate of silver solution of 2 to 3 per 1000, and exposed to bright light (direct sunlight is best) in distilled water. As soon as they have become black or brown they are brought into 1 per cent. acetic acid, where they remain until they have swelled up to their normal dimensions (the swelling induced by the acid serving to make up for the shrinkage caused by the nitrate of silver). They are then examined in a mixture of equal parts of glycerin and water.

This process gives *negative* images, the muscular substance is stained brown, except in the parts where it is protected by the nervous arborescence, which itself remains unstained. The gold-process gives *positive* images, the nervous structures being stained dark violet.

633. Motor Plates, Other Methods.

BREMER (*Arch. f. mik. Anat.*, 1882, p. 195).—Impregnate according to the method of Fischer, *ante*, § 631, and soak for two or three weeks in glycerin containing 20 per cent. of formic acid. Mount in glycerin containing 1 per cent. of formic acid.

CIACCIO (*Journ. de Micrographie*, 1883, p. 38).—Lemon-juice followed by 1 per cent. solution of double chloride of gold and cadmium, and reduction partly in the light and partly in concentrated formic acid in the dark.

WOLFF (*Arch. f. mik. Anat.*, 1881, p. 355).—Study of the living muscle in salt solution.

CARL SACHS, *ibid.*—Dilute acetic acid followed by very weak picric acid for twenty-four hours. After mounting in glycerin, this gives very transparent preparations.

KRAUSE (*Intern. Monatschr. f. Anat. u. Hist.*; *Zeit. f. wiss. Mik.*, 1885, p. 547).—A muscle is put for three or four hours into concentrated solution of oxalic acid, then boiled for two minutes in water, treated for twenty-four hours with 0·1 per cent. osmic acid, and mounted in glycerin.

NEGRO (*Zeit. f. wiss. Mik.*, v, 2, 1888, p. 240).—Make the following solution:

    Concentrated solution of ammonia alum . . 180 parts
    Grübler's saturated alcoholic solution
      of hæmatoxylin . . . . . . 2 ,,

Mix, and let the mixture stand for a week exposed to the air, and add 25 c.c. each of methyl alcohol and glycerin. A drop of this is brought on to a fresh preparation of muscle of *Tropidonotus natrix*, *Lacerta viridis*, or *Rana*, and in a few minutes the stain is washed out and the preparation mounted in balsam.

BOCCARDI (see *Zeit. f. wiss. Mik.*, iv, 4, 1887, p. 492).—Preparations treated according to Ranvier's lemon-juice method or his formic-acid method, are washed in distilled water and put for a couple of hours into oxalic acid of from 0·1 to 0·25 or 0·30 per cent., or, better, into a mixture of

    Pure formic acid . . . . 5 cc.
    1 per cent. oxalic acid . . . . . . 1 ,,
    Distilled water . . . . 25 ,,

Then washed and mounted in glycerin.

KÜHNE (*Zeit. f. Biol.*, xxiii, v, 1887, p. 1; *Zeit. f. wiss. Mik.*, iv, 4, 1887, p. 495). This paper contains a critical review of the different gold-methods, of which the following are the principal conclusions:

(1). The method of Löwit is particularly applicable to thin entire (un-teased) muscles.

(2). Preliminary treatment with weak (0·5 per cent.) formic acid, and reduction in dilute glycerin containing one fourth to one fifth volume of formic acid, in the dark, is a good process for muscles of warm-blooded animals.

(3). The same process, with omission of the preliminary acidification, is to be recommended for cold-blooded animals.

(4). The method of GOLGI (*Mem. delle R. Acad. di Sci. di Torino*, ii, xxxii) is applicable to all classes of objects. It is as follows:—Acidification with 0·5 per cent. arsenic acid, impregnation in 0·5 per cent. solution of double chloride of gold and potassium, and reduction in 1 per cent. arsenic acid in sunlight.

(5). A modification of the last, consisting in impregnation in a mixture containing 0·5 per cent. arsenic acid, 0·25 per cent. chloride of gold and potassium, and 0·1 per cent. osmium, and reduction as before, is stated to be particularly applicable to Reptilia.

Kühne also remarks that teasing ought to be done in the gold solution. Specimens should be removed from the gold every few minutes, and from the reducing medium every hour, so that the right duration both of impregnation and reduction may be hit off. Preliminary acidification is unfavourable to the preservation of the arborescence, and the treatment with acids after impregnation is best abbreviated as much as possible. Mounting in formic-acid glycerin is not favourable for the preservation of detail. This is best studied in arsenic acid. Entire muscles are best mounted in balsam.

634. **Ramifications of Nerves in Muscle** (MAYS, *Zeit. f. Biol.*, 1884, p. 449; *Zeit. f. wiss. Mik.*, 1885, p. 242).—Thin muscles are put into a freshly-prepared mixture of

0·5 per cent. solution of double chloride of gold and potassium	1 gramme.
2 per cent. solution of osmic acid	1 ,,
Water	20 ,,

As soon as the nervous ramifications begin to make their appearance, the muscles are brought into a mixture of

Glycerin	40 grammes.
Water	20 ,,
25 per cent. hydrochloric acid . . .	1 ,,

in which they are left for about a day.

Another method (for *thick* muscles) is as follows:—The fresh muscle is put for twelve hours into 2 per cent. acetic acid, and then into a freshly prepared mixture of

0·5 per cent. solution of double chloride of gold and potassium	1 gramme.
2 per cent. osmic acid solution	1 ,,
2 per cent. acetic acid	50 ,,

The muscle remains in this for two or three hours, until the nerves are impregnated, and then is put for a few hours into the glycerin and hydrochloric acid mixture.

This process colours terminal arborescences. After-blackening is said to be almost entirely absent.

Tendon.

635. Corpuscles of Golgi (RANVIER, *Traité*, p. 929).—Take the tendon of the anterior and superior insertion of the gemini muscles of the rabbit. Free it as far as possible from adherent muscle-fibres. Treat it according to the formic-acid and gold method (§ 208), and after reduction of the gold scrape the tendon with a fine scalpel in order to remove the muscle-fibres that mask the " musculo-tendinous organs."

636. Corpuscles of Golgi (in the tendons of the motores bulbi oculi) (*V.* MARCHI's methods, *Archivio per le Scienze Mediche*, vol. v, No. 15).—The enucleated eyes, together with their muscles, were put for not less than three days into 2 per cent. bichromate of potash. The muscles and tendons were then carefully dissected out, stained with gold chloride and osmic acid (Golgi's method, *supra*, § 633) and by the following methods suggested by MANFREDI.

(1). The muscles and tendons removed from the bichromate solution are put for half an hour into solution of arsenic acid or into a 1 per cent. solution of acetic acid. They are then passed directly into 1 per cent. gold chloride, half an hour; distilled water; then reduced in sunlight (until a deep violet colour is obtained) in 1 per cent. arsenic acid solution, which is changed as fast as it becomes brown.

(2). The muscles and tendons removed from the bichromate solution as before are treated as follows:—Arsenic acid, 1 per cent., half an hour; osmic acid, 1 per cent., five to six hours.

(3). Fresh tissues treated as follows:—Gold chloride, 1 per cent., half an hour; oxalic acid, 0·5 per cent.; warm in a water-bath up to 36°, allow to cool, and examine.

Mount all these preparations in glycerin (balsam clears too greatly). The methods only succeed completely during fine sunny weather.

637. Corpuscles of Golgi (CATTANEO, *Arch. ital. de Biol.*, x, 1888, p. 337).—The method here recommended is the arsenic-acid method of Golgi that has been described above, § 633, *sub voce* " KÜHNE " (4).

Smooth Muscle.

638. Smooth Muscle, Isolation of Fibres (SCHWALBE, *Arch. f. mik. Anat.*, 1868, p. 394).—Maceration in weak chromic-acid solution. (0·02 per cent. proved a generally useful strength.) This is a better reagent than osmic acid, 1 per cent. acetic acid (Moleschott), weak sulphuric acid, pyroligneous acid (Meissner), 20 per cent. nitric acid (Reichert), 32 to 35 per cent. potash solution (Moleschott), as it preserves better than any of these the finer structure of the cells.

GAGE's methods (see *Journ. Roy. Mic. Soc.*, 1887, p. 327).

MÖBIUS, liquid for maceration of the muscle of *Cardium* (see above, § 512).

639. Bladder of Frog, Innervation of (WOLFF, *Arch. f. mik. Anat.*, 1881, p. 362).—A frog is killed and a solution of gold chloride of 1·20,000 injected into the bladder through the anus. (If the injection flows out on removal of the syringe, tie the frog's thighs together.) Now open the frog, dissect away the attachments of the bladder, ligature the intestine above the bladder, and cut away the abdomen of the frog so as to have in one piece bladder, rectum, and hind-legs. (All this time the bladder must be kept moist with weak gold-solution.) The bladder and the rest are now put into gold-solution of 1·2000 for four hours; the bladder is then excised, slit open, and pinned (with hedgehog spines) on to a cork (outside downwards). Place it under running water until all the epithelium is washed away. Use a pencil if necessary. Put for twenty-four hours into gold-solution of 1·6000. Wash in pure water, and put away in the dark "for some time" in acidulated water, and finally reduce in fresh water in common daylight. The muscles should be pale blue-red; medullated nerves dark blue-red; sympathetic nerves and ganglia carmine-red. RANVIER (*Traité*, p. 854) recommends one or the other of his two gold-processes. The bladder of frogs should be carefully distended by injection of the lemon-juice or gold chloride and formic acid through the cloaca.

640. Musculus dilatator pupillæ (DOGIEL, *Arch. f. mik. Anat.*, 1886, p. 403).—An enucleated eye is divided into halves and the anterior one with the iris brought for some days into a mixture of two parts one-third alcohol and part one 0·5 per cent. acetic acid. The iris can then be isolated, and split from the edge into an anterior and posterior plate, and these stained according to the usual methods.

641. Iris (KOGANEI, *Arch. f. mik. Anat.*, 1885, p. 1).—The pigmented epithelium can be removed by brushing with a small brush after prolonged maceration in solution of Müller. The pigment may also be bleached by chlorine water, which however should only be allowed to act for a few hours, until the pigment has become of a light brown; complete decolouration may be obtained by prolonging the reaction for twenty-four hours, but then the tissues suffer. (See *Journ. Roy. Mic. Soc.*, 1886, p. 874).

See also, CANFIELD, in *Arch. f. mik. Anat.*, 1886, p. 121; and DOSTOIEWSKY, *ibid.*, p. 91 (sections stained with macerated oxylin and eosin, or (LIST, *Zeit. f. wiss. Mik.*, iii, 4, 1886, p. 514) with Renaut's hæmatoxylic glycerin.

642. Stomach of Triton (*see* STILLING and PFITZNER, in *Arch. mik. Anat.*, 1886, p. 396).

643. Test for Smooth Muscle (RETTERER, *Comptes Rend. Soc. Biol.*, iv, 1887, p. 645; *Journ. Roy. Mic. Soc.*, 1888, p. 843).— If a specimen of tissue be fixed in a mixture of ten volumes of 90 per cent. alcohol and one volume of formic acid, well washed, and stained for twenty-four to thirty-six hours with alum carmine, the cytoplasm of smooth muscle will be found to be stained red, whilst connective-tissue cells remain unstained, and are swollen.

CHAPTER XXIX.

RETINA, INNER EAR, NERVES.

Retina.

644. Fixation and Hardening.—For section cutting, the retina *of small eyes* is best prepared by fixing the entire unopened bulb with osmium vapour. According to RANVIER (*Traité*, p. 954) you may fix the eye of a triton (without having previously opened the bulb) by exposing it for ten minutes to vapour of osmium. The sclerotic being very thin in this animal, such a duration of exposure is generally sufficient. Then divide it by an equatorial incision and put the posterior pole for a few hours into $\frac{1}{3}$rd alcohol. Stain for some hours in picro-carmine (1·100), treat again with osmic acid " so as to definitively fix the elements," wash with water, and harden in alcohol.

Somewhat larger eyes, such as those of the sheep and calf, may be fixed in solutions without being opened. But it is generally the better practice to make an equatorial incision, and free the posterior hemisphere before putting it into the liquid.

The older practice was to use strong solutions of pure osmic acid; but most of the best recent work has been done with chromic mixtures.

BARRETT (*Quart. Journ. Mic. Sci.*, 1886, p. 607) recommends fixing the entire unopened bulb in a liquid containing 0·2 per cent. of osmium and $\frac{1}{6}$th per cent. of chromic acid in water, for twenty-four to thirty-six hours, with subsequent hardening for fourteen days in alcohol containing 2 per cent. of carbolic acid. He finds that the preservation of specimens so treated is far superior to that of specimens hardened in pure alcohol.

SCHIEFFERDECKER (*Arch. f. mik. Anat.*, 1886, p. 305) used

either Müller's solution, or chromic acid of $\frac{1}{4}$th per cent., or a mixture of one part of wood-vinegar (Acetum pyrolignosum) with three parts of distilled water, which he states gives particularly instructive preparations without any serious alteration of the elements.

CUCCATI (*Mem. R. Accad. Sci. Ist. di Bologna* ; *Zeit. f. wiss. Mik.*, v, 1, 1888, p. 86) fixes large eyes (the lens and vitreous body having been removed) for two days in a solution of

    1 per cent. osmic acid . . . . 14 grammes.
    1 per cent. chromic acid . . . 25 ,,
    Acetic acid . . . . 1 drop,

which should be renewed after the lapse of twenty-four hours. See also DENNISSENKO (*Arch. f. mik. Anat.*, 1881, p. 395).

645. Staining.—May be done in the mass, or sections may be stained.

The best general stains are alum-carmine (several hours) or (FLEMMING) hæmatoxylin of Delafield or Böhmer (one hour in strong solution, or one day in weak). DOSTOIEWSKY and LIST both recommend double staining with hæmatoxylin and eosin (see *Zeit. f. wiss. Mik.*, iii, 4, 1886, p. 514).

BERNHEIMER's hæmatoxylin (*see* S. B. K., *Akad. Wiss. Wien*, 1884; or *Journ. Roy. Mic. Soc.*, 1886, p. 167).

RAMÓN Y CAJAL (*Rev. trim. de Hist. Norm. y Path.*, i, 1888, p. 1; *Anat. Anz.*, 1889, p. 111; *Zeit. f. wiss. Mik.*, v, 3, 1888, p. 373, and vi, 2, 1889, p. 204) applies the silver-nitrate method of Golgi (*infra*, § 652). Fresh retina is hardened for two or three days in a bichromate and osmic acid mixture (for instance, a mixture of four parts 3 per cent. solution of bichromate of potash and 1 part of 1 per cent. osmic acid solution), then brought into 0·75 per cent. solution of silver nitrate for twenty-four to thirty hours.

LENNOX (*Arch. f. Ophthalm.*, xxxii, 1; *Zeit. f. wiss. Mik.*, iii, 3, 1886, p. 408; and *Journ. Roy. Mic. Soc.*, 1887, p. 339) has been applying Weigert's hæmatoxylin method to the retina, with some remarkable results.

CUCCATI (l. c., last §) stained with concentrated aqueous solution of Säurefuchsin, and mounted in balsam.

BARRETT (l. c., last §) stains in the mass either with Kleinenberg's hæmatoxylin, or with a borax carmine that he attributes to Woodward, but the formula of which much more nearly approaches that of Thiersch.

KRAUSE (l. c., § 647) obtains instructive preparations by treating fresh retina with perchloride of iron or of vanadium in 1 per cent. solution, and then with a 2 per cent. solution of tannic or pyrogallic acid. These reagents only stain the granule layers, and the nuclei of the ganglion cells. The elements of the other layers may then be stained with Säurefuchsin, or some other anilin.

646. Sections.—Perhaps the majority of recent workers recommend celloidin; but I see no reason for not employing paraffin. Sections may be mounted in dammar, or (FLEMMING) in glycerin. RAMÓN Y CAJAL (l. c., last §) mounts in dammar or colophonium varnish, without a cover-glass.

647. Dissociation Methods.—For maceration preparations you may use weak solutions (0·2 to 0·5 per cent.) of osmic acid, for fixation, and then macerate in 0·02 per cent. chromic acid (M. SCHULTZE), or in iodised serum (M. SCHULTZE), or in dilute alcohol (LANDOLT), or in Müller's solution, or (RANVIER, *Traité*, p. 957) in pure water, for two or three days. THIN (*Journ. of Anat.*, 1879, p. 139) obtained very good results by fixing for thirty-six to forty-eight hours in one-third alcohol, or in 25 per cent. alcohol, and then staining and teasing.

SCHIEFFERDECKER (l. c., § 644) macerates fresh retina for several days in the methyl mixture, § 520.

KRAUSE (*Intern. Monatsch. f. Anat. u. Hist.*, 1884, p. 225; *Zeit. f. wiss. Mik.*, 1885, pp. 140, 396) recommends treatment for several days with 10 per cent. chloral hydrate solution. Barrett finds that this process preserves the rods and cones admirably.

Inner Ear.

648. Schwalbe's Methods (*Beitr. z. Phys.*, 1887; *Zeit. f. wiss. Mik.*, iv, 1, 1887, p. 90; *Journ. Roy. Mic. Soc.*, 1887, p. 840). —Fix (cochlea of guinea-pig) for eight to ten hours in "Flemming" wash in water, decalcify (twenty-four hours is enough) in 1 per cent. hydrochloric acid, wash the acid out, dehydrate, and imbed in paraffin.

649. Other Methods.—WALDEYER, Stricker's *Handb.*, p. 958 (decalcification either in 0·001 per cent. palladium chloride, containing 10 per cent. of HCl, or in chromic acid of 0·25 to 1 per cent.).

URBAN PRITCHARD (*Journ. Roy. Mic. Soc.*, 1876, p. 211).—(Decalcification in 1 per cent. nitric acid).

LAVDOWSKY (*Arch. f. mik. Anat.*, 1876, p. 497).—Fresh tissues (from the cochlea) are treated with 1 per cent. solution of silver nitrate, then washed for ten minutes in water containing a few drops of 0·5 or 1 per cent. osmic acid solution, and mounted in glycerin.

MAX FLESCH (*Arch. f. mik. Anat.*, 1878, p. 300).

TAFANI (*Arch. Ital. de Biol.*, vi, p. 207).

Nerves.

650. Weigert's Method for Medullated Nerves (*Fortschr. d. Med.*, 1884, pp. 113, 190; 1885, p. 136; *Zeit. f. wiss. Mik.*, 1884, pp. 290, 564; 1885, pp. 399, 484).—The ordinary methods of staining with hæmatoxylin depend on the production of an aluminium lake of hæmatoxylin. Weigert's method depends on the formation of another lake, a chromium or copper lake. In consequence of the formation of these lakes, hæmatoxylin acquires the property of staining the myelin of nerves in a quite specific way.

In Weigert's process the formation of these lakes takes place in the tissue itself. He proceeds now (I pass over the earlier form of the method) as follows:—The tissues are to be hardened in bichromate of potash (the solutions of Müller or Erlicki will do as well, so far as I know). The hardening need only be carried to the point at which the tissues have acquired a *brown*, not a *green*, colouration (but green tissues may be used, provided they have once passed through the brown stage). The preparation is then (but this is not necessary) imbedded by infiltration with celloidin, and the celloidin block fastened on cork and hardened in the usual way. The hardened block is put for one or two days into saturated solution of neutral acetate of copper diluted with one volume of water, the whole being kept at the temperature of an incubating stove. By this treatment the tissues become green, and the celloidin bluish-green. The mordantage of the tissues is now terminated, and the preparation may be kept, till wanted for sectioning, in 80 per cent. alcohol.

Sections are made with a knife wetted with alcohol, and are brought into a stain composed of

Hæmatoxylin	0·75 to 1 part.
Alcohol	10 ,,
Water	90 ,,
Saturated solution of lithium carbonate	1 ,,

(A trace of any other alkali may be added in the place of lithium carbonate. The object of adding a little of some base is to "ripen" the hæmatoxylin solution.)

The sections remain in the stain for a length of time that varies according to the nature of the tissues :—Spinal cord, two hours; medullary layers of brain, two hours; cortical layers, twenty-four hours.

They are then rinsed with water, and brought into a decolourising solution composed of

Borax	2·0 parts.
Ferricyanide of potassium . . .	2·5 ,,
Water	200·0 ,,

They remain in the solution until they are decoloured to the right degree, that is, until complete differentiation of the nerves (half an hour to several hours), and are then rinsed with water, dehydrated with alcohol, and mounted in balsam. They may be previously stained, if desired, with alum-carmine for the demonstration of nuclei.

For very difficult objects, such as pathological nerves, the decolouring solution should be diluted with water, and the immersion in it prolonged. GELPKE (*Zeit. f. wiss. Mik.*, 1885, p. 489) states that for transverse sections of atrophied nerves the solution should be diluted with fifty volumes of water, and the immersion be prolonged to twelve hours at the least; for longitudinal sections it should be diluted with ten volumes of water.

The process is applicable to tissues that have beeen hardened in alcohol or in any other way, provided that they be put into a solution of a chromic salt until they become brown, before mordanting them in the copper solution.

As above stated, it is not necessary that the mordantage be done in the mass with tissues imbedded in celloidin. MAX FLESCH (*Zeit. f. wiss. Mik.*, iii, 1, 1886, p. 50) finds that this practice is unfavorable to subsequent staining with other reagents than hæmatoxylin, and prefers (following LICHTHEIM) to make the sections first, bring them on closet-paper into the mordant, and after mordanting bring them on a spatula into 70 per cent. alcohol, and thence into the stain.

In the process given above, a copper lake is formed in the tissues. In the earlier form of the process, the mordantage with the copper salt was omitted, and the stain depended on the formation in the tissues of a chromic lake. The results were not quite so good, and the process may be taken to be superseded by the copper process. A modified form of, due to MAX FLESCH, has some advantages, and is given in the next §.

If very many large sections have to be prepared, and if the staining solution be thrown away after using, the process may be found somewhat expensive. The following method for regenerating the staining solution is given by FANNY BERLINERBLAU (*Zeit. f. wiss. Mik.*, 1886, p. 50) :—About 2·5 to 5 per cent. of baryta water is added to the used solution; it is well shaken and allowed to stand for twenty-four hours; carbonic acid (obtained from the action of crude hydrochloric acid on marble) is led through it, it is allowed to stand for twenty-four hours more, and then filtered.

PANETH (*ibid.*, 1887, p. 213) makes the stain with extract of logwood, instead of pure hæmatoxylin. One part of commercial extract of logwood is dissolved in 90 parts of water and 10 of alcohol. To the filtered solution is added 8 drops of concentrated solution of lithium carbonate for each 100 c.c. Sections require from eighteen to twenty-four hours in the stain at the normal temperature.

The results obtained by Weigert's method are most splendid. The blue-black nerves stand out with admirable boldness on a golden ground. The method is applicable to the study of peripheral nerves as well as to nerve-centres; and is likely to be of great utility in vertebrate embryology.

Nerve-tissue is not the only tissue stained by the process, which can be usefully applied to lymphatic glands and to skin (see SCHIEFFERDECKER, in *Anat. Anz.*, ii, 1887, p. 680).

651. Modifications of WEIGERT's Method.—MAX FLESCH (*Zeit. f. wiss. Mik.*, 1884, p. 564).—Sections, made either by the celloidin process or otherwise, are put for a few minutes or longer into 0·5 per cent. chromic acid solution; they are then rinsed in water, and brought into the stain, where they take on a sufficient colouration much more rapidly than the tissues will if stained in the mass. Decolouration is done in the usual way. No stove is required in this process, and either brown or green material may be used. Flesch finds (l. c., 1886, p. 51) that this method is to be preferred to Weigert's copper method in all cases in which it is important to produce differential staining in nerve-cells, and especially in the study of peripheral ganglia, and also for producing differential staining of the medulla of central and peripheral nerves, whilst the copper process is most certainly superior for the demonstration of fine nerve-fibres.

BENDA (*Verhandl. Physiol. Ges., Berlin*, 1885–86, Nos. 12 13, 14; *Zeit. f. wiss. Mik.*, iii, 3, 1886, p. 410; *Arch. f. Anat. u. Phys.*, 1886, p. 562).—The tissues are to be fixed and hardened for at least three days in saturated solution of picric acid, washed out for several days in water, hardened in alcohol, imbedded in paraffin, and sections made. The sections are mordanted by treating them for a few minutes or hours with a salt of iron, for which purpose concentrated solution of ammonio-sulphate of iron is recommended. After being well washed in several changes of water, they are stained for about ten minutes in 1 per cent. aqueous solution of hæmatoxylin, then decoloured by treatment for about five minutes in 0·05 per cent. aqueous solution of chromic acid, rinsed in water, dehydrated, cleared, and mounted in balsam.

This method is stated to be peculiarly adapted for the demonstration of very fine fibres and their relations to ganglion-cells.

HAMILTON (*Journ. of Anat. and Physiol.*, xxi, 1887, p. 444; *Journ. Roy. Mic. Soc.*, 1888, p. 1051) mordants his preparations with a copper sulphate solution, and makes sections by the freezing method, after imbedding in collodion.

BEEVER (*Brain*, 1885, p. 227; *Journ. Roy. Mic. Soc.*, 1886, p. 898) gives a very slight modification of Weigert's original process.

PAL (*Wien. med. Jahrb.*, 1886; *Zeit. f. wiss. Mik.*, iv, 1, 1887, p. 92; *Med. Jahrb.*, 1887, p. 589; *Zeit. f. wiss. Mik.*, 1888, p. 88) describes the following process:—After staining in the hæmatoxylin solution the sections are washed in water (if they are not stained of a deep blue, a trace of lithium carbonate must be added to the water). They are then brought for twenty to thirty seconds into 0·25 per cent. solution of permanganate of potash, rinsed in water, and brought into a decolouring solution composed of—

Acid. Oxalic. pur.	1·0
Kalium Sulfurosum (SO_3K_2)	1·0
Aq. dest.	200·0

In a few seconds the grey substance of the sections is decolourised, the white matter remaining blue. The sections should now be well washed out, and may be double-stained with Magdala red or eosin, or (better) with picro-carmine or acetic-acid-carmine.

For further details as to the somewhat elaborate minutiæ of the process see the papers quoted, or BEHRENS, KOSSEL, and SCHIEFFERDECKER's, *Das Mikroskop*, i, p. 199.

652. GOLGI's **Methods for Medullated Nerves.** (1) **Bichromate and Silver-nitrate Process.**—I take the following *résumé* of this method from the interesting paper of Golgi's pupil, REZZONICO, "Sulla struttura delle fibre nervose del midollo spinale," in the *Archivio per le scienze mediche*, iv, No. 4 (1879), p. 85.

1. Take pieces of perfectly fresh spinal cord, and soak them in a 2 per cent. solution of bichromate of potash, for a period of time varying according to temperature. (In summer eight to fifteen days may suffice, in winter about a month is necessary.)

2. Wash them, and put them into a 0·75 per cent. solution of nitrate of silver. The period of immersion therein depends on the temperature: in summer the reaction will be complete throughout the tissues in two or three days; in winter eight, ten, or more days.

3. Dehydrate small pieces with alcohol (make sections if necessary), clear in oil of turpentine, tease in the turpentine, and mount in dammar.

4. The preparations are then left to themselves in order that the secondary impregnation may take place. In direct sunlight, eight to ten days will complete the process; in diffused daylight (or in the dark ?), twenty, thirty, or forty days.

A somewhat greater precision of the reaction is obtained by treating the fresh tissues with osmic acid (by means of interstitial injection) before putting them into the bichromate. In this case a much shorter immersion in the bichromate will suffice (four, six, or eight days).

By this means may be demonstrated in the medullated fibres of the spinal cord a chain of conical funnels, set one within another, and embracing the axis cylinder with their narrow aperture, and the external surface of the following funnel with their greater aperture; and it is seen that they consist of a fine spiral fibre wound into the form of a funnel. (The appearance of rings and strainers is due to imperfect action of the silver.)

(2) **For the Study of Peripheral Nerves** Golgi modifies the process as follows (l. c., 1879, p. 238):

1. Pieces of nerve are immersed in the bichromate solution for from four, six, or eight hours to one day, or at most two days.

2. From time to time pieces are removed into the nitrate of silver; they remain there for from twelve to twenty-four hours.

3. They are washed with several changes of alcohol.

4. Tease in the alcohol, dehydrate, clear with turpentine, mount in dammar.

5. Reduce in direct sunlight; in summer a few days suffice, in cold weather some weeks are necessary.

Does not give quite such fine results as the osmium bichromate silver method next to be described, but the preparations keep indefinitely.

(3) **The Osmium, Bichromate, and Silver Method** (*ibid.*, p. 237). A perfectly fresh piece of nerve is thrown into the following liquid:

2 per cent. solution of bichromate of potash   10 parts. — —
1 per cent. solution of osmic acid   .   .   .   2 parts.

After about an hour's immersion the piece of nerve may be cut into lengths of ½ to 1 cm., which are put back into the liquid.

Four hours after the first immersion of the nerve in the mixture, begin to put the pieces into nitrate of silver solution, transferring a certain number of pieces every three hours, so as to be sure that some of them shall have had a bichromate bath of a proper duration. (This duration may, roughly speaking, be said to lie between six and twenty-four hours.)

The strength of the nitrate of silver solution is 0·50 per cent. The duration of the silver-bath must not be less than eight hours; it may be indefinitely protracted.

Dehydrate, clear with turpentine, mount in dammar.

The method is more expeditious and easier of application than the bichromate and silver-nitrate method, and the results are somewhat more precise, but the preparations do not keep in dammar.

These two methods serve for the demonstration in *peripheral* medullated nerve-fibres of the funnel-shaped coils of sustaining filaments discovered by Rezzonico in the medullated fibres of the spinal cord.

In all methods for the demonstration of the funnels it is important to observe the utmost delicacy of manipulation, and in particular, the fibres *must not be stretched;* their stretching is a weak point in the methods of Ranvier (No. 656).

MONDINO (*Arch. per le Sci. Med.*, viii, p. 45; *Zeit. f. wiss. Mik.*, 1885, p. 547) recommends that the preparations be left in the bichromate mixture longer than directed by Golgi—viz. from one to eight days; and also that they be left for over twenty-four hours in the silver-bath.

For some minutiæ concerning the application of the bichromate and silver method to cerebro-spinal nerves, see PETRONE, in *Internat. Monatschr. f. Anat.*, v, 1, 1888, or *Zeit. f. wiss. Mik.*, v, 2, 1888, p. 238.

653. Stain for Neuroceratin Funnels (GALLI, *Zeit. f. wiss. Mik.*, iii, 1, 1886, p. 467).—An ischiatic nerve is excised, and fixed and hardened for eighteen to twenty days in solution of Müller. Small portions of the nerve are then further treated for one or two days with solution of Müller diluted with two parts of water, then for a quarter of an hour with glycerin containing one or two drops of glacial acetic acid for each cubic centimetre, and finally (without previous washing with water) are stained for fifteen to twenty minutes in aqueous solution of China blue (the best China blue for this purpose is that supplied by the Badische Anilin- und Soda-Fabrik, at Stuttgart). They are washed out in alcohol, cleared in essence of turpentine, and mounted in dammar.

For the results, see the paper and the plate, *l. c.*

PLATNER (*Zeit. f. wiss. Mik.*, vi, 2, 1889, p. 186) obtains a specific stain of the neuroceratin framework of medullated nerve in the following way: small nerves are fixed and hardened for several days in a mixture of one part of Liq. Ferri Perchlor. (Ph. G. Ed. ii) and three to four parts of water or alcohol. They are then to be washed out in water or alcohol till no traces of iron remain in them (the reaction of the washings with rhodanide of potassium being a good test of this), and are stained for several days or weeks in a concentrated solution of "Echtgrün" (Di-nitroresorcin) in 75 per cent. alcohol; after which they are dehydrated, imbedded, and sectioned.

654. Axis Cylinders.

KUPFFER's Method (*Sitzb. math. phys. Kl. k. Bayr. Acad. Wiss.*, 1884, p. 446; *Zeit. f. wiss. Mik.*, 1885, p. 106).—A nerve is stretched on a cork and treated for twenty-four hours with 0·5 per cent. osmic acid. It is then washed in water for two hours and stained for twenty-four to twenty-eight hours in saturated aqueous solution of Säurefuchsin; after which it is washed out for from six to twelve hours (not more in any case) in absolute alcohol, cleared in clove oil, imbedded in paraffin, and cut. Sections are said to show the axis cylinder as a bundle of fibrils (stained red) floating in an albuminous liquid.

NISSL's Method (*Münchener med. Wochenschr.*, 1886, p. 528; *Zeit. f. wiss. Mik.*, iii, 3, 1886, p. 398).—Bichromate objects to be treated as follows:—Alcohol of 95 per cent.; aqueous solution of Congo red, of 5 : 400 strength, seventy-two hours; alcohol of 95 per cent., five to ten minutes; alcoholic solution of nitric acid of 3 per cent. strength, six hours; alcohol, five minutes; clove oil; balsam.

EHRLICH's Methylen Blue Method (*Abh. k. Akad. Wiss. Berlin*, February 25th, 1885).—It has been made out by Ehrlich that by injecting living animals with methylen blue, a specific stain of the axis cylinders of peripheral nerves may be obtained. He found that the axis cylinders that are stained always belong to sensory nerves, motor nerves containing no element that takes the stain.

ARNSTEIN (*Anat. Anz.*, 1887, p. 125), however, found that

motor nerve-endings, and other elements, also stain, though later than the sensory nerves. He proceeded by injecting frogs through the vena cutanea magna with 1 c.c. of saturated solution of the colour. The organ to be investigated is removed after the lapse of an hour or two. The stain generally keeps for only a few minutes, unless it is fixed. This may be done by means of iodine. Arnstein proceeds by injecting a saturated solution of iodine in 1 per cent. solution of iodide of potassium. The organs are then removed and soaked in the iodine solution for from six to twelve hours. They are then washed out with water and put up in acidulated glycerin (see a good abstract of this paper in *Zeit. f. wiss. Mik.*, iv, 1, 1887, p. 84).

In a later paper (*Anat. Anz.*, 1887, p. 551; see *Zeit. f. wiss. Mik.*, 1887, p. 372) Arnstein gives some further details concerning the reaction, and states that the stain can also be fixed, and even better than by iodine, by means of picrocarmine or picrate of ammonia (see also *Journ. Roy. Mic. Soc.*, 1888, p. 515).

BIEDERMANN (*Sitzb. k. Acad. Wiss. Wien.*, *Math. Nat. Cl.*, 1888, p. 8; *Zeit. f. wiss. Mik.*, vi, 1, 1889, p. 65) has been investigating the nerve-endings in the muscles of invertebrates by this method. He injects, for instance, 0·5 to 1 c.c. of a nearly saturated solution of methylen blue in 0·6 per cent. salt solution into the thorax of a crayfish, leaves the animal for from two to four hours in a moist chamber, and then kills it. The animal should then be opened, and the muscles exposed in a roomy, moist chamber for from two to four hours before examining them (in all the forms of this method, the presence of oxygen is necessary to the reaction). Instructions as to the minutiæ of the process as applied to various objects are given in the papers quoted.

As far as I can see, this method, which is at present in its infancy, promises to be a very fruitful one.

655. Peripheral Nerves, Topography and Endings—Other Methods.—It remains here to be noted that the chief methods for the study of these objects are to be found amongst the gold methods given in Chapters XII, in Part I, and XXVII and XXVIII in Part II. The method of Freud for nerve-centres, which will be given in the next chapter, is also a good one for the study of the topography of peripheral nerves.

656. Structure of Medullated Nerve—Other Methods.—In order to demonstrate the axis cylinder and the sheath of Schwann, the myelin may be removed. This may be done by boiling in caustic soda, and then neutralising; by boiling in a mixture of absolute alcohol and ether, and adding caustic soda; by boiling in glacial acetic acid; by boiling in fuming nitric acid and adding caustic potash; or by treating with eau de Javelle.

See also RANVIER, *Traité*, p. 718, *et seq.*; REZZONICO, *Arch. per le Sci. Med.*, 1879, p. 237; TIZZONI, *ibid.*, 1878, p. 4 (a process of boiling in chloroform for an hour or two, then staining and mounting in glycerin); BOVERI, *Zeit. f. wiss. Mik.*, iv, 1, 1887, p. 91; JAKIMOVITCH, *Journ. de l'Anat.*, xxiii, 1888, p. 142, or *Zeit. f. wiss. Mik.*, v, 4, 1888, p. 526 (instructions for impregnating the axis cylinder with silver, followed by reduction in formic acid and amyl alcohol).

CHAPTER XXX.

CENTRAL NERVOUS SYSTEM.

657. Introductory.—For small objects, such as the spinal cord and encephalon of very small Mammalia and of inferior Vertebrata, the ordinary methods of microscopic anatomy are very often sufficient. The cord and encephalon of Batrachia, for instance, may be treated as follows :—Corrosive sublimate, half an hour or so; alcohol of from 50 to 70 per cent., about two hours; then borax-carmine, or Mayer's alcoholic cochineal; paraffin; and sections mounted in balsam. The same organs of somewhat larger animals, such as the cat or rabbit, can also be prepared without recourse to any very special methods. They may be fixed and hardened in liquid of Erlicki, or other bichromate solution, well washed out with water, stained and imbedded, or first imbedded in collodion and stained after sectioning with alum-carmine, ammonia-carmine, or Heidenhain's hæmatoxylin, or, if it be desired to study the topography of nerve-fibres, with Weigert's hæmatoxylin. Even the cord of Man may be treated by these methods ; or rather, could be so treated if it did not almost always come into the hands of the anatomist in a state of *post-mortem* softening that necessitates special precautions in the hardening process.

But the voluminous encephala of Man and the larger Vertebrates cannot be thus simply treated. They require specially modified methods for hardening, for the manipulation of sections, and for staining.

These methods have lately been described with great completeness in the works of BEVAN LEWIS (*The Human Brain; Histological and Coarse Methods of Research*, London (Churchill); and OBERSTEINER (*Anleitung beim Studium des Baues d. nervösen Centralorgane im gesunden u. kranken Zustande*, Leipzig (Toeplitz). These very welcome additions to the literature of the subject relieve me from the obligation of treating the subject with all the minuteness that might be desired by specialists; the more so, as they show that so to treat the subject requires a volume, not a chapter.

Hardening.

658. Hardening by the Freezing Method.—This is in many cases a very good method, and in particular may be of service for the histological study of the cortex.

If it be desired to freeze an organ that has been already hardened by reagents, the freezing may be done by means of a freezing mixture of ice and salt; but in this case the preparation should first be penetrated by a mucilaginous or gelatinous freezing mass (§ 308, *et seq.*), in order to avoid the formation of ice crystals in the tissues. But in the case of fresh tissue, the ether freezing method is to be preferred. This method allows of rapidly producing any desired degrees of hardness, and maintaining them or allowing them to diminish as occasion may require, and is perhaps the only method by which satisfactory sections of unhardened nerve-tissue can be obtained.

The sections should be floated on to water, treated for a minute on the slide with 0.25 per cent. osmic acid solution, and stained or otherwise treated as desired.

For a detailed description of these manipulations *see* BEVAN LEWIS' *The Human Brain*.

659. Generalities on Hardening by Reagents.—If large pieces of nerve-tissue are to be hardened, it is necessary to take special precautions in order to prevent them from becoming deformed by their own weight during the process. Spinal cord or small specimens of any region of the encephalon may be cut into slices of a few millimetres thickness, laid out on cotton-wool, and brought on the wool into a vessel in which they may have the hardening liquid poured over them. The wool performs two functions; it forms an elastic cushion on which the preparations may lie without being distorted by their own weight; and it allows the reagent to penetrate by the lower surfaces of the preparations as well as by their exposed surfaces. A further precaution, which is useful, is to hang up the preparations, lying on or in the cotton-wool, in a glass cylinder, or other tall vessel; by hanging them near the top of the liquid, the processes of diffusion and the penetration of the reagent are greatly facilitated.

If the preparations are placed on the bottom of the vessel, they should never be placed one on another.

If it be desired to harden voluminous organs without dividing them into portions, they should at least be incised as deeply as possible in the less important regions. It is perhaps better in general not to remove the membranes at first (except the dura mater), as they serve to give support to the tissues. The pia mater and arachnoid may be removed partially or entirely later on, when the hardening has already made some progress.

The *spinal cord*, the *medulla oblongata*, and the *pons Varolii* may be hardened *in toto*. The dura mater should be removed at once, and the preparation hung up in a cylinder-glass, with a weight attached to its lower end. The weight has the double function of preventing any part of the preparation from floating above the level of the hardening liquid (a thing that easily happens where somewhat dense liquids, such as Müller's solution, are used), and of preventing the torsions of the tissues that may otherwise be brought about by the elastic fibres of the pia mater and arachnoid.

The *cerebrum* should be very delicately laid out on a layer of cotton wool, or, if possible, hung up in it. Plugs of the wool should be put into the fissure of Sylvius, and between the operculum and the median lobe, and as far as possible between the convolutions. Unless there are special reasons to the contrary, the brain should be divided into two symmetrical halves by a sagittal cut passing through the median plane of the corpus callosum. Betz recommends that after a few hours in the hardening liquid the pia mater should be removed wherever it is accessible, and the choroid plexuses also. I have found this by no means easy, and think it is an operation that can only be recommended for experienced hands.

The cerebellum should be treated after the same manner.

The temperature at which the preparations are kept in the hardening solution is an important point. The hardening action of most solutions is greatly enhanced by heat. Thus WEIGERT (*Centralb. f. d. med. Wiss.*, 1882, p. 819; *Zeit. f. wiss. Mik.*, 1884, p. 388) finds that at a temperature of from 30° to 40° C. preparations may be sufficiently hardened in solution of Müller in eight or ten days, and in solution of Erlicki in four days; whilst at the normal temperature two or three times as long would be required.

But it is not certain that this rapid hardening always gives the best definitive results. SAHLI, who has recently made a detailed study of the hardening action of chrome salts, is of opinion that it does not, and thinks it ought for this reason to be abandoned (see *Zeit. f. wiss. Mik.*, 1885, p. 3).

On the other hand, the slowness of the action of chromic salts at the normal temperature is such that decomposition may easily be set up in the tissues before the hardening and preserving fluid has had time to do its work. For this reason voluminous preparations that are to be hardened in the slow way should be put away in a very cool place—best of all in an ice-safe.

660. The Reagents to be employed.—The hardening agents

most used are the chromic salts. Chromic acid was much used at one time, but most workers now agree that its action, though much more rapid than that of the salts, is much more uneven, and frequently causes a disastrous friability of the tissues. Osmic acid is excellent—according to BEVAN LEWIS, it is the best of hardening agents; but its employment is unfortunately very restricted, as it can hardly be used for objects of more than a cubic centimètre in size.

It has already been noted that the liquid of Erlicki has a more rapid action than the other solutions of chromic salts; for this reason it is one of the most commonly employed solutions. SAHLI, however (l. c.), after having studied the action of the usual solutions, concludes that the best hardening agent for fresh tissue is pure bichromate of potash, in 3 or 4 per cent. solution, the hardening being done in a cold place. And he does not approve of the addition of sodium sulphate (Müller), and rejects the liquid of Erlicki on account of the precipitates it so frequently gives rise to.

OBERSTEINER is of the same opinion, recommending pure bichromate for general hardening purposes; whilst for the study of the most delicate structural relations he recommends fixing in Fol's modification of Flemming's liquid (§ 35, p. 24, *supra*) for twenty-four hours, followed by washing with water and hardening in 80 per cent. alcohol.

In view of the slowness of penetration of chromic salts, it is often advisable to treat preparations for twenty-four hours or more with alcohol of 80 to 90 per cent. before putting them into the hardening liquid, in order to avoid maceration of the deeper layers of tissue.

For the question as to how far certain so-called pathological alterations of ganglion-cells should be attributed to putrefactive changes or to the influence of reagents see *Neurologisches Centralb.* for the years 1884 and 1885.

As to the so-called "pigment spots" produced by the liquid of Erlicki see *supra*, § 79.

661. Strengths of the Reagents.—All hardening reagents (except osmic acid) should at first be taken as weak as is consistent with the preservation of the tissue, and be changed by degrees for stronger.

Osmic acid may be taken of 1 per cent. strength, and will

harden small pieces of tissue sufficiently in 5 to 10 days (EXNER).

Bichromate of potash should be taken at first of not more than 2 per cent. strength; this is then gradually raised to 3 or 4 per cent. for the cord and cerebrum, and as much as 5 per cent. for the cerebellum. Obersteiner begins with 1 per cent. and proceeds gradually during six to eight weeks to 2 or 3 per cent. (This is at the normal temperature; at a temperature of 35° to 45° C. the hardening can be got through in one or two weeks).

Bichromate of ammonia should be taken of half the strength recommended for bichromate of potash, or even weaker at first; it may be raised to as much as 5 per cent. for cerebellum, towards the end of the hardening.

Chromic acid is not much used alone. See *supra*, § 70. It forms part of some of the mixtures mentioned below. A very little chromic acid (say one to two drops of 1 per cent. solution for each ounce) added to bichromate solution will do no harm, and will quicken the hardening.

Nitric acid has been and still is employed in strengths of 10 to 12 per cent., and gives particularly tough preparations. Perhaps a weaker solution might give good results, but I cannot find that any such have been tried.

Neutral acetate of lead in 10 per cent. solution affords an excellent preservation of ganglion-cells, according to ANNA KOTLAREWSKI (see *Zeit. f. wiss. Mik.*, iv, 3, 1887, p. 387).

TRZEDINSKI (*Virchow's Arch.*, 1887, p. 1; *Zeit. f. wiss. Mik.*, iv, 4, 1887, p. 497) finds that as regards the faithful preservation of ganglion-cells (of the spinal cord of the rabbit and dog) the best results are obtained by hardening for eight days in 10 per cent. solution of corrosive sublimate, followed by hardening in alcohol containing 0·5 per cent. of iodine.

DIOMIDOFF (*ibid.* p. 499) also obtained very excellent results by hardening small pieces of brain (as suggested by GAULE, OGATA, and BECHTEREFF) for from five to nine days (not more in any case) in 7 per cent. sublimate solution, and then putting the tissues for twenty-four hours into 50 per cent. alcohol, and for the same time into 70 per cent. and 96 per cent. alcohol successively. (This process produces artificial "pigment spots" similar to those produced by solution of Erlicki; they may be dissolved out by prolonged treatment with warm water, or in five minutes by strong solution of LUGOL.) The tissues are of a good consistence for cutting. Chloride of zinc has been recommended for some purposes, *see* below, § 673.

The next following paragraphs give in detail some methods of hardening recommended by some of the most competent workers.

662. Spinal Cord (KRAUSE, *Arch. f. mik. Anat.*, 1875, p. 226). —Solution of Müller, twenty-four hours. Then chromic acid, 1 per cent. On the fourth day the solution must be changed for fresh. A few days later (when the cord appears hard), the chromic acid is removed by means of water, and the cord put into spirit, followed by absolute alcohol, which must be changed at least twice.

663. **Encephalon** (M. Duval's methods, Robin's *Journal de l'Anatomie*, 1876, p. 497).—*First Method.*—Place the fresh tissues in solution of bichromate of potash 25, water 1000, change the liquid after the first twenty-four hours, and again after three or four days. After two or three weeks place the preparations in chromic acid of 3 per 1000, change the liquid every day for the first week, and after that every eight days until the middle of the second month, after which time it is no longer needful to change the liquid. The preparations must remain at least two months in the chromic acid; the longer they remain in it the better. A few fragments of camphor should be added to the liquid in order to prevent the growth of mould.

Second Method.—Place the fresh tissues in a mixture of equal parts of glycerin and acetic acid; after twenty-four hours remove them to Müller's solution, and after forty-eight hours more to chromic acid. (The strength of the solution is not indicated.) Change the chromic acid once or twice, and the preparations will be fit for cutting in about eight or ten days. (Small encephala (rat, bat) need not be extracted from the cranium provided this be largely opened before immersing them in the glycerin, nor need they be extracted before cutting, as the cranium will be found to be completely decalcified.)

This method is highly expeditious, and furnishes hardened tissues of a singularly homogeneous consistence, quite without fragility, but it should only be employed for the purpose of obtaining general views of structural relations, as the anatomical elements are somewhat changed by it: cells and axis-cylinders swell.

664. **Cerebrum** (Rutherford; from Bevan Lewis).—Place small portions of cerebrum in methylated spirit for twenty-four hours, in a cool place. After that, put them into the following solution:

Chromic acid	1 part.
Bichromate of potash	2 parts.
Water	1200 ,,

Change the liquid at the end of eighteen hours, and then once a week. Should the tissue not be sufficiently tough for cutting at the end of six weeks, place it in a $\frac{1}{6}$th per cent.

solution of chromic acid for a fortnight, and then in rectified spirit.

665. Cerebrum (BEVAN LEWIS, *The Human Brain*, p. 102).—Methylated spirit, twenty-four hours in a cool place. Müller's solution, three days in a cool place. Then change the liquid; and after three days more change it again, or, preferably, substitute a 2 per cent. solution of potassium bichromate. At the end of the second week a solution of double the strength may be added; and if, at the termination of the third week, the mass is still pliable and of the consistence of ordinary rubber, it is as yet unfit for section cutting, and the reagent should be replaced by a solution of chromic acid.

666. Brain (HAMILTON, *Journ. of Anat. and Physiol.*, 1878, p. 254).—Take a fresh brain and make a series of incisions into different parts, still keeping everything *in situ*; or slice it into any number of segments about one inch thick, but of the whole length or breadth of the organ, as may be desired. Do *not* remove the membranes; they form a projection for the superficial layers, and do not interfere with the hardening process. The large segments are placed flat in a large vessel padded with cotton; do not put them one above the other. Cover them with the following fluid:

    Müller's fluid . . . 3 parts.
    Methylated spirit . . 1 part.

(Heat is evolved on mixing these liquids, and the mixture must be allowed to cool before pouring it over the brain tissue.) Put the preparations away in an ice-safe. Turn the segments over next day. Change the solution in a fortnight or three weeks; or if on examining a section of one of the pieces it is found that the hardening reagent has penetrated to the interior, they may be at once removed to the following mixture:

    Bichromate of ammonia . . 1 grm.
    Water . . . . 400 c.c.

in which they remain for one week. Then change the solution to one of 1 per cent. for one week; and let this be followed by a solution of 2 per cent. for another week, or longer if required. The pieces will now be sufficiently hard for cutting; they may be kept permanently in solution of chloral hydrate, twelve grains to the ounce.

Probably the chloral hydrate serves to attenuate the yellow colouration produced by the chromic liquids.

This is a process particularly adapted to the preparation of *large* segments of brain. The consistence is very tough and firm.

667. Entire Encephalon (Deecke, *Journ. Roy. Mic. Soc.*, 1883, p. 449).—"To harden the entire brain so that the inside and the outside shall be hardened equally and properly, Dr. Deecke finally adopted bichromate of ammonia in $\frac{1}{2}$ to 1 per cent. solution, according to the consistence of the brain. When nominally soft he adds say $\frac{1}{8}$th to $\frac{1}{10}$th per cent. of chromic acid to the solution, and always $\frac{1}{6}$th to $\frac{1}{4}$th of the whole volume of alcohol. It is then placed in a refrigerator and the fluid changed frequently. After a month add a little more alcohol from week to week until the alcohol is 90 per cent. This is changed as often as it is discoloured. The treatment requires from twelve to eighteen months."

668. Betz's Methods (*Arch. f. mik. Anat.*, 1873, p. 101).— The *spinal cord, medulla oblongata,* and *pons Varolii* are treated as follows :—The dura mater is removed, and they are hung up in a cylinder containing 75 to 80 per cent. alcohol, to which is added enough iodine to produce a light-brown colouration. After from one to three days the preparation will be found to be somewhat surface-hardened; it is taken down, and the pia mater and arachnoid are removed. If the pia mater does not come away completely enough the preparation is put back for some days into the alcoholic iodine. The membranes having been removed, the preparation is put back into the original fluid, which is found to have become colourless owing to absorption of the iodine by the tissues. Fresh quantities of a strong solution of iodine in alcohol are from time to time added to the liquid in order to keep it at its original strength of iodine (as shown by the colour). If the membranes have been carefully removed it will be found that after about six days the preparation ceases to take up further quantities of iodine. The preliminary hardening may now be considered complete.

The preparation is now brought into a 3 per cent. solution of bichromate of potash. (A small weight is attached to it to prevent any portion of it from floating above the surface of the liquid. After a day or two it will have lost much of its alcohol, and will sink to the bottom of the vessel, which is equally undesirable; this must be watched for, and the preparation hung up or otherwise supported.) The vessel is put away *in a cool place*. As soon as a brown turbidity is seen in the liquid, together with a brown deposit on the preparation, the hardening may be considered to be complete. The preparation must be at once washed with water, and put away until wanted in a ½ to 1 per cent. solution of bichromate.

Cerebellum.—Must be quite fresh, and before placing in the iodine the membranes and vessels must as far as possible be very carefully removed. (If the pia mater does not come away freely, the organ must be macerated for a few hours in iodine solution in which other preparations have been kept, and which is diluted before using for this purpose). The membranes having been removed, the cerebellum is placed (supported on cotton-wool, with which the different organs are so propped up as to preserve their natural position) in solution of iodine for two or three days, and fresh iodine solution frequently added.

The pia mater is now removed from the rest of the preparation, which is put back for seven to fourteen days into the iodine solution. If at the expiration of this time it be found that the cerebellum can be supported on the finger by the vermiculus alone without bending, the preliminary hardening is complete, and it is brought into a 5 per cent. solution of bichromate, where it remains until fit for cutting.

Cerebrum.—The cerebrum is divided into two halves along the median line of the corpus callosum, and put into the iodine solution. After a few hours the pia mater is removed from the fissure of Sylvius and from the corpus callosum, and if possible, the choroid plexus is removed likewise.

The preparation is now put away in the iodine solution in a cool place (in summer in a cool cellar), and fresh iodine added as soon as the liquid is seen to lose colour (which must be watched for). After twenty-four to forty-eight hours the remaining pia mater is carefully removed by means of scissors and forceps from the fissures and convolutions, and one half-volume of fresh iodine solution is added to the liquid. (To facilitate the penetration of the liquid, wads of cotton-wool are stuffed into the fissure of Sylvius, between the operculum and the median (central) lobe, in the direction of the descending cornu, and between the convolutions). After twenty-four to seventy-two hours the brain is brought into fresh solution of iodine in 70 per cent. alcohol, where it remains until the hemispheres are hard enough to be supported on two fingers without bending. (This will not be before ten to fourteen days). It is then put into 4 per cent. solution of bichromate and left to acquire its definitive hardness. If an excessive brown deposit make its appearance, and the brain be found notwithstanding to be not hard enough for cutting, it must be rinsed with water and the bichromate solution changed. When ripe for cutting the brain ought to show an almost equal intensity of yellow-brown stain over the whole surface of a cut made through the total thickness of a hemisphere.

Brains that are not fresh require for hardening longer time and stronger alcohol.

Instead of the iodine solution, it is possible to use for the preliminary hardening a mixture of equal volumes of chloroform and ether; but this mixture is not to be recommended, on account of its solvent action on protoplasm and on the processes of ganglion-cells.

The methods of Betz are particularly adapted to the hardening of voluminous specimens, and of tissues that are in a state of post-mortem softening.

669. Osmic Acid (EXNER, *Sitzb. k. Akad. Wiss. Wien,* 1881, lxxxiii, 3 Abth.; BEVAN LEWIS, *The Human Brain,* p. 105).—A small portion of brain, not exceeding a cubic centimetre in size, is placed in ten times its volume of 1 per cent. osmic acid. The solution should be replaced by fresh after two days, a proceeding which may advantageously be repeated at the end of the fourth day. In from five to ten days the piece is usually hardened throughout, and may be washed with water, treated with alcohol and imbedded. The sections may be treated by a drop of caustic ammonia, which clears up the general mass of the brain substance, leaving medullated fibres black. B. Lewis says that this method exhibits a wealth of structure which no other method displays. The sections may be mounted in soluble glass. The chief value of this method is for tracing the course of medullated fibres.

BELLONCI (*Arch. Ital. de Biol.,* vi, p. 405) has been recently employing this method in his researches on the optic nerve of mammalia. He employs an osmic acid solution of 0·5 to 1 per cent., hardens for only fourteen to twenty hours, makes sections, and treats them for three or four hours with 80 per cent. alcohol, and then with ammonia.

670. GOLGI's Method. See below, § 673, *sub finem.*

671. Nervous Centres of Reptiles and Batrachia (MASON, *Central Nervous System of certain Reptiles*, &c.; WHITMAN's *Methods*, p. 196).
—Iodised alcohol, six to twelve hours; 3 per cent. bichromate, with a piece of camphor in the bottle, and to be changed once a fortnight until the hardening is sufficient (six to ten weeks).

Imbedding and Cutting.

672. The Methods of Imbedding.—The paraffin infiltration method can only be used for the smaller objects of this class. Human spinal cord (which is quite at the upper limit as regards size) can be properly penetrated with paraffin by taking the precaution of first cutting it up into slices of not more than 1 millimetre in thickness. The largest objects of this class, such as entire hemispheres of man, cannot be properly penetrated by any known imbedding mass; and the anatomist must be content with simple imbedding, a proceeding which is here of the greatest service. For intermediate objects— those whose size varies between that of a small nut and a walnut—it appears to me that no hesitation as to the proper course is possible; such objects should be treated by the collodion method, which is at once the safest, the most convenient, and the most advantageous as regards the ulterior treatment of sections.

BEVAN LEWIS recommends the ether freezing method for fresh brain. For hardened brain he recommends some of the old-fashioned wax-and-oil and other fatty mixtures, a doctrine at which I am surprised.

HAMILTON recommends freezing in the gum and syrup mass given above, § 308. He also recommends a method of penetrating with collodion, which is hardened in the usual way, the hardened mass being cut with an ice-and-salt freezing microtome (see *Journ. of Anat. and Physiol.*, 1887, p. 444; or *Journ. Roy. Mic. Soc.*, 1888, p. 1051).

For sections of the entire human brain, DEECKE (l. c.) proceeds as follows: —The brain to be cut is placed upon the piston of the microtome (a Ranvier model) and held *in situ* by several pieces of soft cork. It is then imbedded in a cast of paraffin, olive oil, and tallow, which, after it has become hard, is held in position by a number of small curved rods attached to, and projecting upwards from, the piston to the height of about an inch. Before cutting, and as it proceeds, the cast is carefully removed from around the specimen to the depth of about half an inch (which is easily done by the use of a good-sized carpenter's chisel), so that the knife never comes in contact with the cast.

Cutting is done under alcohol, the entire microtome being immersed in a copper basin. The sections are floated, with the aid of a fine camel-hair brush, on to sheets of glazed writing paper. They are removed thereon successively into staining, washing, and clearing fluids. After clearing, they are brought on the paper on to a slide, and the paper is gently pulled away from them; they are then mounted in chloroform- or benzol-balsam.

It should be noted that the membranes should *not* be removed from the brain; they present no obstacle to cutting if this is done with a slight sawing movement, or with a series of short cuts, instead of one sweep of the knife. By this plan the sections are much more perfect and uniform in thickness, and the loss in a series of from four to five hundred to the inch through the entire cerebrum of man may not amount to more than 2 or 3 per cent.

OSBORN (*Proc. Acad. Nat. Sci. Philadelphia*, 1883, p. 178, and 1884, p. 262; WHITMAN's *Methods*, p. 195) found advantage in employing Ruge's egg-mass (for the brain of Urodela). He recommends that the mass be injected into the ventricles.

The paraffin and collodion methods have already been sufficiently described in Part I.

Staining.

673. Methods for General Stains.—By a "general stain" is here meant one by which it is intended to demonstrate as far as possible all the histological elements of a preparation.

Ammonia-carmine is one of the very best of these stains. Beale's formula is a good one, especially where prolonged staining is required. But in view of the greatest precision of stain, the process of GIERKE (§ 137), or that of BETZ (§ 136), should be preferred.

Picro-carmine is also an excellent reagent. It has much the same action as ammonia-carmine, but gives a better demonstration of non-nervous elements.

BEALE (*Journ. Roy. Mic. Soc.*, 1886, p. 156) recommends that the spinal cord of the ox be stained by immersion *in toto* for twelve hours in picro-carmine. He states that he has found no process that allows of following ganglion-cell processes so far.

Chromic objects stain very slowly in both these media. Sections may, however, be stained with them in a few minutes if they be put into a watch-glass with the stain, and the whole be kept on a wire net over a water-bath heated to boiling point (OBERSTEINER).

UPSON (*Neurolog. Centralb.*, 1888, p. 319; *Zeit. f. wiss. Mik.*, v, 4, 1888, p. 525) employs a stain made by adding to 5 c.c. of strong Grenacher's alum-carmine 1 to 3 drops of phosphomolybdanic acid. In this, sections stain in from two to ten minutes.

He also employs a stain made by saturating alum-carmine with zinc sulphate. Sections stain in this in from half an hour to twelve hours. This is said to give very good differentiations of the elements of peripheral nerves.

A third method of the same author consists in staining with carminic acid,— and afterwards treating with a mordant. *See* the papers quoted.

FREEBORN (*Amer. Mon. Mic. Journ.*, 1888, p. 231; *Journ. Roy. Mic. Soc.*, 1889, p. 305) also recommends carminic acid, followed by treatment with 10 per cent. solution of chloride of iron.

Borax-carmine may be used. It is chiefly useful when employed for double-staining with indigo-carmine or an anilin blue to follow. I have obtained some superb stains with Seiler's borax-carmine and indigo-carmine process (§ 229). Recently (*Zeit. f. wiss. Mik.*, 1884, p. 566, and 1885, p. 349) MERKEL's mixture of borax-carmine and indigo-carmine (§ 230) has been strongly recommended by MAX FLESCH, who says that it gives extremely rich and instructive images.

DUVAL (*Journ. de l'Anat.*, 1876) speaks very highly of the double stain quoted *ante*, § 231.

Alum-carmine (Grenacher's or Csokor's) may be used as a nuclear stain (OBERSTEINER). The stain principally takes effect on non-nervous nuclei.

MAYER's *cochineal* is a very good re-agent for staining in the mass.

Hæmatoxylin may be used as a general stain. Bevan Lewis recommends the formula of Kleinenberg, or a formula that he attributes to Minot, which is essentially the same as Böhmer's. BERNHEIMER's formula, which is quoted in regard to the retina, § 645, is practically the same thing.

Alizarin has been recommended by BENCZUR (*see* THANHOFFER's *Das Mikroskop*; or *Zeit. f. wiss. Mik.*, 1884, p. 97). It is directed that sections be stained for twenty-four hours in a concentrated solution of alizarin in alcohol.

Anilin blue alone is useful for the demonstration of ganglion-cell processes, see ZUPPINGER, in *Arch. f. mik. Anat.*, 1874, p. 255.

Anilin blue-black was first recommended by SANKEY (*Quart. Journ. Mic. Sci.*, 1876, p. 69). He stained in a 0·5 per cent. solution, and, in order to obtain a differential stain, washed out for twenty to thirty minutes in solution of chloral hydrate. BEVAN LEWIS (*Human Brain*, p. 125) considers this to be one of the most valuable stains for nervous centres. He stains sections for an hour in 0·25 per cent. aqueous solution, and clears and mounts (in the case of brain or cord sections); for the cortex of the cerebellum, he washes out for twenty to thirty minutes in 2 per cent. chloral solution. SANKEY and STIRLING have also used anilin blue-black, in a much weaker solution, which Bevan Lewis does not recommend. VEJAS, however (*Arch f. Psychiatrie*, xvi, p. 200), obtained good results by staining from eighteen to twenty-four hours in a solution of 1 in 3000.

GIERKE (*Zeit. f. wiss. Mik.*, 1884, p. 379) was not able to obtain good results with anilin black procured in Germany, and finds that the treatment with chloral is injurious to the preservation of the tissues. MARTINOTTI (*Ibid.*, p. 478) comes to the same conclusion.

LUYS (*Gaz. Méd. de Paris*, 1876, p. 346) greatly recommends the anilin colour known as *Noir Colin*: he stains for three to four minutes in a 0·1 per cent. solution.

JELGERSMA (*Zeit. f. wiss. Mik.*, 1886, p. 39) finds that anilin blue-black gives excellent results, *provided that the English preparation of the colour be alone employed*. He makes solutions of 1·100, 1·800, and 1·2000, of which the first stains sections in a quarter of an hour, the second in five hours, the third in twelve hours. The stain takes effect on ganglion-cells and their processes, and on axis cylinders, but does not demonstrate neuroglia or connective tissue.

MARTINOTTI (l. c., 1884, p. 478) finds that *picro-nigrosin* gives very good results, especially for pathological objects. He stains for two or three hours or days in a saturated solution of nigrosin (§ 110) in saturated solution of picric acid in alcohol, and washes out in a mixture of 1 part of formic acid with 2 parts of alcohol until the grey matter appears clearly differentiated from the white to the naked eye.

Last, but not least, as a general stain we have the *sublimate method of* GOLGI (*Archivio per le Scienze Mediche*, 1878, p. 3). —This method, which may be said to be in principle identical with the bichromate of potash and silver nitrate method of the author, consists, like the latter, of two processes : 1, hardening in bichromate; 2, treatment with bichloride of mercury.

For hardening, use either a solution progressively raised in concentration from 1 per cent. to 2½ per cent., or Müller's solution. Take small pieces of tissue (not more than 1 to 2 c.c), large quantities of liquid, and change the latter frequently so as to have it always clear. Fifteen to twenty days' immersion will suffice, but twenty to thirty should be preferred.

The tissues are then passed direct from the bichromate into the bichloride of mercury. The solutions of the latter employed by Golgi varied from 0·25 per cent. to 0·50 per cent.: he cannot say which strength is to be preferred. The immersion in the bichloride must be much longer than the immersion in the nitrate of silver bath of that process; for the latter, twenty-four to forty-eight hours suffice; but in the bichloride, an immersion of eight to ten days is necessary in order to obtain a complete reaction through the whole thickness of the tissues. During the bath, the bichromate will diffuse out from the tissues into the bichloride, which must be changed every day; at the end of the reaction the preparations will be found decolourised and offering the aspect of fresh tissue. They may be left in the bichloride for any time.

Before mounting, the sections that have been cut must be repeatedly washed with water (if it be wished to mount them permanently), otherwise they will be spoilt by the formation of a black precipitate. Mount in balsam or glycerin; the latter seems the better preservative medium.

The result of this process is not a true stain, but an "apparently black reaction;" the tissues appearing black by transmitted light, *white* by reflected light. Golgi thinks that there is formed in the tissue elements a precipitate of some substance that renders them *opaque*. The elements acted on are (1) the ganglion-cells, with all their processes and ramifications of the processes. These are made more evident than by any other process except the bichromate and silver-nitrate process. An advantage of the mercury process is that it demonstrates nuclei, which is not the case with the silver process. (2) Connective-tissue corpuscles in their characteristic radiate form. But the reaction in this case is far less precise and complete than that obtained by the silver process. (3) The blood-vessels, and particularly their muscular-fibre cells.

The method gives *good* results only with the cortex of the cerebral convolutions, hardly any results at all with the spinal cord, and very scanty results with the cerebellum. And, on the whole, the method shows nothing more than can be demonstrated by the silver-nitrate method, but it is superior to it as regards two points : the reaction can always be obtained with perfect *certainty* in a certain time ; and the preparations can be perfectly preserved by the usual methods.

TAL (*Gazz. degli Ospitali*, 1886, No. 68) finds that if sections made by this process be treated with solution of sodium sulphide, a much darker stain is obtained. Sections may then advantageously be double-stained with Magdala red.

MAGINI (*Boll. Accad. Med. di Roma*, 1886; *Zeit. f. wiss. Mik.*, 1888, p. 87) recommends a development of Golgi's process in which zinc chloride is used in place of sublimate. Portions of tissue of 2 to 3 centimetres cube are hardened for at least two or three months in Müller's solution. They are well washed with distilled water, and brought into a 0·5 to 1 per cent. solution of chloride of zinc. This is changed for fresh every day for seven to ten days (until it does not become yellower than bichromate solution). Sections are then made, washed quickly with alcohol, imperfectly cleared (*see* § 677) with kreasote, and mounted in dammar. This process is said to demonstrate better than Golgi's the finer structure of ganglion-cells and their processes.

Golgi's method may be combined with Weigert's nerve stain (*see* PAL, *Wien. med. Jahrb.*, 1886 ; *Zeit. f. wiss. Mik.*, v, 1, 1887, p. 93).

674. Special Stains.—For the study of medullated nerve-tracts, we have, first and foremost, WEIGERT's *hæmatoxylin* process (§ 650).

PAL's modification of the process (§ 651) is said to be very good.

KULTSCHITZKY (*Anat. Anz.*, 1889, p. 223; *Zeit. f. wiss. Mik.*, vi, 2, 1889, p. 196) proposes the following process, as being simpler than Weigert's and giving similar results :—Specimens are hardened in solution of Müller or Erlicki, and imbedded in celloidin. Sections are made and stained for a few hours (up to twenty-four) in a stain made by adding 1 gramme of hæmatoxylin dissolved in a little alcohol to a mixture of 20 c.c. of saturated aqueous solution of boric acid and 80 c.c. of distilled water. A little acetic acid (two to three drops to a watch-glass) is added to the stain before using. It is well to treat the sections after staining for twenty-four hours with saturated solution of sodium or lithium carbonate. They should then be washed with alcohol and mounted in balsam.

A still simpler hæmatoxylin stain, which is said to give the same results, is composed of 100 c.c. of 2 per cent. acetic acid, with 1 gramme of hæmatoxylin dissolved in a little alcohol.

ROSSI (*Zeit. f. wiss. Mik.*, vi, 2, 1889, p. 182) also recommends a process stated to be simpler than Weigert's. The tissues are hardened in a liquid composed of:—Water, 100 c.c.; chromic acid, 0·75 g. to 1 g.; acetate of copper, 5 g. Human cord requires six to eight days; the cord of the dog, three to four; brain of dog, fifteen to eighteen. They are then dehydrated and imbedded in celloidin. Sections are stained in a liquid made by adding seven to eight drops of a 5 per cent. solution of hæmatoxylin in absolute alcohol to about 30 c.c. of alcohol. After two hours they are brought into a mixture of eight drops of hydrochloric acid with 100 c.c. of absolute alcohol, and washed therein until the white substance is seen to be differentiated from the grey. They are washed out for twenty minutes or more in water, dehydrated, and mounted. The sections may be double-stained with borax-carmine.

FREUD's *gold method* (*Arch. f. Anat. u. Phys.*, 1884, p. 453) may be used for controlling the results obtained by Weigert's process, and for obtaining an impregnation of axis cylinders. This method is based on that of FLECHSIG (*see* § 211), which I suppress, as it gives less certain results. Freud proceeds as follows:—Sections of material hardened in solution of Erlicki (which may be followed by alcohol without hindrance to the impregnation) are washed with water and put for three to five hours into 1 per cent. solution of gold chloride. After again washing with water, they are treated for three minutes with a solution of 1 part of caustic soda in 5 or 6 of water. They are then drained (not washed) and brought into 10 or 12 per cent. solution of iodide of potassium. After from five to fifteen minutes therein, they are washed with water, dehydrated, and mounted in balsam. (In the case of objects that stain easily, it is useful to dilute the gold solution with a volume or two of alcohol—the stain is more highly selective.)

BECKWITH (*Journ. Roy. Mic. Soc.*, 1885, p. 894) modifies this process by intercalating a treatment for thirty minutes with 10 per cent. solution of carbonate of potash between the soda and the iodide.

The gold method is one of the most important methods for the study of the finer nerve fibrils of the spinal cord. The instructions given by GERLACH (*Stricker's Handb.*, p. 678) have been already quoted, *ante*, § 211, p. 117.

BOLL (*Arch. f. Psych. u. Nervenkr.*, iv, p. 42; GIERKE, *Zeit. f. wiss. Mik.*, 1884, p. 403) makes the following observations on Gerlach's process:—The hardening in bichromate ought not to be prolonged further than is absolutely necessary, for after eight days the elective susceptibility of the tissues for gold impregnation begins to diminish, and after fifteen days has almost disappeared. The knife should not be wetted with alcohol; the tissues should never be allowed to come into contact with alcohol. Only a small quantity of gold solution, relatively to the volume of the sections, should be taken. The impregnation should last for twelve hours.

Gierke, l. c., says:—"Gerlach's gold preparations have never had their equal for the demonstration of very fine nerve-fibrils. Weigert's Säurefuchsin preparations are the only ones that can be compared with them in this respect."

SCHIEFFERDECKER (*Arch. f. mik. Anat.*, 1874, p. 472) also recommends gold chloride for demonstrating fine networks in *transverse* sections. Strength 1·5000 or 10,000. Time about one to three hours. After washing in water the sections are put for twenty-four hours into acetic acid of ½ to 1 per cent. then mounted in balsam.

According to the same author, *chloride of palladium* may advantageously be used for the demonstration of the *longitudinal* fibres (and, of course, therefore) for staining longitudinal sections. A solution of 1·10,000 strength is taken; the sections remain in it till light brown (about three to five hours).

For some further methods of the same author, see l. c., 1878, p. 38.

WEIGERT's *Säurefuchsin* method mentioned above appears to be definitively superseded by his hæmatoxylin method. See, however, *Centralb. f. d. med. Wiss.*, 1882, pp. 753, 772; *Fortschr. d. Med.*, 1884, Nos. 4 and 6; *Zeit. f. wiss. Mik.*, 1884, pp. 123, 290.

SAHLI (*Zeit. f. wiss. Mik.*, 1885, p. 1) gives the following method:—Sections of tissue hardened in bichromate to the degree required for Weigert's hæmatoxylin process are washed for not more than five or ten minutes in water, and stained for several hours, until they are of a dark blue colour, in concentrated aqueous solution of methylen blue. They are then rinsed with water and stained for five minutes in saturated aqueous solution of Säurefuchsin. If now they be rinsed with alcohol and brought into a liberal quantity of water, the stain becomes differentiated, axis cylinders being shown coloured red and the myelin sheaths blue. If, instead of rinsing with pure alcohol, alcohol containing from 0·1 to 1 per cent. of caustic potash be taken, the stain differentiated in water, and the sections cleared with cedar oil and mounted in balsam dissolved in cedar oil, still finer images are obtained. Axis cylinders are red as before, but the myelin sheaths are blue in some places, red in others. Sahli thinks that this reaction points to some difference of kind in the nerve-tubes that exhibit it.

The same author (l. c., p 50) also gives a method for obtaining a specific stain of nerve-tubes by means of *methylen blue alone*. Sections of material hardened as before are stained for a few minutes or hours in the following liquid:

Water . . . . 40 parts.
Saturated aqueous solution of methylen blue . 24 „
5 per cent. solution of borax . . . . 16 „
(Mix, let stand a day, and filter).

The sections are then washed either in water or alcohol until the grey matter can be clearly distinguished from the white, are cleared with cedar oil, and mounted in balsam. Nerve-tubes are stained blue, ganglion-cells greenish, nuclei of neuroglia blue. Micrococci are stained, if any be present in the tissues. The preparations are not perfectly permanent.

Safranin followed by methylen blue gives a very special stain of spinal cord. The method is due to ADAMKIEWICS (*Sitzb. k. Acad. Wiss. Wien, Math. Naturw. Kl.*, 1884, p. 245; *Zeit. f. wiss. Mik.*, 1884, p. 587). Sections are washed first with water, then in water acidified with a little nitric acid, and stained in concentrated solution of safranin. They are then treated with alcohol and clove oil till no more colour comes away, and are brought back again into water, washed in water acidified with acetic acid, stained in methylen blue, and cleared as before. The process is said to demonstrate the existence of "chromoleptic zones" which surround the grey matter. Myelin (" erythrophilous substance " of Adamkiewics) is red, nuclei of nerves, of neuroglia, and of vessels, violet. The erythrophilous substance of pathological nerve-tubes does not take the stain, so that the method is valuable for the study of degenerative changes.

NIKIFOROW (*Zeit. f. wiss. Mik.*, v, 3, 1888, p. 338) modifies the foregoing method as follows. The tissue is to be hardened in a chrome salt. It is not to be washed in water afterwards. Sections are brought direct from alcohol into the safranin stain, either concentrated aqueous, or Babes's solution, or a solution in 5 per cent. carbolic acid. Stain for twenty-four hours. Bring the sections into alcohol until the colour is sufficiently extracted for the grey substance to be distinguished from the white. Then put them into a weak (0·1 to 0·2 per cent.) solution of chloride of gold or platinum or any metallic salt. They are to be left in this until the grey substance just begins to get a violet tone (not longer). They are then well washed in water, and put into alcohol until the rosy violet of the grey substance becomes distinctly marked off from the red of the white substance. Clear with clove oil, wash out the clove oil thoroughly with xylol (this is absolutely necessary for the permanence of the preparations), and mount in balsam. Nikiforow says that this method gives much sharper and more certain results than that of Adamkiewics.

Purpurin, according to DUVAL, has a special action on nerve-tissue (especially spinal cord) that has been hardened in bichromate of ammonia. Nerve-cells and their processes, axis cylinders and connective-tissue fibrils remain unstained; whilst the nuclei of connective tissue and of capillaries stain red (*see* § 189).

Mounting.

675. General Mounting Methods.—Under this head it is to be noted that it is often advisable to fix sections to the slide before applying balsam, or other medium, and a cover. For celloidin sections you may use one of the methods described in §§ 325-328, or for large sections the caoutchouc or gutta-percha method of Threlfall or Frenzel, or the gelatin method of Fol, or Minot's shellac method. This is as follows: You arrange the sections on the slide, and cover them with a layer of 10 to 12 per cent. solution of shellac, then warm to 30° or 40° C., until the shellac appears dry, clear with clove oil, and mount in balsam (*Zeit. f. wiss. Mik.*, 1886, p. 175). This process is applicable either to paraffin or collodion sections.

676. Serial Section Mounting.—Besides the ordinary methods for mounting small sections, there are certain processes specially adapted to the mounting of very voluminous sections. One of these is GIACOMINI's collodion-gelatin method, which is certainly a valuable one. But it is hardly more than a macroscopic method, principally adapted for the study of the coarser topography of the brain, and as such may be passed over here, though I would willingly include it did our shortening space allow. Descriptions of the process are to be found in *Gazzetta delle Cliniche*, Nov. 1885; *Zeit. f. wiss. Mik.*, 1885, p. 531; and *Traité des Méth. Techn.*, LEE et HENNEGUY, p. 392.

A further reason for passing over this process is that the same end may be attained, in many cases more effectually, by the beautiful methods of WEIGERT (§ 327), STRASSER (§ 316), and APÁTHY (§§ 326, 326a), which *see*.

A useful method for small sections is that of OSBORN, who arranges his sections, wet with alcohol, in order on the slide, covers them with a cigarette paper, and treats them with the clearing agent through the paper.

For DEECKE's method, *see* above, § 672.

Other Methods for Nervous Centres.

677. The Half-Clearing Method.—Since the days of Lockhart Clarke, to whom the method is due, most observers have been struck by the wealth of minute structure revealed by half-cleared sections. Detail that is invisible in perfectly cleared sections stands out in sculptural relief during the period in which the tissues are only about half cleared. It is of course most important to be able to preserve preparations permanently in this state.

MERKEL (*Arch. f. mik. Anat.*, 1877, p. 622) gives the following instructions:—Sections are dehydrated with alcohol of about 94 per cent. (in which they must remain for at least ten minutes), and then cleared with xylol, in which they are examined.

Certain elements of the tissues retain more obstinately than others the small quantity of water that they bring with them out of the 94 per cent. alcohol. Now, as xylol is absolutely immiscible with water, it can exercise no clearing action on these, and they stand out boldly in the picture by virtue of the difference between the index of refraction of their contained water and that of the xylol. The axis-cylinders are at first all that is visible; after a time the ganglion-cells appear with their processes. Nuclei, vessels, and whatever else may be in the preparation, are totally invisible. Merkel thinks this method superior to all others for the study of the distribution of nerve-fibres. The preparations are not permanent, though they may be kept for some weeks by mounting them in Canada balsam. When a preparation (either in xylol or balsam) has become so transparent as to be of no further use (which will always eventually happen), it may be reprepared by putting it back into the alcohol, and thence again into the xylol.

BEVAN LEWIS (*The Human Brain*, p. 122) proceeds as follows :—A section is placed, saturated with spirit, on a slide. When the spirit has nearly all evaporated, a drop of oil of anise is allowed to flow *over* the section (not to float it up), and the clearing is watched on the stage of the microscope; then, just when the desired image is arrived at, a drop of balsam is allowed to run over the section, and a cover put on.

Instead of oil of anise, Lewis has frequently employed glycerin followed by mounting in glycerin jelly, with the same results.

BYROM BRAMWELL (*Edinb. Med. Journ.* Oct., 1886) also recommends the process. He uses clove oil.

678. GIACOMINI's "Dry" Process for preserving Brains (*Arch. per le Scienze Mediche*, 1878, p. 11).—Although this is in intention a *macroscopic* method, it appears worth while, both on account of its thorough success and on account of its suggestiveness, to give an account of it here.

The object is to make "dry" preparations of the encephalon; by which is meant preparations that can be permanently preserved *in the air*. The methods hitherto employed were not successful because they consisted in making preparations that were "dry" in the literal sense of the word, that is, deprived of their natural water; and since brain-substance contains 88 per cent. of water, such preparations could not of course be obtained without so great an amount of shrinkage as to most seriously diminish the scientific value of the result. The principle of Giacomini's method is, on the contrary, to *retain* the natural water of the tissues, or an equivalent for it, by means of impregnation with a hygroscopic substance,—glycerin.

The process consists of two divisions: 1, hardening; 2, impregnation with glycerin.

1. For hardening may be used, zinc chloride, bichromate of potash, chromic acid, nitric acid, or alcohol.

Chloride of zinc gives the best results. Perfectly fresh brain is put into a saturated aqueous solution of the salt (if there be reason to fear that the tissues are somewhat softened through having been left too long after the death of the subject, it is well first to inject 000 grammes of the solution through the internal carotid arteries). After forty-eight hours' immersion (during which time the floating brain must be turned over three or four times, so that all parts of it may duly come into contact with the liquid) the surface of the brain will have attained a consistency that will allow of the removal of the arachnoid and pia mater. The meninges having been removed, the encephalon is put back into the solution for two or three days more, during which time it will be seen that, increasing in specific gravity, it tends towards the bottom of the vessel containing it. When this is seen to happen, it must be removed into commercial alcohol, as, if allowed to remain longer in the chloride of zinc solution it would take up too much water.

In the alcohol it may remain for an indefinite time, or it may be removed if desired after ten or twelve days. (During the alcohol-bath, it must be frequently turned over in order that no malformation may arise from continuance of pressure on the same part.)

It is then removed into glycerin (either pure or with 1 per cent. of carbolic acid). It floats at first, but gradually sinks as the alcohol evaporates. As soon as it has sunk just below the surface it may be removed and exposed to the air.

It is set aside to "evaporate" in a convenient place for a few days. As soon as the surface has become dry, it is varnished with india rubber or (better) with marine glue varnish diluted with a little alcohol. This completes the process.

If it be desired to make dissected preparations, the necessary dissection should be made on removing the encephalon from the alcohol before putting into glycerin.

Bichromate of potash may be used for hardening in solutions gradually increasing in concentration from 2 to 4 per cent. The liquid must be frequently changed, the immersion must be of not less than a month's duration. Six to eight days will suffice for the alcohol-bath, or this may be altogether omitted.

Nitric acid is used in solutions of from 10 to 12 per cent. for twelve to fifteen days. (Encephala float in this liquid, and must therefore be frequently turned over. It is this reagent that gives the *toughest* preparations.)

Concerning the value of the process, Golgi (from whose abstract I take the foregoing account) states that after a series of experiments he is able to affirm that for preservation of the volume, the colour, the finer relations of the parts, and the physiognomy proper to the organ, the process is far superior to any hitherto known. I am able to add that I saw specimens of Giacomini's preparations at the Milan International Exhibition of 1881, and think it would be hard to overpraise their beauty of aspect.

It should be added that histological detail is preserved to a remarkable extent by this process, and that excellent sections may be cut at any time from the hardened brains. And as the preparations take up as little room as possible, there seems no reason why the process should not be generally adopted in medical schools, lunatic asylums, and similar institutions.

The method may also be applied, with most perfect success, to the preservation of small animals entire, such as Batrachia, Reptilia. It is well to inject them with the zinc solution.

Another "dry" method has been given by MAX FLESCH (*Mt. Naturf. Ges. Bern*, 1887, p. xiii). He hardens in alcohol, and then brings the brains through Calberla's mixture (§ 385) into glycerin containing (1 part to 3000 of) sublimate. He does not appear to varnish his preparations. See *Journ. Roy. Mic. Soc.*, 1888, p. 507.

679. Dissociation methods.—Maceration is a frequently useful process for the study of nervous tissue.

STILLING'S method (*Arch. f. mik. Anat.*, 1880, p. 471;

"*Bau. d. nervösen Centralorgane,*" 1882) is as follows: Tissue hardened in Müller's solution, washed out, and dehydrated, is put for several weeks into pyroligneous acid. The best acid for this purpose is an "artificial pyroligneous acid," composed of—

   Glacial acetic acid . . . . 100 grammes.
   Water 800   „
   Kreasote . . . . . xxx drops.

CARRIÈRE (*Arch. f. mik. Anat.*, 1877, p. 126) hardens spinal cord for ten days in bichromate of potash or chromate of ammonia of 1 to 600, and macerates for from three to five days in strong solution of ammonia-carmine.

SCHIEFFERDECKER (*Ibid.*, 1878, p. 38) gives the following method: Pieces of spinal cord about half a centimetre thick are macerated in a small quantity (just enough to cover them) of Ranvier's alcohol (one-third alcohol) for several days. Small fragments of the grey matter are then taken and well shaken in a test-tube with a small quantity of water. There is then added a little glycerin and a few drops of concentrated solution of picro-carminate of soda, and the whole is set aside for one or two days. Decant, and to the red deposit, which now consists chiefly of stained ganglion-cells, add one or two drops of glycerin, and place the whole for two days in a dessicator with sulphuric acid. (This part of the operation is best performed in a watchglass, or, better, flat-bottomed cell.) The cells are best got on to a slide by pouring a drop of the dehydrated glycerin on to it.

FREEBORN (*Amer. Mon. Mic. Journ.*, 1888, p. 231; *Journ. Roy. Mic. Soc.*, 1889, p. 298) gives the following: Thin slices of spinal cord or cerebellum not over one sixteenth of an inch thick are placed in fifty times their volume of 5 per cent. aqueous solution of potasssium chromate for twenty-four hours. At the end of this time the grey matter has become jelly-like and transparent, and then, having been cut away from the white, is placed in a long narrow tube. Mohr's burette with the lower end plugged with a cork answers the purpose perfectly. The burette is then filled up to within an inch of the top with fresh macerating fluid, and a cork forced in until it comes within half an inch of the surface of the fluid. The burette is then inverted, and this manipulation is repeated at intervals of half an hour until the bits of tissue are reduced to powder. The burette is then placed upright, and when the material has all settled the fluid is poured off. The material is then carefully washed with distilled water by repeated decantation, and finally poured into a conical glass burette. The water is then poured off, and the material stained with picro- or ammonia-carmine. This, which takes from twelve to fifteen hours, is followed by preservation in a mixture of 1 part spirit and 3 parts glycerin. By this method cells from spinal cord and cerebellum may be obtained with their processes attached down to the fourth division.

BEVAN LEWIS's **Compression Method for Fresh Tissue** (*Mon. Mic. Journ.*, 1876, p. 106; *Human Brain*, p. 146) is as follows: Vertical sections, as thin as possible, are made from a piece of a convolution of brain from which the membranes have been removed. The sections are made by free hand by means of a section-knife flooded with spirit. The sections are got on to a slide and treated with Müller's solution for a few seconds. A cover is then put on and steadily pressed down so as to flatten out the sections into an almost transparent film. The slide is then rinsed in water and placed for thirty to forty seconds in a bath of methylated spirit. It is then removed, one edge of the cover-glass steadied with the finger, the blade of a penknife inserted under the opposite edge, and the cover gently lifted. The section, which remains adherent either to the slide or to the cover, is washed with water, stained in any way that may be desired, dehydrated, and mounted in balsam.

For staining the cells of the cortex, Lewis prefers a 1 per cent. solution of anilin-black.

He prefers to dehydrate by drying under a bell-glass in the presence of concentrated sulphuric acid. Clear with chloroform in preference to clove oil. Lewis finds that this process results in far less rupture and tearing of nerve-cells and processes than would be imagined; he considers that "the result is equivalent to the most delicate teasing of tissue, the processes being gradually unravelled from their dense networks, and the structural elements universally displayed to the best advantage."

v. THANHOFFER (*Zeit. f. wiss. Mik.*, iv, 4, 1887, p. 467) describes a similar process.

680. Neuroglia (GIERKE, *Arch. f. mik. Anat.*, 1885, p. 444). —Maceration in the usual media, chromic solutions, iodised serum, &c., but especially in the liquid of Landois (§ 505). For sections, harden for six to ten weeks in bichromate of ammonia, first of 1·5 per cent., and gradually raised to 3 per cent. Avoid paraffin for imbedding, and imbed in celloidin or cut without imbedding. Stain with ammonia-carmine, or picro-carminate of soda, or alum-carmine, or Heidenhain's hæmatoxylin. Other details in this valuable paper.

681. Myelon of Reptilia and Amphibia.—Besides the usual methods which will suggest themselves as being applicable to this class of objects, see MASON's methods, *supra*, § 671; SCHMIDT, *Zeit. f. wiss. Mik.*, 1885, p. 389; v. LENHOSSEK, *Arch. f. mik. Anat.*, 1886, p. 379 (minute directions for dissecting out the spinal ganglia of *Rana*); KORYBUTT-DASZKIEWICS, *Ibid.*, 1889, p. 51; or *Zeit. f. wiss. Mik.*, vi, 2, 1889, p. 203; OSBORN, *Journ. Roy. Mic. Soc.*, 1885, p. 536, or WHITMAN's *Methods*, p. 195).

681 a. APÁTHY's **double stain** for differentiating nervous tissue and connective tissue (*Zeit. f. wiss. Mik.*, vi, 2, 1889, p. 170) consists of staining with Heidenhain's hæmatoxylin, and after-staining with very weak alum hæmatoxylin.

CHAPTER XXXI.

SOME OTHER HISTOLOGICAL METHODS.

Connective tissues.

682. Connective Tissue.—S. MAYER (*Sitzb. k. Akad. Wiss.*, lxxxv, 1882, p. 69) recommends, for staining *fresh* tissue, a solution of 1 gramme of "Violet B." (Bindschedler and Busch, Bâle) in 300 c.c. of 0·5 per cent. salt solution. In this liquid connective tissue cells stain rapidly and energetically. Elastic fibres and smooth muscle also stain, but of different tints.

FREEBORN (*Amer. Mon. Mic. Journ.*, 1888, p. 231 : *Journ. Roy. Mic. Soc.*, 1889, p. 305) recommends (for sections) picro-nigrosin, made by mixing 5 c.c. of 1 per cent. aqueous solution of nigrosin with 45 c.c. of aqueous solution of picric acid. Stain for three to five minutes, wash with water and mount in balsam. Connective-tissue fibres bright blue, nuclei blackish, all the rest greenish-yellow. Sections may be after-stained for five or six minutes in a mixture of 1 c.c. of saturated alcoholic solution of eosin and 49 c.c. of alcohol. The results are as before, except that the yellow colour is replaced by red.

DOGIEL (*Anat. Anz.*, 1887, p. 139 ; *Zeit. f. wiss. Mik.*, 1887, p. 86) puts tendons from the tail of the rat for some hours, days, or weeks into Grenacher's alum-carmine. The tendon-bundles swell up and become transparent, the cells are well stained, and elastic fibres well brought out.

He stains subcutaneous connective tissue with concentrated solution of fuchsin diluted with one volume of water.

For RANVIER's method of artificial œdemata for the study of areolar tissue, see his *Traité*, p. 329.

683. Fat.—DEKHUYSEN (*see* FLEMMING, in *Zeit. f. wiss. Mik.*, 1889, pp. 39, 178) has discovered that fat that has been stained black by treatment with chromo-aceto-osmic acid (*not* with pure osmium)* is dissolved in the course of a few

* The statement made § 28, p. 19, line 9 from bottom, is therefore misleading.

hours in turpentine. It is dissolved also in xylol, ether and kreasote. Flemming finds that very good demonstration-preparations may be made by treating fatty tissue with chromo-aceto-osmic acid, staining with safranin or gentian, and then treating for a few hours with turpentine until all the fat is dissolved. The optical hindrance caused by the high refraction of the fat being thus eliminated, nuclei and cytoplasm may be studied to far greater advantage than in the usual preparations.

684. Granule Cells ("Plasmazellen" and "Mastzellen").—In 1874, there were described by WALDEYER (*Arch. f. mik. Anat.*, Bd. xi) certain special cells existing between the bundles of connective tissue, besides the flat cells, and lymphatic and fat cells. They are large round cells containing large granules; Waldeyer called them *plasma cells* ("Plasmazellen"). Later on, EHRLICH (*ibid.*, xiii, 1877) distinguished in the same tissue and in other places certain cells containing large granules, which have a superficial resemblance to Waldeyer's plasma cells, but which differ from them in staining reaction (*Verhandl. Berl. Physiol. Ges.*, January 17, 1879 : *Reichert u. Du Bois Reymond's Arch.*, 1879, p. 166). Ehrlich named these *food cells* (" Mastzellen "), intending to express thereby the opinion that these cells are derived from fixed connective tissue cells by a transformation brought about by exalted nutrition.

It is a pretty general character of these elements, that they take the stain of anilin colours and retain it on treatment with alcohol with greater energy than other tissue cells. And further (at least so far as regards the true Mastzellen of Ehrlich) that in successful preparations they show the nucleus unstained, the general mass of cytoplasm unstained or but slightly coloured, and, in the cytoplasm, the characteristic granules very intensely stained.

685. EHRLICH'S "**Mastzellen**" (*Arch. f. mik. Anat.*, 1876, p. 263).—*First method.*—Stain with neutral dahlia, and wash out with acidified water, and mount as directed, *supra*, § 103. If there be any Mastzellen in the preparation they will be brought out by the superior energy with which they take the stain.

Second method.—It may be desirable to obtain a *specific* stain of the plasma-cells alone, which may be done as follows :

The tissues must first be well hardened in strong alcohol (chromic acid and its salts must be avoided). They are then placed for at least twelve hours in a staining fluid composed of—

Absolute alcohol	50 c.c.
Aqua	100 c.c.
Acid. acet. glacial	12½ c.c.

to which has been added enough dahlia to give an almost saturated solution. After staining, the preparations are transferred to alcohol, which washes out the stain from all but the plasma-cells, and may then be mounted in the resin-turpentine solution.

The degree to which the colour is removed by this process of washing out depends on the degree of acidity of the staining fluid. "A solution that contained only 7½ c.c. of glacial acetic acid, yet stained the preparation with moderate intensity." Smaller proportions of acetic acid may therefore be taken for preparations in which there is much connective tissue, but for structures in which cells predominate the more strongly acidulated solutions are to be preferred.

Mucus-cells and fat-cells are also sometimes stained by these solutions.

Other media.—In a similar way other soluble anilins may be employed (in the form of a fluid containing 7½ c.c. of acetic acid),—primula, iodine-violet, methyl-violet, purpurin, safranin, fuchsin; of these, methyl-violet gives the best results.

686. Plasma-cells (KOBYBUTT-DASZKIEWICZ's **Methods**) (*ibid.*, xv, 1878, p. 7).—Frogs were kept for two months (in summer) without food, then placed in a reservoir of running water and well fed for four weeks. "Plasma-cells" were then found in abundance. The nerves were first treated with osmic acid of 1·200, and then stained with a slightly acidulated ammonia-carmine. Fuchsin gives the finest stains, but the colour is not permanent.

The best method for permanent preparations is to stain with dahlia, methyl-violet, fuchsin, or other anilin stains, dehydrate in alcohol, and mount in turpentine. Turpentine in which resin is dissolved is a very useful medium for teasing. It is well to first harden the tissues in osmic acid.

687. Plasma-Cells (NORDMANN, *Beitr. z. Kenntniss d. Mastzellen, Inauguraldiss.*, Helmstedt, 1884).—Nordmann finds it useful to employ a solution of vesuvin containing 4 to 5 per cent. of hydrochloric acid. Sections should remain for a few minutes in the solution, and then be dehydrated with absolute alcohol. The paper quoted contains a detailed discussion of the microchemical reactions of granule-cells.

688. Elastic Tissue.—Two of the most salient characters of elastic fibres are that they have a great affinity for osmium, staining with much more rapidity than most other tissue elements, and that they are not changed by caustic soda or potash. A further character is that they have a great affinity for certain anilin dyes, especially Victoria blue.

For a review of the methods of BALZER, UNNA, LUSTGARTEN,

and HERZHEIMER, *see* the paper by MARTINOTTI in *Zeit. f. wiss. Mik.*, iv, 1, 1887, p. 31.

LUSTGARTEN's process, which is simple, and gives very good results, consists in fixing with mixture of Flemming for twenty-four or forty-eight hours, washing with water, hardening with alcohol, making sections, and staining for twenty-four hours in alcoholic solution of Victoria diluted with 2 to 4 parts of water (*see* § 100). The colour used by him was called " Victoriablau 4 B," and this is probably an important detail. The sections are washed, treated with alcohol, and mounted in balsam. I believe it is essential that they should have been fixed in an osmic or chromic mixture; or if fixed with alcohol, they should be mordanted with chromic acid or osmium before staining.

But perhaps the best method is that of MARTINOTTI (l. c.). Fix in a chromic liquid, wash, stain for forty-eight hours in strong (5 per cent. Pfitzner's) solution of safranin, wash, dehydrate, clear, and mount in balsam. Elastic fibres are stained of an intense black, the rest of the preparation showing the usual characters of a safranin stain.

The staining will be performed quicker if it be done at the temperature of an incubating stove (GRIESBACH, *ibid.*, iv, 1887, p. 442). And FERRIA (*ibid.*, v, 3, 1888, p. 342) says that clearer preparations will be obtained if the sections be left for a long time, say twenty-four hours, in the alcohol, or be treated for a short time with very dilute alcoholic solution of caustic potash. This decolourises more completely the ground of the preparations.

For another somewhat complicated method of MARTINOTTI's, with nitrate of silver, which is, I think, of more theoretical than practical interest, see *Zeit. f. wiss. Mik.*, v, 4, 1888, p. 521.

For the methods of UNNA and TAENZER with nitrate of rosanilin salts, see *Monatsch. f. prakt. Dermat.*, vi, 1887; or BEHRENS, KOSSEL, u. SCHIEFFERDECKER, *Das Mikroskop*, 1, p. 204.

689. Bone, Non-Decalcified (RANVIER, *Traité*, p. 297).— Ranvier points out certain precautions that it is necessary to take in the preparation of sections of dry bone. In general, the bones furnished by "naturalists," or procured in anatomical theatres, contain spots of fatty substance that prevent good preparations from being made. Such spots are formed when bones are allowed to dry before being put into water for maceration; when a bone is left to dry the fat of the medullary canals infiltrates its substance as fast as its water evaporates.

Bones should be plunged into water as soon as the surrounding soft parts have been removed, and should be divided into lengths with a saw whilst wet. The medulla should then be driven out from the central canal by means of a jet of water; spongy bones should be submitted to hydrotomy. This may be done as follows:—An epiphysis having been removed, together with a small portion of the diaphysis, a piece of caoutchouc tubing is fixed by ligature on to the cut end of the diaphysis, and the free end of the piece of tubing adapted to a tap through which water flows under pressure.

As soon as the bones, whether compact or spongy, have been freed from their medullary substance they are put to macerate. The maceration should be continued for several months, the liquid being changed from time to time. As soon as all the soft parts are perfectly destroyed, the bones may be left to dry. When dry, they should be of an ivory whiteness, and their surfaces exposed by cutting of a uniform dulness.

Thin sections may then be cut with a saw and prepared by rubbing down with pumice-stone. Compact pumice-stone should be taken and cut in the direction of its fibres. The surface should be moistened with water and the section of bone rubbed down on it with the fingers. When both sides of the sections have been rubbed smooth in this way, another pumice-stone may be taken, the section placed between the two, and the rubbing continued. As soon as the section is thin enough to be almost transparent it is polished by rubbing with water (with the fingers) on a Turkey hone or lithographic stone. Spongy bone should be soaked in gum and dried before rubbing down (but *see* von Koch's copal process, *ante*, § 303, and Ehrenbaum's colophonium process, § 304).

Weil (*Zeit. f. wiss. Mik.*, v, 2, 1888, p. 200) gives a process similar to the foregoing, rough sections or portions of bone and teeth being penetrated with chloroform-balsam, which is hardened in an oven, or over a water-bath, and then ground.

Nealey (*Amer. Mon. Mic. Journ.*, 1884, p. 142; *Journ. Roy. Mic. Soc.*, 1885, p. 348) says that perfectly fresh portions of bone or teeth may be ground with emery on a dentist's lathe, and good sections, with the soft parts *in situ*, obtained in half an hour.

690. Bone, Decalcified (FLEMMING, *Zeit. f. wiss. Mik.*, 1886, p. 47). Sections of decalcified bone are made with the free hand. They are soaked in water, and brought in a drop of water on to a glass plate where they are spread out flat. The excess of water is removed with blotting-paper, and the sections are covered with another glass plate, to prevent them from rolling. The whole is brought into a plate and covered with alcohol. After the lapse of half an hour, the sections have become fixed in the flat position, and may be brought into absolute alcohol without risk of their rolling. To mount them, wash them with fresh alcohol (which may be followed by ether); lay them again flat on glass, and cover them with a double layer of blotting-paper and a somewhat heavy glass plate, and let them dry for a day in the air or in a stove. When they are dry, put a drop of melted balsam on a slide, and let it spread out flat and cool. Prepare a thin glass cover in the same way, put the section on the prepared slide, cover it with the prepared cover, put on a clip, and warm.

By this process, sections can be very expeditiously prepared, which show the lacunar system injected with air in quite as instructive a manner as non-decalcified sections.

KÖLLIKER (*Zeit. f. wiss. Zool.*, xliv, 1886, p. 662) recommends the following process for the demonstration of the "fibres of Sharpey" in decalcified bone. Sections are treated with concentrated acetic acid until they become transparent, and are then put for one quarter to one minute into a concentrated solution of indigo-carmine, then washed in water and mounted in glycerin or balsam. In successful preparations the "fibres of Sharpey" appear stained of a pale or dark *red*, the remaining bone-substance blue.

691. Cartilage (and Decalcified Bone).—For an excellent discussion (especially as regards staining) of the methods that have been recently recommended for these objects, see the exhaustive paper of SCHAFFER in *Zeit. f. wiss. Mik.*, v, 1, 1888, which gives in sufficient detail all the methods in question. The following appear to be the best:—

RANVIER's *purpurin method*. This has been given in detail in § 189.

SCHAFFER's *safranin method*. This method is due in its principle to BOUMA (*Centralb. f. d. med. Wiss.*, 1883, p. 866).

I give it in the form to which it has been brought by the careful study of Schaffer (*Zeit. f. wiss. Mik.*, v, 1, 1888, p. 17). Sections of bone decalcified with nitric acid (chromic acid may be used, but the stain will be less brilliantly contrasted) are stained for half an hour to one hour in 0·05 per cent. aqueous solution of safranin, washed with water, put for two or three hours in 0·1 per cent. solution of corrosive sublimate, and examined in glycerin. In order to make permanent preparations, the sections on removal from the sublimate are rinsed with alcohol, pressed on to a slide with filter paper, cleared *for a long time* in bergamot oil or clove oil, and mounted in xylol balsam.

This is a *double* stain; cartilage, orange; bone, uncoloured (or green in chromic objects); marrow, red.

BAYERL's *method for ossifying cartilage* (*Arch. f. mik. Anat.*, 1885, p. 35) is as follows: Portions of ossified cartilage are decalcified in a mixture of equal parts of 1 per cent. hydrochloric acid and 3 per cent. chromic acid. They are then washed for several days in distilled water, dehydrated, imbedded in paraffin, and cut. The sections are cleaned with turpentine and soaked in absolute alcohol. They are then stained in Merkel's borax-carmine and indigo-carmine mixture (§ 230) washed out with absolute alcohol, cleared with clove oil (or, better, with benzin) and mounted in balsam. In order to ensure a somewhat permanent stain the clove oil must be carefully removed with benzin before mounting, as clove oil oxidises the stain, and causes it to fade. For the characters of the stain, *see* § 230.

MAX FLESCH (*Zeit. f. wiss. Mik.*, 1885, p. 351) particularly recommends this process for the study of the development of dental tissue.

Blood.

692. Fixing Agents for Blood Corpuscles.—The most recent authors (BIONDI, MOSSO, MAX FLESCH) are agreed that by far the most faithful fixing agent for blood corpuscles is osmic acid. A drop or two of blood (Biondi recommends two drops exactly) is mixed with 5 c.c. of osmic acid solution, and allowed to remain in it for from one to twenty-four hours. The exact degree of concentration of the osmium solution is a somewhat important point, and must be made out by experiment for each form. As a rule, it should be strong, 1 to 2 per cent. According to Biondi, 2 per cent. is best. Fixed specimens may be preserved for use in acetate of potash solution (MAX FLESCH, *Zeit. f. wiss. Mik.*, v, 1, 1888, p. 83). The time-honoured process of drying drops of blood over a flame gives rise to great deformation of the elements, and should be abandoned.

693. Stains for Blood.—Blood prepared as above can be satisfactorily stained with many of the usual reagents, such as picro-carmine or hæmatoxylin.

Eosin stains rose-red all parts of blood-corpuscles that contain hæmoglobin (*see* WISSOWSKY, *Arch. f. mik. Anat.*, 1876, p. 479); parts that do not contain hæmoglobin, such as the nucleus, remaining unstained. This suggests double-staining with eosin and hæmatoxylin.

WISSOWSKY (l. c.) stains in a solution of equal parts of eosin and alum in 200 parts of alcohol, and then with hæmatoxylin.

MOORE (*The Microscope*, 1882, p. 73; *Journ. Roy. Mic. Soc.*, 1882, p. 714) stains for three minutes in a similar solution without the alum, washes, and stains for two minutes in a 1 per cent. aqueous solution of methyl-green. Red corpuscles, red; nuclei and white corpuscles, bluish-green.

MERKEL's carmine and indigo-carmine stain has been discussed above, §§ 230 and 691.

HARRIS (*Journ. Roy. Mic. Soc.*, 1885, p. 537) recommends "Spiller's purple" in 1 per cent. solution.

Fresh (unfixed) blood is perhaps best treated as follows (BIZZOZERO and TORRE, *Archivio per le Scienze mediche*, vol. iv, No. 18, 1880, p. 390): Dilute a drop of blood with 0·75 per cent. salt solution in which has been dissolved a little *methyl-violet*. This liquid in no wise affects the form of the elements, stains intensely the nucleus of the red corpuscles, and, in the white, stains the nucleus intensely, and the protoplasm less intensely. May be used for the study of bone-marrow and spleen.

For the staining of the blood-plates of BIZZOZERO, this observer (*Arch. f. path. Anat. u. Phys.*; *Zeit. f. wiss. Mik.*, 1884, p. 389) employs a 0·02 per cent. solution of methyl-violet in salt solution, or a 1 : 3000 solution of gentian-violet.

TOISON (*Journ. Sci. méd. de Lille*, fev., 1885; *Zeit. f. wiss. Mik.*, 1885, p. 398) recommends that blood be mixed with the following fluid:

Distilled water	160 c.c.
Glycerin (neutral, 30° Beaumé)	30 c.c.
Pure sulphate of sodium	8 grammes.
Pure chloride of sodium	1 gramme.
Methyl-violet 5B	0·25 ,,

(The methyl-violet is to be dissolved in the glycerin with one half of the water added to it; the two salts are to be dissolved in the other half of the water; and the two solutions are to be mixed and filtered.) White blood-corpuscles stain in this medium in five or ten minutes; the maximum of coloration is attained in from twenty to thirty minutes. White blood-corpuscles, violet; red blood-corpuscles, greenish.

694. Demonstration of Blood-plates of Bizzozero (KEMP, *Studies fr. the Biol. Lab. Johns Hopkins Univ.*, May, 1886, iii, No. 6; *Nature*, 1886, p. 132).—The mere demonstration of the blood-plates of Bizzozero is easy enough. A somewhat large drop of blood is placed on a slide, and quickly washed with a small stream of normal salt solution. The blood-plates are not washed away, because they have the property of adhering to glass; and on bringing the slide under the microscope they will be seen in large numbers. If it be desired to make permanent preparations of them, they should first be fixed. This is done by putting a drop of osmic acid solution on the finger before pricking it.

695. BIONDI's Section Method for Blood (*Arch. f. mik. Anat.*, xxxi, 1888, p. 103).—None of the foregoing methods are perfectly satisfactory as regards the preservation of the elements of blood without deformation, and at the same time in a perfectly permanent manner. Biondi's ingenious process does this.

Blood is fixed with osmic acid solution as described above, § 692. Four or five drops of the mixture of blood and osmium solution are then mixed with agar-agar jelly melted at 35° to 37° C. The whole is allowed to cool, and the mass is put to harden in alcohol of 85 per cent. After a few days the mass will have attained a consistence that allows of its being imbedded in pith and cut with a microtome. The sections are treated according to the usual methods. The best stains are obtained with methyl-green, methylen-blue, fuchsin, and safranin. Methyl-green and eosin is also a good combination. After staining the sections are cleared and mounted in balsam in the usual way.

Thinner sections can be obtained if the agar-agar mass be imbedded in paraffin by infiltration in the usual way, instead of pith.

Instructions, too long to be abstracted here, are given for the preparation of a suitable agar-agar mass. It may be obtained ready prepared from Herrn König, 29, Dorotheenstrasse, Berlin. Celloidin, and others of the usual imbedding masses, were tried, but without success.

For further details the English reader may consult the *Journ. Roy. Mic. Soc.*, 1888, pp. 313, 659.

This is undoubtedly a valuable method, and is capable of extension to the study of other animal fluids besides blood.

696. Hayem's Methods.—In conclusion, *see* the exhaustive work of Hayem, *Du sang et de ses altérations anatomiques*, pp. 1035, with 126 figures. Paris, Masson, 1889. A report of over twenty pages on this important work is contained in *Zeit. f. wiss. Mik.*, vi, 3, 1889, p. 330. It has reached me too late to be abstracted here.

Glands.

697. Mucus Glands.—It has already been stated that the blue solutions of hæmatoxylin have a special affinity for mucin. For the demonstration of mucus gland-cells the following process is recommended by Flemming (*Zeit. f. wiss. Mik.*, 1885, p. 518) :—Stain sections first with hæmatoxylin of Heidenhain, and afterwards with hæmatoxylin of Delafield or Böhmer. The mucus cells are shown stained violet.

698. "Reticulum" of Mucus Cells.—This was first demonstrated by Schiefferdecker (*Arch. f. mik. Anat.*, 1884, Hft. 3) by means of anilin-green. This reaction has given rise to a long polemic between Schiefferdecker and List, for which see *Zeit. f. wiss. Mik.*, 1885, pp. 51, 222, 223. According to List, the reticula in question stain in an equally specific way in methyl-green, in Bismarck brown, in nitrate of rosanilin (of 0·0001 per cent., ten to fifteen minutes, see *op. cit.*, iii, 3, 1886, p. 393), and in List's double-stains (§§ 253, 254). Schiefferdecker, on the contrary, maintains that the reticula demonstrated by the methods of List are not identical with those demonstrated by anilin-green.

699. Goblet Cells (Flemming, *Zeit. f. wiss. Mik.*, 1885, p. 519). —The contents of certain epithelium cells of the intestine and other organs take a specific dark blue or violet stain on treatment with hæmatoxylin after fixation with osmic acid or an osmium mixture. Staining preparations so fixed with safranin or gentian-violet in the manner described in §§ 102 and 103 also gives rise to an analogous reaction, which is perhaps even more demonstrative.

PAULSEN (l. c., p. 520) confirms these observations, and adds that by staining strongly in hæmatoxylin of Delafield (a quarter of an hour in the undiluted solution, or twelve to fifteen hours in a weak solution) there is obtained, besides the stain of the nuclei, a specific stain of the loose reticulum of mucus cells and goblet cells.

The reactions appear to be different for different animals. Thus PANETH (*Arch. f. mik. Anat.*, xxxi, 1888, p. 113 *et seq.*) found that in the small intestine of the mouse the contents of goblet cells did not stain with Böhmer's hæmatoxylin. And the goblet cells of the small intestine of man did not stain with safranin.

SUSSDORF (*Zeit. f. wiss. Mik.*, vi, 2, 1889, p. 206) mentions special stains of mucus cells obtained with methyl-violet, methyl-blue, and fuchsin.

RANVIER, in a paper too long to be abstracted here (*Comptes rend.*, 1887, 3, p. 145; see also *Zeit. f. wiss. Mik.*, v, 2, 1888, p. 233), describes a specific reaction of perruthenic acid (RuO_4) on goblet cells. By treating the pharyngeal mucosa of the frog first for ten to twelve hours with vapour of osmium, and then for three minutes with vapours of perruthenic acid, the goblet cells are brought out with remarkable distinctness. The contained mucigen is stained black, but the vacuoles are unstained. Since perruthenic acid is very rapidly reduced by organic matter, Ranvier regards this reaction as a proof that the vacuoles do not contain any organic substance, but probably only water and inorganic salts.

For detailed instructions for the study of goblet cells, see LIST, in *Arch. f. mik. Anat.*, xxvii, 1886, p. 481.

See also *ante*, §§ 232, 234, 239, 241, 242.

700. Stomach Glands. — MAX FLESCH (*Zeit. f. wiss. Mik.*, 1885, p. 351) finds that no stain is equal to the carmine and indigo-carmine mixture of Merkel (§ 230) for differentiating the divers cells of stomach-glands.

701. Mucosa of Small Intestine (HEIDENHAIN, *Pflüger's Archiv*, xliii, Supp., 1888, p. 1; *Zeit. f. wiss. Mik.*, v, 4, 1888, p. 519). —Amongst other valuable indications of a special nature, for which I have unfortunately too little space, the following is recommended as an excellent method for differentiating the elements of the stroma of the villi. Portions of small intestine are fixed for twenty-four hours in saturated solution of corrosive sublimate in 0·5 per cent. salt solution. They are washed out with strong alcohol, dehydrated, and imbedded in paraffin and sectioned. The sections are stained on the slide in the mixture quoted § 252, as there described.

702. Salivary Glands (*see* § 239).

703. Nervous Networks of the Liver (IGACUSCHI, *Arch. f. path. Anat.*, xcvii, p. 142; *Zeit. f. wiss. Mik.*, 1885, p. 243).—For the demonstration of those particular networks that are found in the liver, and that have been interpreted as nervous networks by some observers, as elastic tissue by others, Igacuschi proceeds as follows:—Pieces of liver (either fresh, or hardened for some days in solution of Müller) are put for eight or twelve hours into a solution composed of 100 parts of water, 20 parts of grape sugar, and 1 part of common salt. They are then put into 0·5 per cent. solution of gold chloride, and after twelve to twenty-four hours therein are put back again into the grape sugar, where they remain for two or three hours at the temperature of an incubator, or, which is better, twelve to forty-eight hours at the normal temperature. Sections are then made by the freezing method, and mounted in salt solution or glycerin; or by the collodion method and studied in clove oil.

704. Other Glandular Structures.—See, amongst other important papers that want of space precludes me from noticing here, RANVIER, *Les membranes muqueuses et le syst. gland.* (*Journ. de. Microg.*, ix, x, 1885—1886) (principally concerned with the investigation of the liver), and the same author's *Le mécanisme de la sécrétion, ibid.*, x, 1886—1887, which contains a variety of observations on other glandular structures.

CHAPTER XXXII.

SOME ZOOLOGICAL METHODS.

Tunicata.

705. Fixation of Tunicata.—The method of SALVATORE LO BIANCO for killing Ascidians in an extended state has been given above, § 19.

The compound Ascidians with contractile zooids are difficult to manage if one does not go the right way to work. The best process known to me is the following (due to VAN BENEDEN, kindly communicated to me by Dr. C. Maurice). Place the corms in clean sea water, and leave them alone for a few hours, in order that the zooids may become fully extended. Seize the corms with your fingers, and plunge them suddenly into glacial acetic acid. Leave them there for two, four, or six minutes, according to the size of the corms (which of course you will have taken care to select of as small a size as possible). Take them out of the acid with your fingers (or in some manner that may dispense with the employment of steel instruments, which would blacken the tissues) and bring them into 50 per cent. alcohol. Wash them thoroughly in that, and then bring them in the usual way through successively stronger alcohols.

I most strongly recommend this process, which gives admirably preserved preparations quite free from any opacity either in the tissues or the tunic. The acid will not hurt the fingers if they be washed immediately.

Small pelagic tunicates are very easily fixed with osmic acid or acid sublimate solution, with the exception of *Anchinia*. The not very numerous preparations I have made of this exceedingly delicate form have all been unsatisfactory. And some other similar forms may be found difficult. I have just had a striking failure with *Salpa virgula*, which I fixed with

"Flemming," and got a very poor preparation. The very similar *S. pinnata* is fixed perfectly in this medium.

Mollusca.

706. Fixation of Mollusca.—Two groups at least amongst the Mollusca offer considerable difficulties in the way of fixation—Lamellibranchiata and Gastropoda.

If it be attempted to take living and normal Lamellibranchiata from the water they are contained in, in order to throw them into a fixing solution, they invariably withdraw their siphon and foot, shut their valves, and die in a state of contraction. And if it be attempted to open the shell by force after death, the mantle is generally injured and it is impossible to get the foot and siphon into the extended state. DE CASTELLARNAU (*La Estacion Zoolog. de Napoles*, Madrid, 1885) advises that they be killed by the method of EISIG and ANDRES described for Actiniæ in § 13. Before dying, the animals protrude largely their feet, siphons, branchiæ, and tentacles, and die with their shells open. They may be fixed as soon as insensibility has supervened, by bringing them into picro-sulphuric acid, or some other rapidly killing fixing agent.

In order to demonstrate the absence of the supposed aquiferous pores in the root of Lamellibranchiata, FLEISCHMANN (*Zeit. f. wiss. Zool.*, xlii, 1885, p. 376) proceeds as follows :—A mussel is quickly seized at a moment when the foot is fully extended, and the two valves of the shell are forcibly pressed together, so as to prevent any flowing back towards the interior, of the liquid contained in the foot. The foot may then be fixed by holding it for a few minutes in hot sublimate solution.

The same methods recommended for Lamellibranchiata sometimes give good results with Gastropoda. The asphyxiation method has been described in § 18.

The quantity of mucus that exists in the integument of Gastropoda is often a serious obstacle in the way of preparation. MARCHI (*Arch. f. mik. Anat.*, 1867, p. 204) finds that if a living Limax be thrown into moderately concentrated salt-solution it will throw off enormous quantities of mucus, and die in a few hours. The epidermis will be found well preserved. If the animal be thrown into osmic acid or Müller's solution, if I understand the writer justly, no secretion of mucus will occur.

707. Injection of Acephala (FLEMMING, *Arch. f. mik. Anat.*, 1878, p. 252).—To kill the animals, freeze them in a salt-and-ice mixture, and throw them for half an hour into lukewarm water. They will be found dead, and in a fit state for injection. Chloroform and ether are useless. The injection-pipe may be tied in the heart; but when this has been accomplished there remains the problem of occluding cut vessels that it is impossible to tie. To this end, after the pipe has been tied, the entire animal is filled and covered up with plaster of Paris. As soon as the plaster has hardened, the injection may be proceeded with.

708. Foot-glands of Gastropoda (HOUSSAY, *Arch. de Zool. expér.*, 1883, p. 171). Harden the foot for twenty-four to forty-eight hours in 50 per cent. alcohol, make sections, and stain them with picro-carmine. Treat them first with 30 per cent. alcohol, and then stain in a 0·1 to 0·2 per cent. solution of methyl-green in 60 per cent. alcohol. Dehydrate and mount in balsam. The foot-glands alone are coloured green.

709. Eyes of Gastropoda (FLEMMING, *Arch. f. mik. Anat.*, 1870, p. 441).—The first difficulty here is to obtain the excision of an exserted eye. It is impossible to sever the exserted peduncle in a living animal without its retracting at least partially before the cut is completed. Never mind that; make a rapid cut at the base, and throw the organ into very dilute chromic acid, or 4 per cent. bichromate; after a short time it will evaginate, and remain as completely erect as if alive. Harden in 1 per cent. osmium, in alcohol, or in bichromate.

CARRIÈRE (*Zool. Anz.*, 1886, p. 221) gives the following instructions: Remove the eye together with a portion of the tentacle, and fix it by exposing it for some minutes to vapour of osmium. Make sections according to the usual methods, and fix them on a slide with Schällibaum's collodion. Stain them with picro-carmine; or, first depigment them by very careful treatment with *very dilute* eau de Javelle, and then stain with picro-carmine. Mount in dammar. Successful preparations show the tissues perfectly preserved; but Carrière has only been able to make the depigmentation process succeed with *Helix pomatia*; with Prosobranchiata he failed.

710. Eyes of Cephalopoda and Heteropoda (GRENACHER, *Abh. naturf. Ges. Halle-a.-S.*, Bd. xvi; *Zeit. f. wiss. Mik.*, 1885, p. 244).—Fix in picro-sulphuric acid, or in a saturated solution of corrosive sublimate in picro-sulphuric acid (this mixture is especially useful for *Octopus*, *Eledone*, and *Sepia*, but does not succeed with the pelagic forms, such as *Loligo*, *Ommatostrephes*, and *Rossia*). Depigment the specimens with hydrochloric acid (in preference to the nitric acid used by Grenacher in former researches). The mixture § 550 may also be used. The operation of depigmentation may be combined with that of staining; if you stain with borax-carmine and wash out in the last-mentioned mixture, the pigment will be found to be removed quicker than the stain is washed out. But this process is delicate, and requires a practised hand. The operation of depigmentation may be carried out on sections, but it is better to use portions of retina of 2 to 5 mm. in thickness. Grenacher mounted his preparations in castor oil, see § 393.

Similar methods are recommended by the same author for the eyes of **Heteropoda** (see *Abh. naturf. Ges. Halle-a.-S.* 1886; *Zeit. f. wiss Mik.*, 1886, p. 243).

711. Eyes of Chitonidæ (MOSELEY, *Quart. Journ. Mic. Sci.* 1885, p. 40).—Moseley worked by decalcifying the shell and making sections. He places fragments of shell (of which the tissues have previously been hardened in strong alcohol) in 100 to 200 c.c. of distilled water, and adds drop by drop concentrated nitric acid until gas is freely given off, which generally happens when from 3 to 4 per cent. of acid have been added. If the decalcification is not complete at the end of twelve hours, the objects should be removed to fresh distilled water, and the operation repeated. This process is said to give better results than the various processes of slow decalcification.

711 a. Eyes of Pecten and other Forms (*see* PATTEN, in *Mitth. zool. Stat. Neapel*, vi, 4, 1886, p. 733).

712. Maceration Methods for Mollusca.—For the study of

ciliated epithelium the following methods are recommended by ENGELMANN (*Pflüger's Arch.*, xxiii, 1880, p. 505) :

Cyclas cornea (intestine), maceration in osmic acid of 0·2 per cent. (after having warmed the animal for a short time to 45° to 50° C.). Also, concentrated boracic-acid solution.

The intra-cellular processes of the cilia. The entire intra-cellular fibre apparatus may be *isolated* by teasing fresh epithelium from the intestine of a Lamellibranch (*e. g.* Anodonta) in either bichromate of potash of 4 per cent., or salt-solution of 10 per cent. To get good views of the apparatus *in situ* in the body of the cell, macerate for not more than an hour in concentrated solution of boracic or salicylic acid. Very dilute osmic acid (*e. g.* 0·1 per cent.) gives also good results. The "lateral cells" of the gills are best treated with strong boracic-acid solution (5 parts cold saturated aqueous solution to 1 part water).

DROOST (*Morphol. Jahrb.*, xii, 2, 1886, p. 163) recommends, for *Cardium* and *Mya*, a mixture due to MÖBIUS, consisting of—

Chromic acid, 0·25 per cent. ⎫
Osmic acid, 0·1 per cent. ⎬ in sea water.
Acetic acid, 0·1 per cent. ⎭

The animals to remain for a few days in the liquid.

BÉLA HALLER (*Ibid.*, xi, 1885, p. 321) particularly recommends a mixture of 1 part of glacial acetic acid, 1 of glycerin, and 4 of water, for the maceration of nervous centres (of Rhipidoglossa). A sufficient degree of maceration is obtained in from thirty to forty minutes, and there is no shrinkage of the elements.

See also the media recommended by PATTEN (*Mitth. Zool. Stat. Neapel*, vi, 4, 1886, p. 736). Sulphuric acid, 40 drops to 50 grammes of water, is here recommended as a most valuable macerating and preservative agent. Entire molluscs, without the shell, may be kept in it for months.

713. Shell.—Sections of non-decalcified shell are easily obtained by the usual methods of grinding, or, which is often a better plan, by the methods of v. Koch or Ehrenbaum, §§ 303, 304. For sections of decalcified shell, MOSELEY, who has had great experience of this kind of work, particularly recommends the method of decalcification given above, § 711.

Molluscoida.

713 a. Bryozoa.—For the methods of killing and fixing, see §§ 7, 14, and 15.

Arthropoda.

714. General Methods for Arthropoda.—It may safely be stated that, as general methods for the study of chitinous structures, the methods worked out by Paul Mayer (*see* §§ 4 and 5, and also 55 and 168) are superior to all others. It is absolutely necessary that all processes of fixation, washing, and staining should be done with fluids possessing great penetrating power. Hence, picric-acid combinations should be used for fixing, and alcoholic fluids for washing and staining. *Concentrated* picro-sulphuric acid is the most generally useful fixative, 70 per cent. alcohol is the most useful strength for washing out, and tincture of cochineal in alcohol of 70 per cent. (§ 168) is the most generally useful staining fluid. Kleinenberg's hæmatoxylin may sometimes be preferable, and alcoholic-carmine and borax-carmine will occasionally give perfectly satisfactory results.

Some forms are very satisfactorily fixed with sublimate. Such are the Copepoda and the larvæ of Decapoda. Some Copepoda, however (*Copilia, Sapphirina*), are better preserved by means of weak osmic acid, and so are the Ostracoda. In many cases the osmic acid will produce a sufficient differentiation of the tissues, so that further staining may be dispensed with; *Copilia* and *Phyllosoma* are examples of forms that may be prepared in this simple manner.

715. Methods for Clearing and Softening Chitin.—The employment of eau de Javelle or eau de Labarraque, as suggested by Looss, for making chitin transparent and permeable to reagents has been described above § 529.

LIST (*Zeit. f. wiss. Mik.*, 1886, p. 212) has obtained good results with Coccidæ by treating them (after hardening) for eighteen to twenty-four hours with eau de Javelle, diluted with four volumes of water. After washing out with water, the objects may be dehydrated with alcohol and imbedded in paraffin, the chitin being sufficiently softened to allow of their being penetrated and good sections being obtained. You may stain before imbedding, with alum-carmine or picro-carmine (five to six days). I should say Mayer's tincture of cochineal would generally be preferable.

The same methods are applicable to the preparation of the ova of Insecta, for instance *Periplaneta* (*see* MORGAN, *Am. Mon. Mic. Journ.*, ix, 1888, p. 234).

716. Other Depigmentation Methods.—Besides the depigmentation processes discussed in Chap. XXIV, the following methods are available.

SAZEPIN'S **Method for Eyes of Chilognatha** (*Mém. Acad. Imp. St. Petersb.*, xxxii, 9, 1884, pp. 11, 12).—Sazepin treats antennæ that have been dehydrated with alcohol by steeping them in chloroform. The reaction is slow, the chitin becomes gradually less opaque, but the pigment does not entirely disappear. In order to remove the last trace of it, it will be sufficient if a drop of fuming nitric acid be now added to the chloroform. The mixture must be occasionally agitated, in order to prevent the acid from floating on the chloroform. The reaction is complete in twenty-four hours.

Employed in this manner, nitric acid has no injurious action on tissues.

717. Eyes of Arthropods.—LANKESTER and BOURNE (*Quart. Journ. Mic. Sci.*, 1883, p. 180) prepared the eyes of *Limulus* as follows:—Alcohol, turpentine, paraffin; sections made, and carefully depigmented under the microscope with nitric acid of 5 to 10 per cent., then mounted in balsam, some after staining with borax-carmine, others unstained. Non-depigmented sections also mounted in the same manner.

HICKSON (*Ibid.*, 1885, p. 243) prepared the eye of the fly as follows:—Remove the posterior wall of the head, and expose the rest, with the eyes *in situ*, for twenty minutes to vapour of osmium. Wash for a few minutes in 60 per cent. alcohol. Harden in absolute alcohol. Make sections. To depigment them, mount them on a slide with Mayer's albumen, remove the paraffin with turpentine, treat them with absolute alcohol, and invert the slide over a capsule containing 90 per cent. alcohol to which a few drops of strong nitric acid have been added. Nitrous vapours are freely given off, and the pigment dissolves. The reaction may be stopped at any moment by washing with pure alcohol.

For dissociation preparations, put the eye or the optic nerve for twenty-four hours into 5 per cent. solution of chloral hydrate, tease, and mount in glycerin. If the elements of the teased tissues be fixed to the slide by means of Mayer's albumen, they may be washed with alcohol and stained *in situ*, or they may be depigmented before staining.

718. Optic Ganglia (VIALLANES, *Ann. Sci. Nat. Zool.*, 1884, 4, and 1885, 4).—Fix in a mixture of 1 volume of 1 per cent. osmic acid and 2 volumes of 90 per cent. alcohol. After fixation, throw the tissues into absolute alcohol. After twenty-four hours, stain with alum-carmine or Kleinenberg's hæmatoxylin, make sections and mount them in balsam or glycerin.

CUCCATI (*Zeit. f. wiss. Zool.*, xlvi, 1888, p. 241) fixes the brain of the blow-fly (*Somomya erythrocephala*) (after having carefully incised the integuments of the head in several places) for twenty-four hours in liquid of Flemming or Rabl's mixture. The object is imbedded in paraffin and sectioned in the usual way. The sections are fixed to the slide with Mayer's albumen and there freed from the paraffin and carefully brought through successive alcohols into water. They are then stained for half an hour in a solution composed of 3 grammes of Säurefuchsin, 1 of chloral hydrate, and 100 c.c. of water. They are washed for ten minutes in water, rapidly dehydrated with alcohol, cleared with clove oil and mounted in balsam.

719. Nerve and Muscle of Arctiscoida (DOYÈRE, *Arch. f. mik. Anat.*, 1865, p. 105).—A score or so of *Milnesium tardigradum* are collected (it is well to have a large number, as the process by no means succeeds with all individuals) and put into a test-tube with water that has been deprived of its air by boiling. A drop of oil is run on to the surface of the water, so as thoroughly to exclude the air. After twenty-four to forty-eight hours the animals will be found, not dead, but fixed and extended in a cataleptic state; the circulation of the perivisceral fluid has ceased, the pigment of the cuticle has disappeared or collected into patches that are no hindrance to observation, the entire animal has gained in transparency, and the nervous and muscular systems stand boldly out. The animals are examined in boiled water, unless it be wished to study the phenomena of resuscitation, in which case spring-water should be used.

720. Sarcolemma of Insecta (THANHOFFER, *ibid.*, 1882, p. 27).
—In order to demonstrate the two plates of the sarcolemma, digest muscle (of an insect) either in the stomach of a living animal (by wrapping it in gauze and introducing it through a fistula) or in artificial gastric juice (in the former case several hours, in. the latter half to one hour, at the temperature of the room in summer). Examine in gastric juice.

721. Phalangida (RÖSSLER, *Zeit. f. wiss. Zool.*, xxxvi, 1882, p. 672). The animals are killed in boiling water; the water is allowed to boil up several times, so that the albumen of the tissues may be coagulated; they are then brought into alcohol, first of 70, then 90 per cent., then absolute, until all water is removed from them. They are then imbedded in soap. The soap is remelted, and allowed to cool once or twice, in order to get the objects thoroughly penetrated. Sections are then made, and stained on the slide with some colouring matter dissolved in absolute alcohol.

Paraffin was tried for imbedding, but gave no good results, on account of the brittleness of the tissues caused by the preliminary treatment with turpentine or oil of cloves. I fancy that this difficulty would be easily overcome by clearing with cedar oil instead of clove oil, as I have constantly recommended for the special purpose of avoiding brittleness in tissues.

722. Macrotoma Plumbea (SOMMER, *Inaug. Diss.*, 1884, p. 4; *Zeit. f. wiss. Mik.*, 1885, p. 234).—Fix with boiling water according to the method of Rössler, last §, and harden for several hours in picro-sulphuric acid diluted with 5 volumes of water. Wash out with alcohol; stain with alum-carmine, or Hamann's carmine (§ 156), or borax-carmine. Penetrate with chloroform and imbed in paraffin.

723. Aphidæ (see *ante*, § 586).

724. Other Methods for Arthropoda.—For Embryological methods, *see* Chap. XXV, §§ 581 to 593.

For Spermatological methods, *see* Chap. XXVI, § 612.

Vermes.

725. Cestodes.—This group must of course be chiefly studied by the usual section methods. It is only necessary here to remind the reader that, as pointed out by VOGT and YUNG (*Trait. d'Anat. Comp. prat.*, p. 204), the observation of the living animal may be of service, especially in the study of the excretory system. And, as shown by PINTNER, tæniæ may be preserved alive for several days in common water to which a little white of egg has been added.

726. Trematodes (FISCHER, *Zeit. f. wiss. Zool.*, 1884, p. 1).—*Opisthotrema cochleare* may be mounted entire in balsam, after treatment with absolute alcohol, picro-carmine, or hæmatoxylin or ammonia-carmine, and clearing with clove oil. For sectioning, Fischer recommends imbedding in a mass made by dissolving 15 parts of soap in 17·5 parts of 96 per cent. alcohol. This mass melts at about 60° C., penetrates very rapidly, and solidifies very quickly. The sections should be studied in glycerin.

WRIGHT and MACALLUM (*Journ. of Morph.*, i, 1887, p. 1) find that *Sphyranura* is for most purposes best fixed in liquid of Flemming, and stained with alum cochineal.

Cercariæ.—SCHWARZE (*Zeit. f. wiss. Zool.*, xliii, 1886, p. 45) found that the only fixing agent that would preserve the histological detail of these forms was cold-saturated sublimate solution warmed to 35° to 40° C.

727. Turbellaria.—For *Rhabdocœla* BRAUN (*Zeit. f. wiss. Mik.*, iii, 1886, p. 398) proceeds as follows:—For preparing entire animals the animals are got on to a slide, lightly flattened out with a cover, and killed by running under the cover a mixture of three parts of liquid of Lang with 1 per cent. osmic acid solution. After due fixation therein, they are treated for some time with successive alcohols, the cover is removed, the objects are stained with alum-carmine (two to three minutes is enough), washed, and brought into balsam in the usual way. Other fixing media than that described were not satisfactory. (BÖHMIG, however, commenting on this, says that for some of the tissues, such as muscle and body parenchyma, nitric acid and picro-sulphuric acid are very useful.) Sections may be made by the usual paraffin method.

DELAGE (*Arch. de Zool. exp. et gén.*, iv, 2, 1886; *Zeit. f. wiss. Mik.*, iii, 2, 1886, p. 239) strongly recommends fixation (of Rhabdocœla) by the osmium-carmine mixture, § 161. Concentrated solution of sulphate of iron is also an excellent fixing medium. The animals (*Convoluta*) die in it fully extended. Liquid of Lang was not successful.

For staining, he recommends either the osmium-carmine stain or impregnation with gold (one-third formic acid, two minutes; 1 per cent. gold chloride, ten minutes; 2 per cent. formic acid, two or three days in the dark. It is well to allow an excessive reduction to take place, and then lighten the stain by means of 1 per cent. solution of cyanide of potassium).

BÖHMIG, commenting on the above, says that he has obtained very instructive images with Plagiostomidæ fixed with sublimate and stained with osmium-carmine.

For *Dendrocœla* (*Polycladidea*) LANG (*Fauna u. Flora d. Golfes v. Neapel*, 1884, p. 30) recommends the following procedure :—Fix with one of the mixtures, § 46, or with hot alcohol. After washing out, stain for eight to fourteen days in picro-carmine, wash with 70 per cent. alcohol until the greater part of the picrin is extracted, stain for from one to fourteen days in borax carmine, and wash out in the usual way with acidulated alcohol. The result is a sharp stain of nuclei by the borax-carmine, a stain of extra-nuclear parts by the picro-carmine, and a slight maceration of the tissues by the prolonged action of the picro-carmine, which serves to make the limits of cells much more evident than they would be otherwise. Mayer's cochineal is useful for the study of glands. Sections made by the paraffin method.

728. Nemertina.—After considerable experience of this difficult group I have to say that I know of no method of fixation that will certainly give good results. My best results have always been obtained with cold saturated sublimate solution, acidified with acetic acid. I have tried most of the energetically hardening fixing agents, such as the osmic and chromic mixtures, and do not recommend them for this group; for they seem (the chromic mixtures and perchloride of iron in particular) to act as irritants, and provoke such violent muscular contractions that the whole of the tissues are crushed out of shape by them. And, besides, they do not kill as quickly as sublimate.

I have found it a good plan to decapitate the animals (in

the larger forms), cut them up quickly into lengths (not too long), and throw these sharply into the sublimate, the muscular contractions being less energetic in segments that are no longer in connection with the cerebral ganglia.

Perhaps a better method than this will be found in the simple process, suggested to me by Prof. DUPLESIS, of fixing with hot (almost boiling) water. On the few occasions on which I have tried it, the animals have died in extension, without vomiting their proboscis; and I think it is certainly worth trial, especially for the larger forms.

I have tried FOETTINGER's chloral hydrate method (§ 14); my specimens died fairly extended, but vomited their proboscides.

DE CASTELLARNAU (*Estacion Zool. de Napoles*, p. 137) says that Nemertians can be successfully narcotised by Eisig's alcohol method, described § 13.

For staining *in toto* I hold that it is absolutely necessary to employ alcoholic stains, for even the most delicate species are not satisfactorily penetrated by watery stains in any reasonable lapse of time. Borax-carmine or Mayer's alcoholic-carmine may be recommended; not so cochineal or hæmatoxylin stains, on account of the energy with which they are held by the mucin which in general exists in such great abundance in the skin of these animals.

Sections by the paraffin method, after penetration with oil of cedar (chloroform will fail to penetrate sometimes after the lapse of weeks).

729. Nematodes.—The extremely impermeable cuticle of these animals is a great obstacle to preparation. According to LOOSS (*Zool. Anz.*, 1885, p. 318) this difficulty may be overcome by treating the animals (or their ova, which are in the same case) with eau de Javelle or eau de Labarraque, in the manner described in § 529.

BRAUN (see *Journ. Roy. Mic. Soc.*, 1885, p. 897) recommends that small unstained Nematodes be mounted in a mixture of 20 parts gelatin, 100 parts glycerin, 120 parts water, and 2 parts carbolic acid, which is melted at the moment of using.

730. Acanthocephali.—It is very difficult to kill Echinorhynci so as to have the animals duly extended and the tissues well preserved. Neither corrosive sublimate nor strong osmic acid attains this end, even after preliminary intoxication with tobacco smoke or chloroform; the animal thus treated dying contracted.

SAEFFTIGEN (*Morphol. Jahrb.*, x, 1884, p. 120; *Journ. Roy. Mic. Soc.* (N.S.), v, 1885, p. 147) obtained the best results by killing gradually with 0·1 per cent. osmic acid; the animals placed in this contract during the first hours, but stretch out again and die fully extended.

Another method of killing is treatment with 0·1 per cent. chromic acid; Echinorhynci live for days in it, but eventually die fully extended.

731. Gephyrea.—DE CASTELLARNAU (*La Est. Zool. de Napoles*, p. 137) says that *Phascolosoma, Phoronis hippocrepis, Sipunculus nudus*, and *S. tesselatus* should be killed and fixed with chromic acid; and *Aspidosiphon Mülleri, Bonellia viridis*, and *B. fuliginosa* with picro-sulphuric acid.

VOGT and YUNG (*Anat. comp. prat.*, p. 373) direct that *Sipunculus nudus* be kept for some days in perfectly clean basins of sea water, in order that the intestine of the animals may be got free from sand, which would be an obstacle to section cutting, and then anæsthetised with chloroform, under which treatment they die extended, and may be fixed as desired.

APEL (*Zeit. f. wiss. Zool.*, xlii, 1885, p. 461) says that *Priapulus* and *Halicryptus* can only be satisfactorily killed by heat. The animals may either be put into a vessel with sea water and be heated on a water-bath to 40° C.; or they may be thrown as rapidly as possible into boiling water, which paralyses them so that they can be quickly cut open and thrown into one third per cent. chromic acid, or picro-sulphuric acid.

732. Rotatoria.—By far the most important method for the study of this group consists in the observation of the living animals. Great difficulty exists in the way of getting them to keep sufficiently quiet. VOGT and YUNG (*Anat comp. prat.*, p. 420) say that a drop of solution of any of the soluble salts of strychnin run under the cover sometimes renders service. WEBER (*Arch. de Biol.*, viii, iv, 1888, p. 713) finds that strychnin, prussic acid, and curare act too strongly; of all the reagents he tried, 2 per cent. solution of hydrochlorate of cocain gave the best results. Warm water gave him good results for large species, such as those of *Hydatina* and *Brachionus*.

HARDY (*Journ. Roy. Mic. Soc.*, 1889, p. 475) recommends thick syrup added drop by drop to the water. HUDSON (*Ibid.*, p. 476) mentions weak solution of salicylic acid.

Annelida.

733. Cleansing Intestine of *Lumbricus* (KÜKENTHAL, *Journ. Roy. Mic. Soc.*, 1888, p. 1044).—Put the animals into a tall glass vessel which has been filled up with bits of moistened blotting paper. They gradually evacuate the earthy particles from the gut and fill it instead with paper.

734. Killing Annelida.—*Lumbricus* may be anæsthetised by putting the animals in water with a few drops of chloroform. In order to kill *Criodrilus lacuum*, COLLIN (*Zeit. f. wiss. Zool.*, xlvi, 1888, p. 474) puts the animals into a closed vessel with a little water, and hangs up in it a strip of blotting paper soaked in chloroform. KÜKENTHAL (*Die mik. Technik*, 1885; *Zeit. f. wiss. Mik.*, 1886, p. 61) puts annelids into a glass cylinder filled with water to the height of 10 centimetres, and then pours 70 per cent. alcohol to a depth of 1 to 2 centimetres on to the water. The animals will be found sufficiently narcotised for fixation in from four to eight hours. For *Opheliadæ* he also employs 0·1 per cent. of chloral hydrate in sea water.

The *Polychæta sedentaria* offer the difficulty of a complex and very contractile branchial apparatus. They may sometimes be satisfactorily fixed by bringing them rapidly into corrosive sublimate. De Castellarnau (l. c., p. 139) says that SALVATORE LO BIANCO obtains magnificent preparations of *Spirographis* by this means. The species of *Polychæta errantia* that offer a contractile branchial apparatus, as *Eunice* and *Onuphis*, may be treated in the same way.

The greater part of the remaining forms of Polychæta and other marine annelids may be satisfactorily killed, according to De Castellarnau, by the alcohol method of Eisig, § 13.

Very delicate pelagic forms, such as *Alciope*, *Alciopina*, *Vanadis*, may be killed and fixed by iodised alcohol. SALVATORE LO BIANCO brings them for one or two minutes into a mixture of 100 parts of 70 per cent. alcohol with 3 parts of iodine; and after fixation washes out with 70 per cent. alcohol until all the colouration due to the iodine has disappeared.

I can recommend as a good fixing and hardening mixture for Anuelids in general, the following fluid due to EHLERS (I do not know whether it has been published) :—To 100 c.c. of chromic acid of 0·5 to 1 per cent. add from 1 to 5 drops of glacial acetic acid. The proportion of acetic acid indicated is sufficient to counteract any tendency to shrinkage due to the chromic acid.

735. Blood-vessels of Annelids (KÜKENTHAL, *Zeit. f. wiss. Mik.*, 1886, p. 61).—The animals should be laid open and put for two or three hours into *aqua regia* (4 parts of nitric acid to two of hydrochloric acid). The ramifications of the vessels will then be found to be stained black, the rest of the preparation yellow.

736. Hirudinea.—WHITMAN (*Meth. in Mic. Anat.*, p. 27) recommends that they be killed with sublimate. This reagent kills leeches with such rapidity that they die in general without having time to change the attitude in which they were found at the moment when the liquid came into contact with them.

Injection.—WHITMAN (*Amer. Natural.*, 1886, p. 318) states that very perfect natural injections may often be obtained from leeches that have been hardened in weak chromic acid or other chromic liquid. He considers that these injections are the best for the purpose of the study of the circulatory system by means of sections.

Of course Hirudinea (or any other Annelids) on which it is desired to make artificial injections must be killed by some procedure that leaves the tissues in a state that will allow the injection to run freely.

JAQUET (*Mitth. Zool. Stat. Neapel*, 1885, p. 298) advises that leeches be put into water with a very small quantity of chloroform; they soon fall to the bottom of the vessel and remain motionless. They should be allowed to remain a day or two in the water before injecting them.

Echinodermata.

737. Holothuroidea.—These animals are difficult to fix, on account of their contracting with such violence under the influence of irritating reagents as to expel their viscera through the oral or cloacal aperture. It has been recommended that they be seized by the middle of the body and firmly squeezed in the hand, and so plunged in a fixing liquid (acetic acid, for instance); or that they be anæsthetised (in the case of *Synapta* and *Cucumaria*) by adding ether to the water in which they are contained. So far as my experience goes, I am bound to say that I know no better way of killing them than that of simply putting them into fresh water, in which they generally die without contraction and with their tentacles extended.

Vogt and Yung (*Anat. Comp. Prat.*, p. 641) say that *Cucumaria Planci* (*C. doliolum*, Marenzeller) is free from the vice of expelling its intestines under irritation; but they recommend that it be killed with fresh water, or by slow intoxication with alcohol, chromic acid, or sublimate added to the sea-water in which it is contained.

738. Asteroidea.—There are great difficulties in the way of fixation here too. It is quite possible to obtain a fixation of the ambulacral feet, branchiæ and tentacles in the extended state, by throwing the animals into boiling water, and then bringing them into a fixing liquid. But this method has the fault that the fixing liquid so employed only penetrates extremely slowly into the interior of the animal, and therefore does not give a good fixation of internal organs.

Hamann (*Beitr. z. Hist. d. Echinodermen*, ii, 1885, p. 2) finds it preferable to inject the living animal with a fixing liquid. The canula should be introduced under the integument at the extremity of a ray, and the liquid injected into the body-cavity. The ambulacral feet and the branchiæ are soon distended by the fluid; and as soon as it seems to have penetrated sufficiently, the animal is thrown into a quantity of the same reagent.

The study of the *eyes* present points of special difficulty. In order to study them in sections, with the pigment preserved *in situ*, the eye should be removed by dissection, should be hardened in a mixture of equal parts of 1 per cent. osmic acid and 1 per cent. acetic acid, and imbedded in a glycerin gum mass, or some other mass that does not necessitate treatment with alcohol (which dissolves out the pigment, leaving the pigmented cells perfectly hyaline). For maceration, use one-third alcohol; the aceto-osmic mixture failing to preserve the rods of the pigmented cells.

739. Ophiuroidea.—Should be killed in fresh water if it be desired to avoid rupture of the rays (DE CASTELLARNAU, *La Est. Zool. de Napoles*, p. 135).

740. Larvæ of Echinodermata.—I am greatly obliged to my able friend Dr. BARROIS for kindly writing down for me (for the *Traité des Méth. Techn.*, from which they are translated) the following instructions, which are the outcome of a prolonged and minute study of the metamorphoses of the Echinodermata.

Pluteus.—In order to a fruitful study of the metamorphoses of the Echinoidea and Ophiuroidea, it is necessary to obtain preparations that offer the advantages presented by the study of the living larvæ; and especially such as give distinct images of the different organs, and show the calcareous skeleton preserved intact (a point of considerable importance, since this skeleton frequently affords landmarks of the greatest value). These preparations should further possess the following points:—They should give clear views of the region of formation of the young Echinoderm (which is generally opaque in the living larva). And they should possess sufficient stiffness to allow of the larva being turned about in any desired way and placed in any position under the microscope.

It is not very easy to obtain preparations fulfilling these conditions, on account of the difficulty of obtaining a selective stain whilst preserving the integrity of the calcareous skeleton. The following method is recommended:—*Pluteus* larvæ are fixed in a cold saturated solution of corrosive sublimate, in which they remain not more than two or three minutes. They are then washed with water, and brought into dilute Mayer's cochineal. This should be so dilute as to possess a barely perceptible tinge of colour. The objects should remain in the stain for from twelve to twenty-four hours, being carefully watched the while, and removed from the stain at the right moment and mounted in balsam, or, which is frequently better, in oil of cloves or cedar wood. This method is perfectly satisfactory for the study of the chief phases of metamorphosis.

Auricularia and *Bipinnaria*.—The method described above is equally applicable to these forms, and seems to be altogether the best method for the study of the metamorphosis of Bipinnaria. The earlier stages of the metamorphosis of Auricularia are better studied by fixing with osmic acid, staining with Beale's carmine, and mounting in glycerin.

Larvæ of Comatula.—The best method for the study of the embryonal development of *Comatula* consists in fixing with liquid of Lang, and staining with dilute borax carmine. It is important (for preparations that are not destined to be sectioned) to use only *dilute* borax carmine, as the strong solution produces an over-stain that cannot easily be reduced.

Narcotisation by chloral hydrate before fixing is useful, especially for the study of *Pentacrinus* larvæ and of the young *Synaptæ* formed from Auricularia. Without this precaution, you generally get preparations of larvæ either shut up (*Pentacrinus*), or entirely deformed by contraction (young *Synaptæ*).

Coelenterata.

741. Anthozoa: Fixation.—HERTWIG's method for killing Actiniaria has been given, § 9. The methods employed by ANDRES have been given in great part in §§ 10 and 13. In *Le Attinie, Fauna u. Flora d. Golfes v. Neapel*, the following are also given:—Hot corrosive sublimate often gives good results. In the case of the larger forms, the solution should be injected into the gastric cavity, and a further quantity of the liquid be poured over the animals.

Freezing sometimes gives good results. A vessel containing Actiniæ is put into a recipient containing an ice-and-salt freezing mixture and surrounded by cotton wool. After freezing, the block of ice containing the animals is thawed in alcohol or some other fixing liquid.

The Zoantharia with calcareous skeletons are also difficult to deal with on account of the great contractility of the polyps. Sublimate solution, which ought very often to be taken boiling, sometimes gives good results. DE CASTELLARNAU (*La Est. Zool. de Napoles*, p. 132) says that this process succeeds well with *Dendrophyllia, Antipathes, Astroides, Cladocora,* and *Caryophyllia.*

For preparing sections, besides the usual methods for sectioning decalcified specimens, we have the valuable methods of von Koch and Ehrenbaum, §§ 303 and 304, which, being applicable to undecalcified specimens and furnishing preparations showing at one and the same time soft parts and hard parts *in situ*, render most inestimable services.

The Alcyonaria have also extremely contractile polyps. I suggest for their fixation either hot sublimate solution, or glacial acetic acid (§ 51). GARBINI (*Manuale*, p. 151) says that the polyps may be fixed in the state of extension by drenching them with ether, and then bringing them into strong alcohol.

WILSON (*Mitth. Zool. Stat. Neapel*, 1884, p. 3) kills Alcyonaria with a mixture of 1 part of strong acetic acid and 2 parts of concentrated solution of corrosive sublimate, the animals being removed as soon as dead and hardened for two or three hours in concentrated sublimate solution.

SCHULTZE (*Biol. Centralb.*, 1887, p. 760) says that for Pennatulida with large polyps the gradual addition of fresh water is a good plan.

BRAUN (*Zool. Anz.*, 1886, p. 458) recommends that for both Zoantharia and Alcyonaria a little osmic acid be added to the sublimate employed for fixation. For *Alconium palmatum, Sympodium coralloides, Gorgonia verrucosa, Caryophyllia cyathus,* and *Palythoa axinellæ* he proceeds as follows:—The animals are left for a day or two in a glass vessel, so that the polyps may become thoroughly extended. They are then suddenly drenched with a mixture of 20 to 25 c.c. of concentrated solution of sublimate in sea-water with four to five drops of 1 per cent. osmic acid. This is allowed to act for five minutes.

(This method also gives good results with *Hydra* and some Bryozoa and Rotifers.)

742. Anthozoa: Maceration.—The HERTWIGS (*Jen. Zeitschr.*, 1879, p. 457) treat the tissues of Actiniæ for two or three minutes with a mixture of:

0·04 per cent. osmic acid in sea-water .	1 part.
0·2 per cent. acetic acid . . .	1 ,,

and then wash in acetic acid of the same strength till all traces of osmium are removed from the tissues. They are then left to macerate in the acid for a day; after which they are stained in picro-carmine if the degree of maceration seems sufficient; if not, in Beale's carmine. Finally, teased preparations are made in glycerin. This method has become classical.

LIST (*Zeit. f. wiss. Mik.*, iv, 2, 1887, p. 211) recommends dilute liquid of Flemming. Tentacles of *Anthea cereus* and *Sagartia parasitica* treated for ten minutes with a mixture of 100 c.c. of sea-water with 30 c.c. of Flemming's liquid (the strong solution, § 36), then washed out for two or three hours in 0·2 per cent. acetic acid, and teased in dilute glycerin, give fine dissociations of the connective, sensory, and urticant cells of the exoderm, and after removal of the epidermis allow of the demonstration of ganglion-cells and the supporting lamella. Picro-carmine may be used for staining.

743. Hydroidea, Polypoid Forms: Fixation.—In general the polyps may be very well killed in boiling sublimate solution, in which they should be plunged for an instant merely, and be brought into alcohol. Ether attentively administered gives good results with Campanulariadæ. *Hydra* is very easily killed by treatment with a drop of osmic acid on a slide. BRECKENFELD (*Amer. Mon. Mic. Journ.*, 1884, p. 49) obtains good results by heating the animals in a drop of water on a slide for from three to five seconds over a petroleum lamp. The methods for sections are the usual ones.

744. Medusæ: Fixation.—There is some difficulty in properly fixing the forms with contractile tentacles, which easily roll up on contact with reagents. The best results I have had with these forms have been obtained by means of VAN BENEDEN'S acetic-acid method, § 51, followed by alcohol. A similar method, with the difference that a mixture of chromic acid and alcohol is used for washing out instead of pure alcohol, is recommended by DE CASTELLARNAU (*La Est. Zool. de Napoles*, p. 133) for *Oceania, Lizzia, Bougainvillia, Podocoryne, Syncoryne, &c.* I have seen good results obtained by etherisation, and also (for large forms) by poisoning with solid corrosive sublimate added little by little to the water containing the medusæ.

Medusæ that have not very contractile tentacles can be satisfactorily killed by solution of sublimate or a chromic or osmic liquid in the usual way.

According to De Castellarnau (*l. c.*) *Cassiopeia borbonica* requires a special treatment. The animals should be treated with osmic acid until they begin to change colour, and then be put for two or three days into 5 per cent. solution of bichromate of potash, and then into alcohol.

745. Medusæ: Sections.—I am not acquainted with any perfectly satisfactory method of sectioning these extremely watery organisms. Paraffin and collodion will afford good sections of some organs, but are certainly not satisfactory as all-round methods for this group. Some modification of the method employed by the HERTWIGS (*Nervensystem der Medusen,* 1878, p. 5) might be successful. They imbedded in liver with the aid of glycerin gum, and hardened the objects and the mass in alcohol. I should think better results would be obtained by one of the freezing methods given in §§ 308 to 312.

746. Medusæ: Maceration.—The methods of the HERTWIGS (*Das Nervensystem u. die Sinnesorgane der Medusen,* Leipzig, 1878, p. 5) have deservedly become classical for the study of the tissues of this group. The objects are treated for two or three minutes, according to their size, with a mixture of equal volumes of 0·2 per cent. acetic acid and 0·5 per cent. osmic acid, and then washed in repeated changes of 0·1 per cent. acetic acid until all traces of free osmic acid are removed. They then remain for a day in 0·1 per cent. acetic acid, are then washed with pure water, stained with Beale's carmine, and preserved in glycerin.

Amongst other advantages of this mixture it is noted that the reduction of osmic acid by albuminates is greatly hastened by the presence of acetic acid, which in the case of animals so transparent and poor in cells as medusæ is an advantage for the study of the nervous system. For ganglion-cells and nerve-fibrils reduce osmium quicker than common epithelium-cells. They become greenish brown, and are easily distinguished from surrounding tissues.

The isolation of the elements of the macerated tissues is best done by gently tapping the cover-glass (which may be supported on wax feet). This gives far better results than teasing with needles. A camel-hair pencil also sometimes renders good service.

747. Siphonophora.—This group contains some of the most difficult forms to preserve that are to be found in the whole range of the animal kingdom. You have not only to deal with the very great contractility of the zooids, but with the tendency to general disarticulation of the swimming-bells and prehensile polyps.

The recent method of BEDOT has been given in § 44A. Bedot states that the most important point in this process is the bringing into alcohol of gradually increased strength. The liquid of Flemming for hardening ought to be added to the solution of sulphate containing the Siphonophore, and about two volumes of it should be taken for one of the sulphate solution. After hardening in the mixture a few drops of 25 per cent. alcohol should be added to the fluid, with a pipette, being dropped in as far as possible from the colony, which should be disturbed as little as possible; and further alcohol of gradually increasing strength should be added so gradually that the strength of 70 per cent. be not attained under fifteen days at least. 90 per cent. alcohol should be used for definitive preservation.

KOROTNEFF's method of paralysing with chloroform has been given in § 11. I would only add that I have seen *Physophora* very successfully killed by the careful administration of ether.

748. Ctenophora : Fixation.—The small forms are very easily prepared by means of osmic acid. The large forms are for the most part difficult to deal with on account of the extreme delicacy of the tissues. I suppose nobody has ever successfully prepared a large *Eucharis*. DE CASTELLARNAU recommends fixing in a mixture of 200 c.c. of 1 per cent. chromic acid with four to five drops of 1 per cent. osmic acid. According to GARBINI (*Manuale*, p. 154) SALVATORE LO BIANCO employs for *Callianira* a mixture of "1 part chromic acid, 1 part pyroligneous acid, and 2 parts sublimate," but, unfortunately, Garbini does not give the strength of the solutions.

Porifera.

749. Spongiæ: Fixation.—The smaller forms (Calcispongiæ) can be fairly well fixed by the usual reagents, osmic acid being one of the best. For the larger forms no satisfactory fixing agent has yet been discovered, so far as I can ascertain. The tissues of this group are very watery, very delicate, very friable after hardening, and macerate with the greatest facility. For all but very small specimens, absolute alcohol is apparently the best fixing agent. If any watery fluid be preferred, care should at all events be taken to get the sponges into strong alcohol as soon as possible after fixation, on account of the rapidity with which maceration sets in in watery fluids. FIEDLER (*Zeit. f. wiss. Zool.*, xlvii, 1888, p. 87) has been using (for *Spongilla*) besides absolute alcohol an alcoholic sublimate solution, and the liquids of Kleinenberg and Flemming, with good effect.

Staining.—On account of the great tendency to maceration above referred to, I hold (notwithstanding many recommendations of watery stains that are to be found in the literature of the subject) that alcoholic stains be alone employed for staining sponges, and I particularly recommend Mayer's tincture of cochineal as giving the best results known to me. It is better than Kleinenberg's hæmatoxylin for this purpose, because hæmatoxylin tends to make the tissues brittle.

Sectioning.—Calcareous sponges may be decalcified in alcohol slightly acidified with hydrochloric acid, and then imbedded in the usual way. Siliceous sponges may be desilicified by Mayer's hydrofluoric acid method mentioned, *ante*, § 541. But in view of the really dangerous nature of this operation, I feel bound to recommend that it be avoided. Fair sections may be obtained from sponge tissues well imbedded in paraffin without previous removal of the spicula. The spicula appear to be cut; probably they break very sharply when touched by the knife. Of course you will not use your best knives for cutting such sections.

Preparation of Hard Parts.—Siliceous spicules are easily cleaned for mounting by treating them on a slide with hot concentrated nitric or hydrochloric acid, or solution of potash or soda. The acids mentioned are very efficient, but it must be pointed out that they will attack the silex of some delicate spicules. Thus DEZSÖ found that the small stellate spicules of the cortex of *Tethya lyncurium* are completely dissolved by boiling hydrochloric acid. Potash solution is therefore frequently to be preferred, notwithstanding that, in my experience, it does not give such clean preparations.

According to NOLL, eau de Javelle is preferable to any of these reagents (*see* § 528).

Impregnation with Silver (*see* § 202).

Larvæ of Spongiæ. SCHULTZE (*Zeit. f. wiss. Zool.*, xxxi, p. 295) places the ova and larvæ of *Sycandra raphanus* in hanging-drop moist chambers, oxygenated by means of a few fronds of green algæ. He also (*ibid.*, xxxiv, 1880, p. 416) found that the best sections of the more advanced *sessile* larvæ of *Plakina* were obtained by selecting larvæ that had settled down on thin fronds of algæ and treating them, together with the fronds, with osmic acid, then alcohol of 52 per cent., alum-carmine, Aq. dest., alcohol of 52 per cent., then 70 per cent., 95 per cent., then absolute alcohol, turpentine, and finally paraffin.

Protozoa.

750. Introductory.—Since the Protozoa may be considered as free cells, and their peculiar organs known as "nucleus" and "nucleolus," "macronucleus" and "micronucleus," &c., present in the main the same reactions as cell-nuclei, it is evident that the reagents and methods of cytology are in great part applicable to this group. One of the most generally useful of these reagents will be found in the acid solution of *Methyl green*; it is the reagent that allows of the readiest and best demonstration of the presence and form of the nucleus and nucleolus (BALBIANI et HENNEGUY, *Compt. rend. Soc. de Biol.*, 1881, p. 131).

Amongst useful reagents not mentioned in the following descriptions of the methods employed by different authors, I call attention to the weak solutions of alum, potash, and borax, which serve to demonstrate the striations of the cuticle and the insertions of the cilia of Infusoria.

751. Staining *intra vitam.*—The possibility of staining Infusoria *intra vitam* was discovered independently and almost simultaneously by BRANDT (*Verh. d. physiol. Ges. Berlin*, 1878), by CERTES (*Soc. Zool.*, 25 janv., 1881), and by HENNEGUY (*Soc. philom.*, 12 fév., 1881).

CERTES found that living Infusoria stain, while continuing in life for a certain time, in weak solutions of cyanin, Bismarck brown, dahlia, violet 5 B, chrysoidin, nigrosin, methylen blue, malachite green, iodine green, and other tar colours, and hæmatoxylin. The solutions should be made with the liquid that constitutes the natural habitat of the organisms. They should be very weak, that is, of strengths varying between 1 : 10,000 and 1 : 100,000. For cyanin, 1 : 500,000 is strong enough.

The "nucleus" may be stained in the living organism by dahlia and malachite green. Bismarck brown only colours the "nucleus" of certain species (*Nychtoterus, Opalina*, HENNEGUY). The "nucleus" frequently behaves differently in allied species.

A double stain of the nucleus (green) and protoplasm (violet) may be obtained by the simultaneous employment of dahlia and malachite green.

752. PFITZNER's **Method** (*Morph. Jahrb.*, xi, 1885, p. 454).— For *Opalina* Pfitzner proceeds as follows: The animals are got on to a slide in a drop of water and covered. A drop of concentrated solution of picric acid is run in under the cover and the whole is put away in a moist chamber for several days. The preparation is then washed out with water and stained for a day or more with alum carmine or for a few hours with Delafield's hæmatoxylin. The preparation is then well washed out with water, dehydrated, cleared with clove oil followed by xylol, and mounted in balsam.

All the reagents employed must be very carefully run in under the cover, by placing a drop at the edge of the cover and allowing it to penetrate gradually; the drops must not be drawn in by means of blotting-paper applied at the other side of the cover, as this occasions a too rapid change of liquids and produces shrinkage or swelling in the Infusoria. The water used for washing out the picric acid may, however, be used in this way, and should be plentifully employed.

753. KORSCHELT's **Methods** (*Zool. Anz.*, 1882, p. 217).— Infusoria are fixed on the slide by means of a drop of 1 per cent. osmic acid placed at the edge of the cover and

drawn in by means of blotting-paper applied at the opposite edge. They are washed by the same process with alcohol of 70 per cent., 90 per cent., and then water. The objects are then stained by a drop of Weigert's picro-carmine, which is allowed to act for from half an hour to two hours in a moist chamber. They are then washed out with alcohol of 70 per cent., followed by 90 per cent. and absolute alcohol, are cleared with clove oil and mounted in balsam.

For Amœbæ a drop of 2 per cent. solution of chromic acid, which is allowed to act for two or three minutes, gives better results than osmic acid.

LANDSBERG (*ibid.*,p. 336) treats the organisms with the same reagents, but operates by taking up the organisms separately with a capillary tube and bringing them separately into a drop of the respective reagents. Care must be taken to have a small drop of liquid in the lower end of the tube before bringing it near to the object to be taken up, in order to check the violent rush up of the liquid into the tube. Landsberg recommends that *Actinosphærium* be mounted in glycerin instead of balsam.

754. BLANC (*ibid.*, 1883, p. 22) employs for fixing Infusoria a picro-sulphuric acid of the following composition :

Saturated picric acid solution . . 100 vols.
„ sulphuric acid . . . 2 „
Water . . . . . . 600 „

This liquid is for larvæ of Echinodermata, of Medusæ, and of Porifera; for Rhizopoda and Infusoria add two or three drops of 1 per cent. acetic acid for every 15 c.c. of the liquid. The acetic acid is added in order to bring out the nuclei and "nucleoli." Blanc fixes under a cover-glass, notwithstanding the objections of Korschelt. Wash out with 80 per cent. alcohol, followed by 90 per cent. and absolute. Stain with saffron solution (§ 192), wash out with 80 per cent. alcohol until the colour is sufficiently extracted, and pass through absolute alcohol into clove oil.

Blanc recommends the method for the preservation of most microscopic organisms, and in particular for marine *Nematodes*, the stain being sufficiently penetrating to pass through their thick chitinous integument.

755. CATTANEO'S **Methods** (*Bollettino Scientifico*, iii and iv;

Journ. Roy. Mic. Soc., 1885, p. 538).—Fix for a few minutes with ⅓ per cent. aqueous solution of chloride of palladium. This is the best fixing agent, as it hardens in a few minutes without blackening the structures. Double chloride of gold and cadmium also fixes well, and brings out nuclei even better than the palladium. Solution of iodide of mercury and potash (½ per cent.) is useful for bringing out protoplasmic networks, as it stains the granules of protoplasm black. Corrosive sublimate in 5 per cent. solution gives good results. Osmic acid causes darkness and opacity in the preparations. Nitrate of silver (in ½ to 1 per cent.) solution may be used, the objects being washed out with solution of acid sulphate of soda. Chromic, picric, and picro-sulphuric acids and bichromate of potash are only second-rate fixing agents for this purpose.

The best stains are carmine and picro-carmine. Magenta-red and fuchsin both give good results; nigrosin and hæmatoxylin (Kleinenberg's) still better. These should be used in weak solutions and allowed to act for a long time.

756. BRASS (*Zeit. f. wiss. Mik.*, 1, 1884, p. 39) employs for fixing unicellular organisms the following liquid :

Chromic acid	1 part.
Platinum chloride	1 ,,
Acetic acid	1 ,,
Water	400 to 1000 ,,

For protozoa that are opaque through accumulation of nutritive material, he proceeds as follows : The organisms are treated for three or four minutes with liquid of Kleinenberg, and then for some time with boiling water. They are then brought into water containing a small proportion of ammonia, in which they reassume their natural forms and dimensions. The ammonia is then neutralised by addition of a little acetic acid, and the preparation is stained with borax carmine or ammonia carmine. After washing, the objects are mounted in dilute glycerin. This treatment is said to afford extremely transparent preparations.

Brass also obtained good results with sublimate solution.

757. CERTES (*Comptes rend.*, 1879, 1 sem., p. 433) makes permanent preparations as follows : Fix with osmic acid of 2 per cent. (In the case of very contractile Infusoria, place a drop of the solution on the cover-glass, and place it on the drop of water that contains them. But generally speaking it is best to employ only the vapour of the solution, exposing the organisms to its action for not more than from ten to thirty minutes).

The objects having been covered, the excess of liquid is removed by means of blotting paper, and the following stain is allowed to flow in:

Glycerin . . . . 1 part.
Water . . . . 1 „
Picro-carmine . . . . . 1 „

(Eosin may also be used. Soluble anilin-blue does not give such good results.) The stain should be placed at the edge of the cover and the slide put away in a moist chamber in order that the water may evaporate very slowly and be changed very gradually for the glycerin-mixture; if this precaution is not taken, shrinkage may occur. When the exchange has taken place, strong glycerin may be added and gradually substituted for the dilute glycerin.

Certes states that the organisms thus prepared are fixed perfectly in their natural form, and allow of the study of the minutest detail of cilia, flagella, and the like, with the highest powers; the green colouration of Euglenæ and Paramœcia is preserved. The nuclear structures are sharply brought out by the picro-carmine.

758. SAVILLE KENT and BERTHOLD (*Manual of the Infusoria; Journ. Roy. Mic. Soc.*, 1883, p. 451) prefer a brownish-yellow solution of potassium iodide to osmic acid for fixing.

759. **Demonstration of Cilia** (WADDINGTON, *Journ. Roy. Mic. Soc.*, 1883, p. 185).—Solution of tannin, or a trace of alcoholic solution of sulphurous acid.

760. Other Methods for Protozoa.
MAUPAS, *Comptes rend.*, 1879, I^r sem., p. 1276, and 2me sem., p. 251.—(Fixation with alcohol followed by picro-carmine, glacial acetic acid, and glycerin).

DU PLESSIS, VOGT et YUNG, *Trait. Anat. Comp. Prat.*, p. 92.—(Fixation with 0·2 per cent. solution of corrosive sublimate. Let the preparation dry up, and if the organisms have preserved their shape, stain, and mount in balsam).

GÉZA ENTZ, *Zool. Anz.*, 1881, p. 575.—(Fixation of the organisms in a watch-glass with liquid of Kleinenberg. Wash out with alcohol, stain for ten to twenty minutes with picro-carmine, wash with water and mount in equal parts of glycerin and water).

FOL (*Lehrb.*, p. 102) fixes delicate marine Infusoria (*Tintinnodea*) with the perchloride of iron solution § 50 added to the water containing them, and stains with gallic acid as directed § 214, and states that this is the only method that has given him good results, especially as regards the preservation of cilia.

KÜNSTLER (*Journ. de Microgr.*, 1886, pp. 17 and 58). For Monadina. Fixation by means of a drop of very concentrated osmic acid solution (a gramme of osmium dissolved in only a few c.c. of distilled water).

SCHEWIAKOFF (*Morphol. Jahrb.*, xiii, 1887, p. 193).—Fixation with liquid of Flemming allowed to act for only a very short time and very thoroughly washed out afterwards. Staining with alum carmine (or picro-carmine, to be carefully watched, as it easily overstains).

BRANDT'S **Methods for Sphærozoa,** *Fauna u. Flora d. Golfes v. Neapel*, 1885; see also *Journ. Roy. Mic. Soc.*, 1888, p. 665. For some forms, fixation for fifteen to thirty minutes in a mixture of 1 part 70 per cent. alcohol, 1 part sea-water, and as much tincture of iodine as will impart a distinctly yellow colour to the mixture. This is followed by washing out with alcohol. For other forms, 0·4 to 1 per cent. chromic acid. For others, 5 to 15 per cent. solution of sublimate in sea-water. The best stain was found to be aqueous hæmatoxylin.

APPENDIX.

761. Green Light.—The employment of green light in microscopy has been alluded to above (§ 606) as recommended by Rabl. The suggestion is, I believe, due to ENGELMANN (*Pflüger's Arch.*, 1880, p. 550). He strongly recommends the use of green light for delicate observations as giving sharper definition, allowing finer detail to be seen, and tiring the eyes less than white light. Green glass of sufficiently good quality is found in commerce. The glass is best put between the mirror and the object, *e. g.* on the diaphragm. Blue glass (cobalt or ammonio-sulphate of copper) is also useful, but less so than green. Red light is most hurtful. " The explanation of these points, so important in practice, may be found in the results obtained by Lamansky in his researches on the " Limits of sensibility of the eye to the different colours of the spectrum" (*Arch. f. Ophthalm.*, xvii, p. 123, 1871)."

762. Cleaning Slides and Covers (HANAMAN, *Journ. Roy. Mic. Soc.*, i, 1878, p. 295; *American Naturalist*, xii, p. 573).—To a cold saturated solution of bichromate of potash, add $\frac{1}{8}$ of its bulk of strong sulphuric acid (care must be taken on account of the heat and vapours evolved).

(HENEAGE GIBBES, *ibid.*, iii, 1880, p. 392).—Place the cover-glasses in strong sulphuric acid for an hour or two, wash well until the drainings give no acid reaction; wash first with methylated spirit, and then with absolute alcohol, and wipe carefully with an old silk handkerchief.

(SEILER, *ibid.*, p. 508).—*New* slides and covers are placed for a few hours in the following solution:

Bichromate of potash	2 ounces.
Sulphuric acid	3 fluid ounces.
Water	25 ,,

Wash with water. The slides may be simply drained dry; the covers may be wiped dry with a linen rag.

Slides and covers that have been used for mounting either with balsam or a watery medium are treated as follows:—The covers are pushed into a mixture of equal parts of alcohol and hydrochloric acid, and after a few days are put into the bichromate solution and treated like new ones. The slides are scraped free of the mounting medium with a knife and put directly into the bichromate solution.

FOL (*Lehrb.*, p. 132) recommends either a solution containing 3 parts of bichromate, 3 of sulphuric acid, and 40 of water; or simply dilute nitric acid.

GARBINI (*Manuale*, p. 31) puts slides for a day into 10 per cent. sulphuric acid, then washes, first with water and then with alcohol.

BEHRENS (*Zeit. f. wiss. Mik.*, 1885, p. 55) treats slides first with concentrated nitric acid, then with water, alcohol, and ether.

JAMES (*Journ. Roy. Mic. Soc.*, 1886, p. 548) treats used slides with a mixture of equal parts of benzin, spirit of turpentine, and alcohol.

The readiest way known to me of freeing slides from balsam, damar, and cement is to wet with water and scrape with an old knife; using afterwards, if necessary, one of the solvents mentioned above.

763. Gum Mucilage, for Labels, &c.—The *Journ. of the Chemical Soc.* says that the adhesive qualities of gum may be very much exalted by the addition of aluminium sulphate (the so-called "patent" alum) to the mucilage. "Two grammes of crystallised aluminium sulphate dissolved in 20 grammes of water, is added to 250 grammes strong gum arabic solution (2 grammes in 5 grammes water). Ordinary solutions of gum arabic, however concentrated, fail in their adhesive power in many cases, such as the joining together of wood, glass, or porcelain; prepared, however, according to the above receipt, the solution meets all requirements" (from *Public Opinion*, Feb. 19th, 1886).

FOL (*Lehrb.*, p. 148) advises that slides be prepared for labelling by spreading over one end a layer of aluminium-chloride gelatin dissolved in acetic acid, and allowing it to dry before putting on the label.

Why do not the glass makers furnish slides with roughened (ground) end-surfaces for the reception of labels?

For four other receipts for gums and pastes for labels, see ELIEL, in *Engl. Mechan.*, 1887, p. 535; *Amer. Mon. Mic. Journ.*, 1887, p. 93; *Zeit. f. wiss. Mik.*, v, 1, 1888, p. 69.

764. Lubricating Medium for the Thoma Microtome.—According to my experience, it is extremely important that sliding microtomes of this class should be so oiled as to reduce friction to the greatest extent possible. I used for a long time the finest watch-maker's oil I could procure. But a mixture, for the knowledge of which I have to thank Prof. v. KOROTNEFF, is greatly superior. It is simply the usual pharmaceutical mixture of equal parts of castor oil and oil of almonds. It has not only the advantage of being a better lubricant (for this purpose) than the most highly refined animal oil, but also those of having a less oxydizing action on bronze, of not becoming thick by exposure to air, and of not becoming charged with dust to the same extent.

765. Levulose as a Mounting Medium.—Levulose is recommended as a mounting medium by BEHRENS, KOSSEL, u. SCHIEFFERDECKER (*Das Mikroskop. u. d. Meth. d. mik. Unters.*, Braunschweig, 1889). It is uncrystallisable, and preserves well carmine and coal-tar stains (hæmatoxylin stains fade somewhat in it). The index of refraction is somewhat higher than that of glycerin. Objects may be brought into it out of water.

766. Methylen-Blue Impregnation Method (DOGIEL, *Arch. f. mik. Anat.*, xxxiii, 4, 1889, p. 440, *et seq.*). Suitable pieces of tissue (thin membranes by preference) are brought fresh into a 4 per cent. solution of methylen blue in physiological salt solution. After a few minutes therein they are brought into saturated solution of picrate of ammonia, soaked therein for half an hour or more, then washed in fresh picrate of ammonia solution, and examined in dilute glycerin.

If it be wished only to demonstrate the outlines of endothelium cells, the bath in the stain should be a short one, not longer than ten minutes in general. Whilst if it be desired to obtain an impregnation of ground-substance of tissue so as to have a negative image of juice-canals or other spaces, the staining should be prolonged to fifteen or thirty minutes, and it is advisable to remove the endothelial covering of the objects operated on before putting them into the stain.

If it be desired to preserve the preparations permanently, they had better be mounted in glycerin saturated with picrate of ammonia.

The effect is practically identical (except as regards the colour) with that of a negative impregnation with silver nitrate. If the process bears out all that is claimed for it, it will certainly be valuable; for there are many objects to which metallic impregnation cannot be readily applied. Marine animals furnish many cases in point.

767. Venice Turpentine for Mounting (VOSSELER, *Zeit. f. wiss. Mik.*, vi, 3, 1889, p. 292, *et seq.*).—Vosseler strongly recommends this medium as having considerable advantages over Canada balsam or damar. Commercial Venice turpentine is mixed in a tall cylinder glass with an equal volume of 96 per cent. alcohol, allowed to stand in a warm place for three or four weeks, and decanted. It is stated that preparations may be mounted in this medium without previous clearing with essential oils or the like. The index of refraction being lower than that of the above-named balsams delicate details are more distinctly brought out. Stains keep well in the medium, and Vosseler states that he possesses preparations made fifteen years ago that are perfectly well preserved.

768. Gelatin-Soap Imbedding Mass (GODFRIN, *Journ. de Bot.*, 1889, 5, p. 87; abstract in *Zeit. f. wiss. Mik.*, vi, 3, 1889, p. 317). Very complicated and, as far as I can judge from the report, rather of the nature of an emulsion than of a homogeneous mass.

769. Platino-Aceto-Osmic Mixture (HERMANN, *Arch. f. mik. Anat.*, xxxiv, 1889, p. 58).—The author obtained excellent results by substituting 1 per cent. platinum chloride for the chromic acid in Flemming's strong formula, the other ingredients either remaining as before, or the osmium being diminished one half. Thus—1 per cent. platinum chloride 15 parts, glacial acetic acid 1 part, and 2 per cent. osmic acid either 4 parts or only 2 parts. Hermann found that protoplasmic structures are thus better preserved than with the chromic mixture, which appears to me very likely.

770. Iodine Hæmatoxylin (SANFELICE, *Journ. de Microgr.*, xiii, 1889, p. 335; *Journ. Roy. Mic. Soc.*, 1889, p. 837).—Dissolve 0·70 g. hæmatoxylin in 20 g. absolute alcohol, and 0·20 g. alum in 60 c. c. distilled water. Add the first solution, drop by drop, to the second. Expose the mixture to the light for three or four days, add 10 to 15 drops of tincture of iodine, agitate, and allow to stand for some days. Stain for twelve to twenty-four hours, and wash out for the same time in 90 per cent. alcohol, acidulated with acetic acid.

771. Carbonate of Soda Carmine (CUCCATI, *Zeit. f. wiss. Mik.*, iv, 1, 1887, p. 50).—This carmine is stated to have a bleaching action on the retina of Arthropods. Apart from that, it appears to me superfluous.

772. The Laboratory Table.—The following list is intended for a memorandum of the reagents required for *ordinary* zoological work, and is given in the belief that it may be useful as a reminder to those whose duty it is to furnish tables for students in public laboratories, and as advice to any beginner who may desire to set up a table for himself. The list comprises reagents alone, and of these only such as I apprehend are really necessary for the general scope of zoological work. It gives the material necessary for one student only, and states the minimum quantity of each reagent required. The items marked with asterisks are considered to be not strictly indispensable, and may be omitted if it be desired to simplify the table. By "the table" is of course meant the actual table supplemented by shelf or cupboard space.

The prices affixed to the several reagents are approximate only, as the market price of these commodities is variable. They are taken from recent notes kindly furnished me by Dr. Grübler, and indicate the prices both of the raw reagents and of reagents made up for use by him—these latter being, of course, somewhat in excess of the market value of the raw chemicals.

Quantity required.	Description.	No. in this book.	Price.
			s. d.
1 lit.	96 per cent. alcohol.		
1 lit.	70 ,, ,,		
1 lit.	50 ,, ,,		
1 lit.	30 ,, ,,		
1 lit.	Distilled water.		
*1 g.	Nicotin	10	0 9
100 g.	Chloroform	11, 12	1 0
*100 g.	Ether	13	10
*50 g.	Hydrate of chloral	14	2 6
*1 g.	Cocain	15	1 4
1 g.	Osmium	26	4 0
50 c.c.	2 per cent. sol. of osmium	26, 35	} 4 6
	in 1 per cent. chromic acid sol.	36	
*100 g.	Chromic anhydride	30	1 0
50 g.	Pure formic acid	32	0 4
1 lit.	Chromo-aceto-osmic solution (strong formula)	36	3 10
*2 g.	Argentum nitricum	38	0 6
500 c.c.	1 per cent. chromic acid solution	—	0 3
50 c.c.	1 per cent. solution of silver nitrate	38, 198	—
100 g.	Bichromate of potash	44, 77–9	0 5
*500 g.	Pure cupric sulphate	44, 44a	0 4
50 g.	Neutral chromate of potash	178	0 4

APPENDIX.

Quantity required.	Description.	No. in this book.	Price.
			s. d.
50 g.	Pure concentrated nitric acid	37, 39, 56	0 11
50 g.	Potash alum, fresh	—	—
10 g.	Pure permanganate of potash	44c	½
1 lit.	Solution of Erlicki or Müller	78, 79	2 0
*50 g.	Sulphate of soda	78	—
50 c.c.	Strong sol. of permanganate of potash	44c	0 6
200 g.	Corrosive sublimate	45	2 0
1 lit.	Strong sol. of sublimate in 5 per cent. acetic acid, or liquid of Lang	45, 46	2 6
1 kilo.	Glacial acetic acid	51	4 0
1 lit.	Concentrated picrosulphuric acid	55	2 6
50 g.	Picric acid	54	0 7
50 c.c.	Pure sulphuric acid (conc.)	55	—
100 c.c.	Ripart and Petit's liquid	64	—
10 g.	Cupric acetate	64	0 1½
10 g.	Cupric chloride	64	0 2
30 c.c.	Saturated solution of iodine in iodide of potassium	45, 66, 89	0 3½
*10 g.	Victoria blue	100	0 8
10 g.	Gentian violet	102	0 5
30 g.	Pure anilin	102, 104	0 4
*10 g.	Dahlia	103	0 5
10 g.	Safranin	104	0 9½
*10 g.	Methyl violet	107	0 7
10 g.	Methyl green	105	0 8
10 g.	Bismarck brown	106	0 4
*50 c.c.	Sat. aqueous sol. of Victoria	100	0 6
50 c.c.	Bizzozero's gentian fluid	102	—
30 c.c.	Sol. of gentian in chloral hydrate salt sol.	93	—
50 c.c.	Babes's anilin safranin liquid	104	—
50 c.c.	Acetic acid methyl green sol.	105	0 4
*50 c.c.	Sol. of Bismarck brown in dilute glycerin	106	0 5
10 g.	Methylen blue	127	0 7
*10 g.	Bleu de Lyon	109, 231	0 7
10 g.	Eosin (wasserlöslich)	238, 245	0 6½
50 c.c.	Aqueous solution of eosin	116	0 6
*10 g.	Anilin blue black	673	0 6½
50 g.	Nakaret carmine	130	4 0
*50 c.c.	Ranvier's picro-carmine (solution)	145	0 7½
10 g.	,, ,, (in powder)	145	4 6
100 c.c.	Grenacher's alum carmine	149	0 5
*50 c.c.	Orth's lithium carmine	153	0 6
50 c.c.	Schneider's aceto-carmine	154	0 6
100 c.c.	Grenacher's borax carmine	163	0 11
100 c.c	70 per cent. alcohol acidulated with 4 drops of HCl	163	—
100 c.c.	Mayer's alcoholic carmine	164	1 0
50 c.c.	Pure fuming hydrochloric acid	—	—
100 c.c.	Mayer's cochineal	168	1 0
30 c.c.	Delafield's hæmatoxylin	174	0 4
20 g.	Hæmatox. crist.	—	4 5
*100 c.c.	3 per cent. sol. of hæmat. crist. in distilled water	178	1 0
*200 c.c.	0·5 per cent. sol. of neutral chromate of potash	178	0 7
100 c.c.	Calcium chloride solution	184	0 4

Quantity required.	Description.	No. in this book.	Price.
			s. d.
30 c.c.	Conc. sol. of hæmatox. crist. in absolute alcohol	184	1 6
10 g.	Nitrate of potash...	202	—
1 g.	Double chloride of gold and potassium...	211	2 0
*2 or 3	Lemons ...	—	—
*30 c.c.	Tinctura ferri perchlor. ...	214	0 4
*10 g.	Pyrogallic acid ...	214	0 4½
30 c.c.	Renaut's hæmatoxylic eosin	239	0 11
250 g.	Paraffin, 45° C. melting point	273	1 2½
250 g.	„ 52° C. „	273	1 2½
20 c.c.	Mark's thin collodion (in small bottle fitted with camel-hair brush in cork)	275	0 7
100 c.c.	Pure naphtha, or toluol ...	276	—
1	Tablet of celloidin	290	3 0
100 c.c.	Thin celloidin solution	292	0 10
100 c.c.	Thick celloidin solution ...	292	1 3
50 c.c.	Ol. origani cretici	298	1 6
*20 g.	Gum copal	303	0 11
15 c.c.	Mayer's albumen...	318	0 3
50 c.c.	Clove oil ...	332	1 2½
100 c.c.	Cedar oil ...	331	2 2½
50 c.c.	Bergamot oil	334	2 0
*30 c.c.	Conc. sol. of carbolic acid	338	0 2½
*30 g.	Crystallised carbolic acid	338	0 3¼
1 lit.	Normal salt solution	345	—
30 c.c.	Iodised serum	346	—
*50 c.c.	5 per cent. chloral hydrate solution	361	—
100 c.c.	Liquid of Calberla	385	—
50 c.c.	Pure glycerin	—	0 2½
*50 c.c.	„ with 1 per cent. acetic acid	—	0 2½
30 g.	Glycerin jelly (any formula)	386 to 391	0 4
*20 c.c.	Xylol balsam	397	0 2½
20 c.c.	Xylol damar	399	0 3½
	or the colophonium sol.	401	—
1	Small bottle Bell's cement	411	1 0
*50 g.	Hardened turpentine	420	—
*50 g.	Marine glue	416	—
50 g.	Fol's carmine gelatin mass	446	0 5
50 g.	Fol's Berlin blue gelatin mass ...	453	0 4
50 c.c.	Beale's Prussian blue glycerin mass	477	0 5
*50 c.c.	40 per cent. sol. of caustic potash	502	0 2½
*30 c.c.	Glycerin solution of pepsin	521, 522	1 0
50 c.c.	Eau de Javelle ...	528	0 1¼
	or Eau de Labarraque...	529	0 2
*100 c.c.	Weigert's hæmatoxylin ...	650	1 0
*100	Weigert's borax-ferricyanide liquid	650	0 6
50 c.c.	Liquor ammoniæ fortissimus	—	0 1½
5 g.	Thymol ...	—	0 5

773. **The Zoologist's Travelling Case.**—I suggest the following as a portable selection of reagents not easily obtainable of country druggists. It is not intended to include all that is necessary for travellers beyond the limits of even those resources of civilisation.

50 g. each of

50 per cent. solution of chromic acid.
Mayer's alcoholic carmine.
Grenacher's borax carmine.
Delafield's hæmatoxylin.
Solution of Ripart and Petit.
Toluol.
Cedar oil.
Clove oil.

10 g. each of

Alum carmine in powder.
Pure carmine.
Bismarck brown.
Eosin (wasserlöslich).
Renaut's hæmatoxylic eosin.
Gentian violet.
Victoria blue.
Safranin.
Methylen blue.
Methyl green.
Picric acid.
Soluble Berlin blue.
Bichromate of potash.
Iodised serum.
Pure anilin.

Origanum oil.
Mayer's albumen.
Glycerin jelly.
Chloral hydrate.

5 g. each of

Hæmatox. crist.
Bleu de Lyon.
Picro-carmine, dry.
Permanganate of potash.
Methyl violet, 1 B.

In a drawer.

1 g. nicotin.
1 g. gold chloride.
2 g. silver nitrate.
1 g. cocain.
2 half-tubes osmium.
1 tube Canada balsam.
¼ plate celloidin.
50 to 100 g. corrosive sublimate.
Paraffin, 45° C. fusion.
Paraffin, 52° C. fusion.
3 corked flat-bottomed tubes, to take 3 × 1 in. slides, for staining on the slide.

Such a selection, in appropriate bottles, fitted into a case measuring 1 ft. 4 in. × 5½ in. × 4½ in., may be obtained from Grübler, at the price of about £2 5s.

Grübler also makes a somewhat larger case, containing, in addition to the above, sixteen more 50-g. bottles, with the following reagents:—Strong solution of Flemming (two bottles), glacial acetic acid, concentrated nitric acid, corrosive sublimate, solution of Perenyi, Schneider's aceto-carmine, thin celloidin solution, thick celloidin solution, pure ether, chloroform, glycerin, Bell's cement or asphalt, eau de Javelle, stick potash, and Beale's Prussian blue injection. This is not too heavy to be carried in the hand. The price is £3.

The address is, Dr. G. Grübler, 12, Bayersche Strasse, Leipzig. *See* also *ante*, § 94.

INDEX.

The numbers refer to the Paragraphs, not to the Pages.

A.

Absolute alcohol, preparation and uses, 62.
Acanthocephali, 730.
Acephala, 706, 707, and *see* Mollusca.
Acetate, of alumina for mounting, 358; of copper, 64; of lead, 87, 661; of potash, 359; of uranium, 65.
Acetic acid for fixing, 51, 52.
Aceto-carmine, 154—156.
Achromatic figure, 603, 604, 607.
Acids, *see* Acetic, Chromic, Formic, Hydrochloric, &c.
Acidulated alcohol, 63.
Actiniaria, fixation, 9, 10, 12—14; maceration, 514, 741.
Actinosphærium, 753.
ADAMKIEWICS, stain for nervous centres, 674.
Adipose tissue, 36, 683.
AGASSIZ and WHITMAN, pelagic ova, 575.
Agelena, embryology, 588.
Albumen, imbedding masses, 306, 312; injection mass, 471; section-fixing process, 318.
Alciope, Alciopina, 734.
Alcohol, for narcotising, 13; fixing, 60—63; hardening, 88, 89; macerating, 498; preserving, 362; clearing, 341, 398.
Alcoholic carmines, 133, 163—166.
Alcoholic cochineal, 167, 168, 171.
Alcoholic hæmatoxylins, 179, 184.
Alcyonaria, 7, 741.
Alcyonella, 14.

ALFEROW, silver staining, 199.
Alizarin, 188, 673.
ALTMANN, fixing methods, 37; impregnation and corrosion methods, 530.
Alum, for fixing, 44b; preserving, 357.
Alum-carmine, 149—152; ditto, with acetic acid, 151; with osmic acid, 150; with picric acid, 152.
Alum-cochineal, 169—171.
Alumina, acetate, 358.
Amarœcium, embryology, 577a.
Amber varnish, 422, 423.
Ammonia, bichromate, 80; neutral chromate, 80, 81; vanadate, 484; molybdate, 221.
Ammonia-carmine, 134—139, 673.
Ammonium, *see* Ammonia.
Amœba, 753.
Amphibia, for cytological study, 599; embryology, 563 *et seq.*
Amphioxus, 630.
Amphipoda, embryology, 592.
Amyl-nitrite, 473.
Amyloid matter, 105, 107.
Anchinia, 705.
ANDEER, phoroglucin, 540.
ANDRES, Actiniæ, 10, 13, 741.
ANDREWS, imbedding apparatus, 266.
Anilin dyes, generalities, 92, 95; classified list, 95; Hermann-Böttcher or Flemming staining process, 95—104.
Anilin black, 95, 129.
Anilin blue, 95, 125, 673.
Anilin blue-black, 95, 129, 673.
Anilin green, 95, 101, 123.
Anilin violet, 107.

The numbers refer to the Paragraphs, not to the Pages.

Anilin oil, anilin water, 298 ; p. 64, note.
Annelida, 733—736; blood-vessels, 735;
    cleansing intestine, 733; fixing, 734.
Anthea, 742.
Anthozoa, 741, 742.
Antipathes, 741.
APÁTHY, hæmatoxylin, 179; celloidin imbedding, 292; preserving celloidin blocks, 295; serial sections, 326, 326a; cement, 430a; double stain for nervous tissue, 681.
APEL, *Sipunculus* and *Halicryptus*, 731.
Aphides, preparation and embryology, 586.
Appendicularia, 202.
Aqua Javelli, 528, 548, 563.
Aquiferous pores of molluscs, 706.
Arabin, 319.
Arancina, embryology, 587.
ARCANGELI, carmines, 162.
Arctiscoida, 719.
Areolar tissue, 682.
ARNSTEIN, methylen-blue for nerve-endings, 654.
Arthropoda, fixation, 55, 62, 65, 714; general methods, 714; clearing and softening chitin, 715; depigmentation, 716, 717; eyes, 717; optic ganglia, 718; nerve and muscle, 719; sarcolemma, 720; embryology, 581—593; spermatology, 612.
Artificial œdemata, 682.
Artificial pigment spots, 79, 661.
Ascaris, ova, 52, 611.
Ascidians, 19, 705.
Asphalt varnish, 412.
Asphyxiation of Gastropods, 18.
Aspidosiphon, 731.
Astacus, embryology, 591.
Asteracanthion, 14.
Asteroidea, 738.
Astroides, 741.
AUBERT, tenacity of cements, 410.
Auréoline, 547.
Auricularia, 740.
Aves, embryology of, 556—559a.
Axis cylinder, 654.
Axolotl, ova, 564.

B.

BABES, picro-carmine, 147.
BABES (IN), cytological methods, 604, 610; safranin stain, 103; supersaturated do., 604.
BALBIANI, "noyau vitellin," 608; living ova, 552; ova of Phalangida, 589; Protozoa, 750; spermatological methods, 612.
BALFOUR, embryology of Aves, 556, 557; of Araneina, 587.
Balsam, 396—398, 400.
BALZAR, elastic tissue, 688.
BARFF, boroglyceride, 384.
BARRETT, retina, 644, 645, 647.
BARROIS, embryology of Echinodermata, 740.
BASTIAN, gold method, 211.
BAUMGARTEN, cytological methods, 610; fuchsin and methylen blue stain, 246; picro-borax carmine, 168; triple stain, 247.
Bayberry tallow, 281.
BAYERL, stain for cartilage, 691.
BEALE, carmine stain, 134; digestion fluid, 521; glycerin jelly, 388; injections, 474, 476, 477; preservative fluids, 355; staining spinal cord, 673.
BECHTEREFF, hardening brain, 661.
BECKWITH, gold-staining nervous centres, 674.
BEDOT, fixing process, 44a, 747.
BEEVER, Weigert's hæmatoxylin, 651.
BEHRENS, cements, tenacity of, 410; amber varnish, 422; cleaning slides, 762.
BÉLA HALLER, macerating mixture, 515, 712.
BELLONCI, preparation of brain, 669.
BELL'S cement, 411.
BENCZUR, staining nervous centres, 673.
BENDA, hæmatoxylin, 612; spermatological methods, 612; modification of Weigert's hæmatoxylin, 651.
BENEDEN, VAN, acetic alcohol, 52; Ascidians, 705; cytological methods, 611; embryology of Mam-

INDEX. 893

The numbers refer to the Paragraphs, not to the Pages.

malia, 555; of Tænia, 594; glacial acetic acid, 51.
BENEDEN, VAN, and NEYT, cytological methods, 603, 607, 611.
Bengal rose, 95, 117.
Bengalin, 110.
Benzol, for clearing, 540.
Benzo-purpurin, 95, 114.
Bergamot oil, 334.
BERLINERBLAU, F., regeneration of Weigert's hæmatoxylin, 650.
BERNHEIMER, hæmatoxylin, 645.
BERRY'S Hard Finish, 403.
BERTHOLD, Infusoria, 758.
BETZ, carmine, 136; brain methods, 668.
BEVAN LEWIS, central nervous system, 657, 658, 660, 665, 672, 673, 677, 679.
BIANCO, S. LO, *Alciope, Alciopina,* 734; Ascidians, 19; *Callianira,* 748; narcotising mixture, 13; *Spirographis,* 734; *Vanadis,* 734.
Bichloride of mercury, *see* Corrosive Sublimate.
Bichromate of ammonia, 80, 661.
Bichromate of potash, for fixing and hardening, 40, 43, 44, 77—79; 661; for maceration, 507.
BICKFALVI, digestion fluid, 523.
Biebricher Scharlach, 95, 115.
BIEDERMANN, methylen blue for nerve-endings, 654.
Bilberry juice stain, 196.
Biniodide of mercury medium, 394.
BIONDI, triple stain, 252; blood, 692, 695.
Bismarck brown, 93, 95, 104, 106, 751.
Bitumen, 412.
BIZZOZERO, cytological methods, 610; gentian violet, 102; picro-carmine, 147; maceration, 613; blood, 693, 694.
BJELOUSSOW, injection, 472.
BLACKBURN, vegetable wax, 281.
Black currant stain, 196.
Blackley blue, 110.
Bladder of frog, 639.
BLANC, saffron stain, 192; picro-sulphuric acid, 754; Nematodes, 754.

Blattida, embryology of, 584.
Bleaching, 524 *et seq.*
Bleu de Lyon, 95, 109.
Bleu de nuit, 109.
Bleu lumière, 95, 108.
BLOCHMANN, ova of Amphibia, 563.
Blood, 692—696.
Blood-vessels of Vermes, 735, 736.
BOBRETZKY, embryology of Lepidoptera, 583.
BOCCARDI, gold method, 211; motor plates, 633.
BÖHM, gold method, 211.
BÖHMER, hæmatoxylin, 175.
BÖHMIG, Rhabdocœla, 727.
BÖHN, carmine, 142.
BOLL, gold-staining nervous centres, 674.
Bone, decalcification, 531, 532; preparation, 689—691.
Bonellia, 731.
Boracic preservative medium, 384.
Borax-carmine, GRENACHER'S, 158, 163; GIBBES', 160; NIKIFOROW'S, 157; WOODWARD'S, 159; BAUMGARTEN'S, 163; DUTILLEUL'S, 166; ditto and indigo, 229, 230, 673.
BORDEN, cochineal, 171.
Boric-acid carmine, 162, 165.
BORN, paraffin imbedding, 266; section-fixing method, 321; plastic reconstruction, 554.
Boroglyceride, 384.
BÖTTCHER-HERMANN, staining process, 96; p. 57, note.
Bougainvillia, 744.
BOUMA, stain for cartilage, 691.
BOVERI, ovum of Ascaris, 607, 611 medullated nerve, 656.
Brachionus, 732.
BRADY, chloral hydrate medium, 361.
Brain, methods of, BETZ, 668; BEVAN LEWIS, 657, 658, 660, 665, 672, 673, 677, 679; DUVAL, 663; DEECKE, 667, 672; GIACOMINI, 676, 678; GOLGI, 673; HAMILTON, 666, 672; MERKEL, 677; of insects, 718; and *see* Central nervous system.

The numbers refer to the Paragraphs, not to the Pages.

BRAMWELL, study of brain, 677.
BRANDT, glycerin jelly, 389; Protozoa, 751, 760.
BRASS, alcoholic carmine, 164; paraffin imbedding, 268, 279, 280; Protozoa, 756.
BRAUN, Alcyonaria, &c., 741; Rhabdocœla, 727; Nematodes, 729.
BRECKENFELD, *Hydra*, 743.
BREMER, motor plates, 633.
BROCK, macerating medium, 511.
BRÖSICKE, staining method, 222.
BRÜCKE, digestion fluid, 522; injection masses, 451.
Brunswick black, 413.
Bryozoa, 7, 14, 15, 713a, 741.
BUDGE, injection mass, 488.
BUSCH, eosin and hæmatoxylin, 238; decalcification, 531, 532.
BÜTSCHLI, paraffin imbedding, 269.

C.

Cadmium chloride, 211.
CALBERLA, methyl green, 105; methyl green and eosin, 250; glycerin liquid, 385; macerating mixture, 504; imbedding, 306.
Calcispongiæ, 749.
Calcium chloride mounting medium, 360.
CALDWELL, serial sections, 317.
Campanulariadæ, 743.
Canada balsam, 396—398, 400; for imbedding, 305.
CANFIELD, iris, 641.
Cannel oil, 333.
Caoutchouc for section fixing, 324.
CAPPARELLI, dichroism of methyl violet, 107.
Carbolic acid, 338, 350, 352.
Carbolised alcohol, 644.
Carbolised syrup, 350.
Carbonic acid narcotisation method, 17.
Carmine, acetic, SCHNEIDER'S, 154; SCHWEIGGER-SEIDEL'S, 155; alcoholic, GRENACHER'S, 163; MAYER'S, 164; BRASS'S, 164; FRANCOTTE'S, 165; DUTILLEUL'S, 166; alum, 149 —152; ammonia, 134—139; ARCANGELI'S, 162; BÖHM'S neutral, 142; borax, *see* Borax-carmine; boric acid, 162, 165; CUCCATI'S soda, 781; DELAGE'S osmium, 161; HEIDENHAIN'S neutral, 143; HOYER'S neutral, 141; ORTH'S lithium, 153; PERL'S soluble, 162; picro-carmine, 144—148; RANVIER'S picro-carmine, 144, 145; his neutral, 140; salicylic, 162.
Carmine double stains, 225—236.
Carmine, generalities on, 130—133; the sorts of, 130; use of, 131.
Carminic acid, 162.
Carminroth, 162.
CARNOY, acetic alcohol, 52; chromo-aceto-osmic acid, 36; cement, 426; cytological methods, 601, 602, 611; spermatology of Arthropods, 612.
CARRIÈRE, corpuscles of Herbst and Grandry (gold method), 618; eyes of Gastropods, 709; maceration of nervous tissue, 679.
CARTER, injection mass, 443.
Cartilage, 691.
Caryophyllia, 741.
Cassiopeia, 744.
CASTELLARNAU, de, Actiniæ, 741; Eucharis, 748; Gephyrea, 731; Lamellibranchiata, 706; Medusæ, 744; Nemertina, 728; Ophiuroidea, 739; Polychæta, 734; Zoantharia, 741.
Castor oil for mounting, 393, 605.
CATTANEO, corpuscles of Golgi, 637; Infusoria, 755.
Cedar oil for imbedding, 269; for clearing, 331; for mounting, 396.
Cell researches, *see* Cytological methods.
Celloidin, 290 *et seq.*; clearing sections, 298.
Cements, 407 *et seq.*; generalities, 407; comparative tenacity of, 410; APÁTHY'S, 430a; BELL'S, 411; colophonium, 421; French, 418; HARTING'S gutta-percha, 417; KITTON'S, 429; Knotting, 419;

INDEX. 395

The numbers refer to the Paragraphs, not to the Pages.

LOVETT'S, 430; MARSH'S gelatin, 408; STIEDA'S, 427; Tolu balsam, 426; turpentine, 420; ZIEGLER'S, 428.

Central nervous system, introduction, 657; hardening, 658—669, 673, 678; imbedding and cutting, 672; general stains, 673; special stains, 674; mounting, 675, 676; the half-clearing method, 677; dry processes for preserving brains, 678; dissociation methods, 679; neuroglia, 680; double stain, 681a; myelon of reptiles, &c., 681; methods of ADAMKIEWICS, 674; APÁTHY, 681; BEALE, 673; BECHTEREFF, 661; BECKWITH, 674; BENCZUR, 673; BETZ, 136, 668; BEVAN LEWIS, 657, 658, 660, 665, 672, 673, 677, 679; BOLL, 674; BYROM BRAMWELL, 677; CARRIÈRE, 679; DEECKE, 667, 672; DIOMIDOFF, 661; DUVAL, 231, 603; EXNER, 661, 669; FLESCH, MAX, 651, 673, 678; FREEBORN, 673, 679; FREUD, 674; GAULE, 661; GERLACH, 674; GIACOMINI, 674, 678; GIERKE, 505, 673, 680; GOLGI, 633, 645, 652, 673; HAMILTON, 666, 672; JELGERSMA, 673; KORYBUTT-DASZKIEWICS, 681; KOTLAREW-SKY, 661; KRAUSE, 662; KULT-SCHITZKY, 674; v. LENHOSSEK, 681; LUYS, 673; MAGINI, 673; MARTINOTTI, 673; MASON, 671; MERKEL, 673, 674; NIKIFOROW, 674; OBERSTEINER, 657, 660, 661, 673; OGATA, 661; OSBORN, 667, 672, 681; PAL, 673; ROSSI, 674; RUTHERFORD, 664; SAHLI, 659, 660, 674; SANKEY, 673; SCHIEF-FERDECKER, 674, 679; SCHMIDT, 681; STILLING, 679; STIRLING, 673; TAL, 673; v. THANHOFFER, 679; TRZEBINSKI, 661; UPSON, 673; WEIGERT, 650, 659, 674; ZUPPINGER, 673.

Cephalopoda, eyes, 710; ova, 578.
Cercaria, 726.

CERTES, Protozoa, 751, 757.
Cestodes, 725.
Chain section-cutting, 274.
Chilognatha, 716.
China blue, 653.
Chinolin blue, 111.
Chironomus, salivary gland, 45.
Chitin, clearing and softening, 529, 715, 716.
Chitonidæ, 711.
Chloral hydrate for narcotising, 14, 717; preservative solutions, 361; for preserving injection masses, 432; for maceration, 501, 647, 717.
Chloride of cadmium, 211.
Chloride of calcium mounting medium, 360.
Chloride of copper fluid, 64.
Chloride of gold, *see* Gold chloride.
Chloride of palladium, 84, 755.
Chloride of platinum, 42, 47, 88.
Chloride of zinc, 85; for hardening brain, 678.
Chlorine solution, 544; *see* also Bleaching.
Chloroform, for paraffin imbedding, 269; for celloidin imbedding, 294; for bleaching, 549, 716; for narcotisation, 11, 12.
Chondrosia, 202.
Chromate of ammonia, 43, 80, 81.
Chromate of lead stain, 219.
Chromates as fixing agents, 43.
Chromatin, reactions of, 602.
Chromic acid, generalities, 30; fixing, 30; hardening, 70, 660, 661; maceration, 507; decalcification, 534, 535; chromic acid with acetic acid, 31; with alcohol, 33, 71; with nitric acid, 39, 40, 535; with platinum chloride, 42; with osmic acid, 34; with acetic and osmic acid, 35, 36.
Chromic objects, action of light on, 30.
Chromo-acetic acid, 31; chromo-aceto-osmic, 35, 36; chromo-formic, 32; chromo-nitric, 39, 40; chromo-osmic, 34, 72; chromo-picric, 41, 74; chromo-platinic, 42, 73.

The numbers refer to the Paragraphs, not to the Pages.

CHRSCHTSCHONOWIC, gold method, 211.
Chrysaurein, 95, 104.
CHUN, imbedding Siphonophora, 282.
CIACCIO, gold process, 211; motor plates, 633.
Cilia of Infusoria, 759, 760.
Ciliated epithelium, 712, 761.
Cladocora, 741.
Cleaning slides and covers, 762.
Clearing, generalities, 2.
Clearing agents, 329 *et seq.* STIEDA's experiments on 330; NEELSEN and SCHIEFFERDECKER'S, 330—336; celloidin sections, 298.
Clove oil, 4, 332.
Cocain for narcotising, 15.
Coccidæ, 715.
Cochineal, use of, 167; BORDEN'S, 171; CZOKOR'S, 170; KLEIN'S, 170; MAYER'S, 168; PARTSCH'S, 169.
Cochlea, 648, 649.
Cœlenterata, 741—748.
COHNHEIM, gold staining, 206; silver staining, 632.
COLE, freezing method, 309; gum medium, 381.
COLLIN, *Criodrilus*, 734.
Collodion imbedding, 290 *et seq.*; section-fixing process, 315, 316, 327.
Collodionising paraffin sections, 275.
Colophonium, cement, 421; imbedding mass, 304.
Comatula, 740.
Combination stains, 224 *et seq.*; and *see* Stains, combination.
Congelation imbedding methods, 307—312.
Congo red, 95, 113.
Conjunctiva, 621.
Connective tissue, 682 *et seq.*
Convoluta, 727.
Copal, imbedding process, 303; varnish for mounting, 403.
Copepoda, 714.
Copilia, 714.
COOK, hæmatoxylin, 177.
Coral, 303.
Corallin, 95, 104.
Cornea, 506, 624, 625.
CORNIL, preservative media, 365.

Corpuscles, tactile, 616, 620; corpuscles of Golgi, 635—637; of Herbst and Graudry, 618; of Krause, 621; of Pacini, 619.
Corrosion, 527 *et seq*; ALTMANN'S methods, 530.
Corrosive sublimate, fixing liquids, 45, 46; LANG's liquids, 46; preservative liquids, 363—369.
Covers and slides, cleaning, 762.
Creosote, *see* Kreasote.
Criodrilus, 734.
Cristatella, 14.
Croccin, 95, 120.
Crystalline, 626, 627.
Ctenophora, killing, 44b; imbedding, 300; preparation, 748.
CUCCATI, picro-carmine, 147; hæmatoxylin, 187; brain of blow-fly, 718; retina, 644, 645; soda-carmine, 781; Säurefuchsin stain, 718.
Cucumaria, 737.
Cupric fixing mixtures, 44, 44a, 64, 727; hardening ditto, 82; preservative ditto, 64; impregnation process, 218.
Curare for narcotisation, 16.
CURSCHMANN, amyloid matter, 105.
Cyanin, 95, 111, 751.
CYBULSKY, gold method, 212; muzzle of ox, 615.
Cytological methods, 598 *et seq.*; methods of study, 598; observation of living cells, 599; staining ditto, 600; fresh cells, 601; microchemical reactions, 602; fixing, 603; cytological stains, 604; mounting, 605; synthetic review, 606; Nebenkern, Archoplasmakugel, sphère attractive, 607; nucleus of Balbiani, 608; cytology of the ovum, 609, 611; spermatology, 612; methods of BABES, 604, 610; BALBIANI, 608; BAUMGARTEN, 610; BENEDEN, van, 611; BENEDEN, van, and NEYT, 603, 607, 611; BIZZOZERO, 610; BOVERI, 607, 611; CARNOY, 601, 602, 611; FLEMMING, 599, 601,

The numbers refer to the Paragraphs, not to the Pages.

603, 606, 609; GEHUCHTEN, van, 603, 611; GILSON, 603; GRENACHER, 605; KULTSCHITZKY, 611; NISSEN, 610; PEREMESCHKO, 599; PFITZNER, 606; PLATNER, 607; RABL, 603, 606; SCHOTTLÄNDER, 610; STRASBURGER, 606; TIZZONI, 610; USKOFF, 606; v. la VALETTE ST. GEORGE, 600, 601; ZACHARIAS, 611; ZWAARDEMAKER, 610.
CZERNY, macerating mixture, 504.
CZOKOR, cochineal, 170; turpentine cement, 420.

D.

Dahlia, 93, 95, 103, 685, 686.
Dammar, 396, 399, 400.
DAVIDOFF, ova of *Distaplia*, 577.
DAVIES, injection, 444.
DEANE, mounting medium, 377; glycerin jelly, 386.
Decalcification, 531 *et seq.*
Decapoda, 714.
DECKER, section stretcher, 273.
DEECKE, hardening brain, 667; imbedding and cutting, 672.
Dehydration, 2.
DEKHUYSEN, fat, 683.
DELAFIELD, hæmatoxylin, 174.
DELAGE, osmium-carmine, 161; sulphate of iron for fixing, 727; Rhabdocœla, 727.
Deltapurpurin, 95, 114.
Dendrocœla, 727.
Dendrophyllia, 741.
DENNISSENKO, retina, 644.
Depigmentation, 542 *et seq.*
Desilicification, 541.
DEZSÖ, *Tethya*, 749.
Digestion, 521 *et seq.*
DIMMOCK, carminic acid, 162.
DIOMIDOFF, hardening brain, 661.
DIPPEL, hæmatoxylin, 185.
Diptera, embryology, 585.
Dissections, minute, 4.
Dissociation, 495 *et seq.*; of nervous tissue, 679.
Distaplia, 577.

DOGIEL, olfactive organs, 623a; iris, 640; methylen-blue impregnation method, 706; connective-tissue, tendon, 682.
DOSTOIEWSKY, iris, 641; retina, 645.
Double imbedding in celloidin and paraffin, 299.
DOWDESWELL, stain for spermatozoa, 612.
DOYÈRE, nerve and muscle of Arctiscoida, 719.
DRASCH, gold staining, 205; tactile hairs, 628.
DROOST, Mollusca, 712.
DUNHAM, clearing celloidin sections, 298.
DU PLESSIS, fixing method, 44c; Nemertians, 728; Protozoa, 760.
DUTILLEUL, picro-borax-carmine, 166.
DUVAL, silver staining, 199, 200a; carmine and anilin blue, 231; collodion imbedding, 290 *et seq.*; embryology of Aves, 559; hardening encephalon, 663.

E.

Ear, inner, 648, 649.
Eau de Javelle, 528, 548, 563.
Eau de Labarraque, 529, 548, 563, 582.
Echinodermata, 737—740.
Echinorhyncus, 730.
Echtgelb, 95, 120.
Echtgrün, 653.
Echtroth, 95, 104.
EDINGER, nerve-centres, 79.
Egg-emulsion imbedding masses, 306, 312.
EHLERS, fixing fluid for Annelids, 734.
EHRENBAUM, imbedding method, 304.
EHRLICH, dahlia, 103; hæmatoxylin, 182; methylen blue for nerves, 654; plasma-cells, 684, 685.
EISIG, fixing mixture, 83; narcotising, 13, 728, 734.
Elastic tissue, 100, 688.
ELIEL, gum for labels, 763.
ELOUI, eosin, 116.
Embryology, general methods, 551—554; of *Agelena*, 588; *Amarœcium*,

The numbers refer to the Paragraphs, not to the Pages.

577a; Amphibia, 551, 563—570; Amphipoda, 592; Aphides, 586; Araneina, 587; Arthropoda, 581 *et seq.*; *Ascaris*, 52, 611; *Astacus*, 591; Aves, 556—559a; *Axolotl*, 564; Blattida, 584, 715; Cephalopoda, 578; *Distaplia*, 577; Diptera, 585; Echinodermata, 740; Gastropoda, 579, 580; *Lacerta*, 561; Lepidoptera, 583; *Limax*, 580; *Lumbricus*, 596; Mammalia, 555; Mollusca, 578 *et seq.*; *Orchestia*, 592; Phalangida, 589, 590; Phryganida, 584; *Planaria*, 595; Porifera, 749; rabbit, 555; *Rana*, 568; Reptilia, 560—562; *Salamandra*, 567; Salmonidæ, 551; *Tænia*, 594; Teleostea, 40, 551, 571—576; *Triton*, 565, 566; Tunicata, 577, 577a; Vermes, 594 *et seq.*

Embryos, reconstruction from sections, 554.

EMERY, embryological methods, 55; injection, 483.

Encephalon, 657 *et seq.*; and *see* Nervous centres.

Endosmosis, to avoid, 4.

ENGELMANN, ciliated epithelium, 712; green light in microscopy, 761.

ENTZ, GÉZA, Protozoa, 760.

Eosin, 95, 116; ELOUI'S formulæ, 116; FISCHER'S, 116; LAVDOWSKY'S, 116; eosin and anilin green, 256; — and dahlia, 257; — and gentian, 245; — and hæmatoxylin, 116, 238, 239; — and iodine green, 259; — and methyl green, 250, 251; — and methyl violet, 258; — and picro-carmine, 235; and silver nitrate, 262.

Epidermis of Mollusca, 706, 712.

Epithelium, 613 *et seq.*; ciliated, 712, 761; glandular, 613.

ERLICKI, hardening fluid, 79.

Erythrosin, 116.

Essence, *see* Oil, and Clearing agents.

ETERNOD, histological rings, 201.

Etherisation, 13.

Eucharis, 748.

EULENSTEIN, cement, 414.

Eunice, 734.

Examination media, 342 *et seq.*, 767.

EXNER, hardening brain tissue, 661, 669.

Exosmosis, to avoid, 4.

Eyes, of Arthropoda, 717, 718; of Cephalopoda and Heteropoda, 710; of Chitonidæ, 711; of Gastropoda, 709; of *Pecten*, 711a; of Vertebrates, 644 *et seq.*

F.

FABRE-DOMERGUE, glucose preservative medium, 382.

FARRANT, do., 376, 379, 380.

Fat, 36, 683.

Fecundation, artificial, 551.

FEBRIA, elastic tissue, 688.

FIEDLER, *Spongilla*, 749.

FISCHER, eosin, 116; soap imbedding mass, 726; Trematodes, 726; tactile corpuscles, 616; motor plates, 631.

Fixing, generalities, 2, 6 *et seq.*, 22 *et seq.*

Fixing agents, 26 *et seq.*; action of, 23; choice of, 24; methods of using, 25; washing out, 23; acetic acid, 51, 52; ALTMANN'S methods, 37; alum, 44b; bichloride of mercury, 45, 46; bichromate, 43; CARNOY'S, 36, 52; chromates, 43; chromic mixtures, 31—36, 39—42; cytological fixing agents, 603; DU PLESSIS'S method, 44c; FLEMMING'S, *see* FLEMMING; MAX FLESCH'S, 34; FOL'S, 41, 50; gold chloride, 49; heat, 7, 25; HERMANN'S, 769; iodine, 66; KLEINENBERG'S method, 55; KULTSCHITZKY'S, 44; LANG'S, 46; MAYER'S, 56, 57, 63; MERKEL'S fluid, 42; nitric acid, 37; nitric and chromic acid, 39, 40; nitrate of silver, 38; osmic acid, 26—29; PERENYI'S liquid, 39; picric acid fluids, 41, 54—59; platino-aceto-osmic mixture, 769; RIPART and PETIT'S solution, 64.

INDEX.

The numbers refer to the Paragraphs, not to the Pages.

FLECHSIG, gold method, 211.
FLEISCHMANN, *Limax*, 706.
FLEMMING, chromo-aceto-osmic acid, weak solution, 35; strong ditto, 36; cytological methods, 599, 601, 603, 606, 609; damar solution, 399; double stains, 225; eyes of Gastropods, 709; fixing liquids, 31, 35—37, 51, 59; imbedding mass, 285; injection of Molluscs, 707; retina, 645, 646; spermatological methods, 612; fatty tissue, 683; bone, 690; mucus cells, 697; goblet cells, 699.
FLESCH, MAX, chromo-osmic mixture, 34, 72; double stains, 230, 233; inner ear, 649; modification of Weigert's hæmatoxylin stain, 650, 651; staining nervous centres, 673; dry preservative process, 678; dental tissue, 691; ditto blood, 692; stomach-glands, 700.
FLÖGEL, serial sections, 319.
FLORMAN, celloidin imbedding, 294.
FOETTINGER, chloral hydrate narcotisation method, 14, 728; prepared paraffin, 279.
FOL, albumen fixative, 318; chromo-aceto-osmic acid, 35; cleaning slides, 762; gelatin fixative, 322; glycerin jelly, 392; gumming labels, 763; imbedding *in vacuo*, 271; injection masses, 446, 453, 458, 466, 470; Infusoria, 50, 214, 760; narcotisation method, 17; paraffin oven, 270; perchloride of iron fixing and staining process, 50, 214; picro-carmine, 147; picro-chromic acid, 41; reconstructing sections, 554; ribesin, 196.
Foot-glands of Gastropoda, 708.
Formic acid, for fixing, 53.
FOSTER and BALFOUR, embryology of Aves, 556, 557.
Fowl, embryology of, 556—559a.
FRANCOTTE, alcoholic carmine, 165; section-stretcher, 273; vacuum imbedding, 271; vegetable wax ditto, 281.
FREEBORN, staining nervous centres, 673; dissociation of ditto, 679; connective tissue, 682.
FRENZEL, serial section methods, 320, 324.
FREUD, gold process for nervous centres, 674.
FREY, ammonia-carmine, 139; injection mass, 464; iodised serum, 347; Parma blue, 108.
Frog, bladder, 639.
Fuchsin, 95, 104, 107a, 246.
Fuchsin S., 112; and *see* Säurefuchsin.

G.

GAGE, picro-carmine, 147; section-stretcher, 273; preservative fluid, 369; injection, 491; smooth muscle, 638.
GALLI, stain for neuroceratin funnels, 653.
GANNAL, mounting medium, 358.
GARBINI, safranin staining, 103; double stain, 248; Alcyonaria, 741; cleaning slides, 762.
Gastropoda, embryology, 579, 580; foot-glands, 708; eyes, 709; preparation, 18, 706.
GAULE, section-fixing process, 314; hardening brain, 661.
GEHUCHTEN, VAN, cytological methods, 603, 611.
Gelatin, imbedding masses, 286—289, 310, 311, 768; injection masses, 432 *et seq.*; cement, 408.
GELPKE, Weigert's hæmatoxylin, 650.
Gentian violet, 95, 102, 245.
Gephyrea, 731.
GERLACH, gold method, 211, 674; gelatin imbedding, 289; injection, 441; embryology of Aves, 556a.
GÉZA ENTZ, Protozoa, 760.
GIACOMINI, gelatin process for serial sections, 674; dry process for preserving brains, 678.
GIBBES, HENEAGE, borax-carmine, 100, 229; combination stains, 234, 261; cleaning slides, 762.
GIERKE, ammonia-carmine, 137; mace-

The numbers refer to the Paragraphs, not to the Pages.

ration of nerve-tissue, 505; staining nervous centres, 673; neuroglia, 680.
GIESBRECHT, serial section method, 317; paraffin imbedding, 269.
GIESON, VAN, clearing colloidin sections, 298.
GILSON, preservative fluid, 368; cytological methods, 603, 612.
Glandular structures, 697 et seq.
Glucose preservative medium, 382.
Glue, marine, 416.
Glycerin, 383—385; glycerin mounts, to close, 407, 409; glycerin and gum, 379, 380; glycerin jelly, 386—392, 729.
GOADBY, preservative fluids, 366.
Goblet cells, 699.
GODFRIN, imbedding mass, 768.
Gold chloride, for fixing, 49; for impregnation, generalities, 204, 205; methods of BASTIAN, 211; BECKWITH, 674; BOCCARDI, 211; BÖHM, 211; CARRIÈRE, 618; CHRSCHTSHONOWIC, 211; CIACCIO, 211; COHNHEIM, 206; CYBULSKY, 212; DRASCH, 205; FLECHSIG, 211; FREUD, 674; GERLACH, 211, 674; GOLGI, 633; HÉNOCQUE, 211; HOYER, 211; IGACUSCHI, 703; KOLOSSOW, 211; LÖWIT, 207; MANFREDI, 211, 636; MAYS, 634; NESTEROFFSKY, 211; RANVIER, 208, 209; REDDING, 211; VIALLANES, 210.
Gold orange, 95, 120.
Gold size, 415.
GOLGI, corpuscles of, 635, 636, 637; gold process, 633; nerve-tissue, 633, 645, 652, 673; stain for nervous centres, with mercury, 673.
Gorgonia, 741.
GORONOWITSCH, embryology of Teleostea, 576.
GRAM, stain for bacteria, 102.
Granule cells, 684.
GRASER, staining process, 107.
Green light in microscopy, 606, 761.
GRENACHER, alum-carmine, 149—152; alcoholic carmine, 163, 164; bleaching mixture, 550; borax-carmine, aqueous, 158; alcoholic, 163; castor oil for mounting, 393, 605; hæmatoxylin, 174; purpurin, 189.
GRIESBACH, methyl green, 105, note; Bengal rose, 117; iodine green, 121; double stain, 260; elastic tissue, 688.
GRÜBLER, his histological reagents, 94, 103, 782, 783.
Grünstichblau, 109.
GUIGNET, soluble Prussian blue, 450.
Gum arabic, pure, 319; preservative media, 376—381; imbedding masses, 308—311; injection mass, 472; mucilage for labels, 763.
Gum damar, 399.
Gutta-percha, cement, 417; section-fixing process, 324.

H.

Hæmatoxylic eosin, 239.
Hæmatoxylin, acid, anonymous, 186; APÁTHY'S, 179; ARNOLD'S, 177; BENDA'S, 612, 651; BÖHMER'S, 175; COOK'S, 177; CUCCATI'S, 187; DELAFIELD'S, 174; DIPPEL'S, 185; EHRLICH'S, 178; GRENACHER'S, 174; HEIDENHAIN'S, 178; HICKSON'S, 177; KLEINENBERG'S, 184; KULTSCHITZKY'S, 674; MINOT'S, 181; MITCHELL'S, 177; PRUDDEN'S, 174; RANVIER'S, 176; RENAUT'S, 183; SANFELICE'S, 780; WEIGERT'S, 180, 650, 651; generalities, 172, 173; double stains, 237—244.
Hairs, tactile and others, 613, 628.
Halicryptus, 731.
HALLER, BÉLA, macerating fluid, 515, 712.
HAMANN, carmine, 150; Echinoderms, 738, 739.
HAMILTON, freezing process, 308; WEIGERT'S hæmatoxylin stain, 651; hardening brain, 666; freezing ditto, 672.

INDEX.

The numbers refer to the Paragraphs, not to the Pages.

HANAMAN, cleaning slides, 762.
HÄNTSCH, glycerin medium, 385.
Hardening agents, 67 *et seq.*; alcohol, 88; bichromate of ammonia, 80; of potash, 77 ditto with cupric sulphate, 79; chloride of palladium, 84; of platinum, 83; chromate of ammonia, 81; chromic acid fluids, 70—74; nitric acid, 76; osmic acid, 75, 661, 669; ERLICKI's solution, 79; MERKEL's, solution, 83; MÜLLER's solution, 78; generalities, 67 to 69; hardening nerve-centres, 659 *et seq.*; PERÉNYI's solution, 39; picric acid, 86.
HARDY, Rotifers, 732.
HARMER, silver staining marine animals, 202.
HARRIS, stain for blood, 693.
HARTIG, white injection mass, 463.
HARTING, preservative fluid, 363; cement, 417.
HAYEM, blood, 696.
Heat, killing by, 7; fixing by, 25.
HEIDENHAIN, carmine, 143; hæmatoxylin, 173; anilin blue, 125; triple stain, 252; small intestine, 701.
Helix, eyes, 709; ova, 579; sexual gland, 612.
HENKING, collodionising sections, 275; embryology of Diptera, 585; of Phalangida, 590.
HENNEGUY, alum-carmine, 151; embryology of Aves, 557; of *Axolotl*, 564; of Gastropoda, 579; of Mammalia, 555; of Teleostea, 571; sexual gland of *Helix*, 612; Protozoa, 750, 751.
HÉNOCQUE, gold process, 211.
HERMANN, silver staining, 199; papillæ foliatæ of rabbit, 623; platino-aceto-osmic mixture, 769.
HERMANN-BÖTTCHER staining process, p. 57, note.
HERTWIG, silver staining, 199, 202; macerating methods, 514; Actiniæ, 9, 742; Medusæ, 745, 746; embryology of *Rana*, 568, 611; of

Triton, 566; of *Strongylocentrotus*, 611.
HERXHEIMER, elastic tissue, 688.
HESCHL, amyloid matter, 105.
Heteropoda, eyes, 710.
HEYDENREICH, amber varnish, 423.
HEYS, balsam, 397.
HICKSON, hæmatoxylin, 177; maceration, 501; eye of *Musca*, 717.
Hirudinea, 737.
HIS, fixing, 37; cornea, 624.
HOCHSTETTER, injection, 487.
HOFFMANN, vacuum imbedding, 271.
"Hofmann's Grün," 121.
HOGGAN, silver staining, 199; histological rings, 201; perchloride of iron stain, 214.
HOLL, toluol for imbedding, 268.
Holothuroidea, 737.
Horn, 613.
HOUSSAY, foot-glands of Gastropods, 708.
HOYER, ammonia-carmine, 138; neutral ditto, 141; picro-carmine, 141; silver staining, 199; gold ditto, 211; gum arabic mounting medium, 378; gelatin injection masses—carmine, 445; Berlin blue, 454; lead chromate, 457; silver nitrate, 459; green, 460; shellac mass, 489; oil-colour masses, 490.
HUDSON, Rotifers, 732.
HUXLEY and MARTIN, ammonia-carmine, 139.
HYATT, imbedding method, 302.
Hydatina, 732.
Hydra, 741, 743.
Hydrate of chloral, for narcotising, 14; and *see* Chloral.
Hydrochloric acid for decalcification, 536, 537.
Hydrofluoric acid for desilicification, 541.
Hydrogen peroxide, for bleaching, 30, 547.
Hydroidea, 7, 743.
Hypochlorite of potassium, *see* Eau de Javelle.
Hypochlorite of sodium, *see* Eau de Labarraque.

The numbers refer to the Paragraphs, not to the Pages.

I.

IGACUSCHI, gold process for liver, 703.
IIJIMA, embryology of *Planaria*, 595.
Imbedding, defined, 2; generalities on, 263 *et seq.*; imbedding *in vacuo*, 271; and *see* also Celloidin, Paraffin, &c.
Imbedding masses, albumen, 306, 312; bayberry tallow, 281; celloidin, 290 *et seq.*; collodion, *ibid.*; colophonium, 304; copal, 303; egg emulsion, 306; freezing masses, 294, 307—312, 672; gelatin, 286—289, 310, 311; gelatin soap, 768; gum, 300; gum and syrup, 308, 309; paper, 266; paraffin, 276—280; shellac, 302; soap, 283—285, 726; spermaceti, 280; vegetable wax, 281; wax and oil, 280.
Impregnation methods, 197 *et seq.*, 530, 766; and *see* Gold chloride, Nitrate of silver, &c.
Inchiostro di Leonardi, 107.
Indian ink injection, 485.
Indifferent liquids, 342 *et seq.*
Indigo-carmine, 190; THIERSCH'S staining method, 191; SEILER'S, 229; MERKEL'S, 230.
Indirect staining method, 96 *et seq.*
Indulin, 95, 110.
Infusoria, 750 *et seq.* and *see* Protozoa.
Injection of leeches, 736.
Injection of Mollusca, 707.
Injections, 431 *et seq.*; injection-masses, *albuminous*, JOSEPH'S, 471 : *aqueous*, BJELOUSSOW'S gum, 472; EMERY'S carmine, 483; HOYER'S silver nitrate, 459; LETELLIER'S vanadate of ammonia, 484; MAYER'S blue, 482; MÜLLER'S blue, 481; RANVIER'S blue, 480; TAGUCHI'S Indian ink, 485 : *collodion*, HOCHSTETTER'S, 487; SCHIEFFERDECKER'S, 486 : *fatty*, HOYER'S oil colour, 490; TEICHMANN'S linseed oil, 492: *gelatinous*, BRÜCKE'S Berlin blue, 451; CARTER'S carmine, 443; DAVIES'S carmine, 444; FOL'S carmine, 446; FOL'S black (or brown), 466; FOL'
blue, 453; FREY'S white, 464; GERLACH'S carmine, 441; GUIGNET'S Prussian blue, 450; HARTIG'S white, 463; HOYER'S carmine, 445; HOYER'S Berlin-blue, 454; HOYER'S chrome-yellow, 457; HOYER'S silver-yellow, 459; HOYER'S green, 460; MILLER'S purple, 467; RANVIER'S carmine, 439; RANVIER'S Prussian blue, 448, 449; RANVIER'S silver nitrate, 469; ROBIN'S vehicles, 432, 433; ROBIN'S anilin masses, 438a; ROBIN'S cadmium, 437; ROBIN'S carmine, 434; ROBIN'S mahogony, 435; ROBIN'S Prussian blue, 436; ROBIN'S Scheele's green, 438; TEICHMANN'S white, 465; THIERSCH'S carmine, 442; THIERSCH'S Prussian blue, 452; THIERSCH'S lead chromate, 456; THIERSCH'S transparent green, 461; VILLE'S carmine, 440 : *glycerin*, BEALE'S carmine, 474; BEALE'S Prussian blue, 476, 477; RANVIER'S Prussian blue, 478; ROBIN'S carmine, 475; ROBIN'S cadmium, 437; ROBIN'S mahogany, 435; ROBIN'S Prussian blue, 436; ROBIN'S Scheele's green, 438: *resinous*, BUDGE'S asphaltum, 488; HOYER'S shellac, 489; HOYER'S oil-colour, 490 : *starch*, PANSCH'S, 491.
Injections, natural, 494.
Insecta, *see* Arthropoda.
Insecta, brain of, 718.
Intestine, 81; mucosa of, 701.
Intra-epidermic nerve-fibres, 614.
Iodine, for fixing, 66; for hardening, 89; LUGOL'S solution, 66.
Iodine green, 95, 105, 121.
Iodine hæmatoxylin, 187, 780.
Iodised serum, 346, 347; ditto for macerations, 496, 497.
Iris, 640, 641.
Iron, perchloride, for fixing, 50; for staining, 50, 214, 645; pyrogallate of, 215; sulphate of, 727.
ISRAEL, orcin, 194.

The numbers refer to the Paragraphs, not to the Pages.

J.

JACOBS, freezing mass, 311.
JACQUET, injection of leeches, 736.
JÄGER, glycerin medium, 385.
JAKIMOVITCH, silver staining, 200; medullated nerves, 656.
JAMES, cleaning slides, 762; damar solutions, 399.
Japan wax, 281.
JELGERSMA, staining nervous centres, 673.
JENSON, spermatological methods, 612.
JOLIET, imbedding method, 300.
JOSEPH, injection-mass, 471.

K.

KADYI, imbedding mass, 284.
KAISER, glycerin jelly, 286, 288, 390.
KASTSCHENKO, reconstruction from sections, 554.
KEMP, blood-plates, 694.
KENT, S., Infusoria, 66, 758.
Kernschwarz, 195, 607.
Kidney, 704.
Killing tissues and organisms, 6—21; killing by heat, 7; by rapid fixing agents, 8; by narcotisation, 9; by nicotin, 10; by chloroform, 11, 12; by ether and alcohol, 13; by chloral, 14; by cocain, 15; by carbonic acid, 17; by asphyxiation, 18; by chromic acid, osmic acid, "Kleinenberg," or sublimate, 19; by fresh water, 28; by warm water, 21.
KINGSLEY, imbedding method, 266.
KITTON, asphalt, 412; cement, 429.
KLEBS, glycerin jelly, 287.
KLEIN, fixing mixture, 33; hardening intestine, 81; cornea, 624.
KLEINENBERG, colophonium for mounting, 401; embryology of *Lumbricus*, 596; picro-sulphuric acid, 55; hæmatoxylin, 184.
KLEMENSIEWICS, picro-carmine, 147.
Knotting, for cement, 419.
KOCH, VON, coral method, 303.

KOGANEI, iris, 641.
KÖLLER, embryology of Aves, 559a.
KÖLLIKER, ovum of rabbit, 555; bone, 690.
KOLLMANN, fixing mixture, 40.
KOLOSSOW, gold method, 211; osmium mixture, 28, 222.
KOROTNEFF, killing Siphonophora, 11; paraffin imbedding, 272; lubricating microtomes, 764.
KORSCHELT, Infusoria, 753.
KORYBUTT-DASZKIEWICS, myelon of reptiles, 681; plasma-cells, 686.
KOSSINSKI, double stains, 249.
KOTLAREWSKY, central nervous system, 661.
KRAUSE, corpuscles of, 621; molybdate of ammonium, 221; motor plates, 633; silver staining, 199; retina, 645, 647; spinal cord, 662.
Kreasote, for bleaching, 545; for clearing, 298, 339; preservative fluids, 353—355.
KRONECKER, artificial serum, 348.
KRÖNIG, cement, 421.
KÜHNE, digestion fluids, 526; macerating mixture, 517; motor plates, 633.
KÜKENTHAL, Annelids, 14, 733, 734.
KULTSCHITZKY, fixing mixture, 44; double imbedding, 298; preservative fluids, 362; ovum of *Ascaris*, 611; tactile corpuscles, 620; stain for nervous centres, 674.
KÜNSTLER, Protozoa, 760.
KUPFFER, embryological methods, 561; axis cylinder, 654.
KUSKOW, digestion fluid, 524.
KUTSCHIN, clearing method, 330.

L.

Labelling slides, 763.
Labyrinth of ear, 648, 649.
LAMANSKY, sensibility of the eye, 761.
Lamellibranchiata, aquiferous pores, 706; epithelium, 712; eyes, 711a; injection, 707; sexual gland, 612; shell, 303, 304, 713.

INDEX.

The numbers refer to the Paragraphs, not to the Pages.

LANDOIS, impregnation methods, 220; macerating mixture, 505.
LANDOLT, retina, 647.
LANDSBERG, Infusoria, 753.
LANG, fixing liquids, 46; methods for Platyhelmia, 46, 235, 727.
LANGERHANS, *Amphioxus*, 630; gum and glycerin medium, 380; tactile corpuscles, 617.
LANGHANS, dichroism of hæmatoxylin stains, 173.
LANKESTER, eyes of *Limulus*, 717.
LAVDOWSKY, bilberry juice stain, 196; eosin, 116; myrtillus, 196; chloral hydrate medium, 361; macerating fluid, 501; inner ear, 649.
LAWRENCE, glycerin jelly, 387.
Lead, acetate, 87, 661; chromate, 219.
LEBER, impregnation methods, 217, 219.
LEGAL, alum-carmine, 152.
LEGROS, silver staining, 200a.
LENHOSSEK, myelon of reptiles, 681.
LENNOX, retina, 645.
Lens, crystalline, 626, 627.
LÉON, nucina, 196.
Lepidoptera, embryology, 583.
LEUCKHART, imbedding boxes, 266.
LEVEN, saffron stain, 192.
Levulose for mounting, 765.
LEWIS, BEVAN, *see* BEVAN LEWIS.
Lichtblau, 108, 126.
Lichtgrün, 105.
LICHTHEIM, staining with Weigert's hæmatoxylin, 650.
Light, action of, on alcohol containing chromic objects, 30; green, in microscopy, 606, 761.
Limax, 706; embryology, 580.
Limulus, 717.
Liquidambar, 405.
LIST, hæmatoxylin and eosin, 238; ditto and nitrate of rosanilin, 242; other combination stains, 251—256; macerations, 498; Anthozoa, 742; Coccidæ, 715; retina, 645; mucus cells, 698; goblet cells, 699.
Lithium-carmine, 153.
Litmus stain, 196.
Liver, nervous networks of, 703.

Lizzia, 744.
LOCY, embryology of *Agelena*, 588.
LOEWE, crystalline, 626.
LOEWENTHAL, nerve-centres, 79; picro-carmine, 148.
LONGWORTH, corpuscles of Krause, 621.
LOOSS, softening chitin, 529; Nematodes, 729.
LOVETT, cement, 430.
LÖWIT, gold method, 207, 631.
Loxosoma, 202.
Lubricant for microtomes, 764.
LUGOL, solution of, 66.
Lumbricus, 733, 734; embryology, 580.
LUSTGARTEN, elastic tissue, 100, 688.
LUYS, noir colin, 673.
Lymphatics, 650; and *see* Injections.

M.

MACALLUM, *Sphyranura*, 726.
Maceration, 495—520, 712; nervous tissue, 679, 680.
Macrotoma, 722.
Magdala red, 95, 104.
Magenta, 95, 104.
MAGINI, stain for nervous centres, 673.
Mammalia, embryology, 555.
Manchester brown, 106.
MANFREDI, gold method, 211, 636.
MARCHI, VON, *Limax*, 706; corpuscles of Golgi, 636.
Marine animals, killing, 20, 21; fixing, 25.
Marine glue, 416.
MARK, collodionising sections, 275; embryology of *Limax*, 580.
MARSH, bleaching, 543; gelatin cement, 408.
MARTINOTTI, safranin staining, 103; damar solutions, 399; anilin black, 673; picro-nigrosin, 673; elastic tissue, 688.
MASON, nervous system of Reptiles, 671.
Mastzellen, 684—687.
Maturation of ovum, 609, 611; and *see* Cytological methods.
MAUPAS, Protozoa, 760.

INDEX.

The numbers refer to the Paragraphs, not to the Pages.

MAURICE and SCHULGIN, double stain, 231; embryology of *Amarœcium*, 577a.

Mauvein, 95, 104.

MAX FLESCH, *see* FLESCH.

MAYER, PAUL, acidulated alcohol, 63; alcoholic carmine, 164; prep. of Arthropods, 5, 55, 168, 714; albumen fixative for sections, 318; bleaching, 542; Bismarck brown, 106; cochineal, 168, 673; desilicification, 541; injection, 482; picro-carmine, 147; picro-sulphuric acid, 55; picro-hydrochloric acid, 57; picro-nitric acid, 56; section-stretcher, 273; section fixing methods, 317, 318; shellac fixative, 317.

MAYER, S., "violet B.," for connective tissue, 682.

MAYS, nerve-endings (gold process), 634.

MAYSEL, Bismarck brown, 106.

Medullated nerve, 650—656.

Medusæ, 12, 19, 44b, 514, 744—746.

Mercury, bichloride of, 45, 46, 363 *et seq.*; and see Corrosive sublimate.

MERK, making up chromo-aceto-osmic acid, 36.

MERKEL, chromo-platinic liquid, 42, 73, 83; impregnation method, 221; carmine and indigo stain, 230, 673, 693; celloidin imbedding, 290; half-clearing brain sections, 674.

Metallic stains, 197 *et seq.*

Metanil yellow, 95, 119.

Methanilin green, 105; methanilin violet, 107.

Methyl green, 95, 105, 602, 750; combination stains with, 250—254.

Methyl violet, 95, 104, 107.

Methylen blue, 93, 95, 127; for nerve-endings, 654; for central nervous system, 674; for impregnation, 766.

MEYER, salicylic solutions, 375, 376.

MICHELSON, corpuscles of Pacini, 619.

Microtomes, 263; oiling, 764.

MILLER, vegetable wax, 281.

MILLON'S test, 602.

Milnesium, 719.

MINOT, hæmatoxylin stains, 181; clearing celloidin sections, 298; macerating skin, 613.

MITCHELL, hæmatoxylin, 177.

MITROPHANOW, gold process, 613; maceration, *ibid.*; sense organs of Amphibia, 622.

MÖBIUS, macerating media, 512, 712.

MOLESCHOTT, potash or soda solution, 502; MOLESCHOTT and PISO BORME'S macerating fluid, 500.

Mollusca, 706—713; aquiferous pores, 706; embryology, 578 *et seq.*; eyes, 709—712; fixation, 706; foot-glands, 708; injection, 707; maceration, 712; removing mucus from, 706; shell, 303, 304, 713.

Molluscoida, 7, 14, 15; and *see* Bryozoa.

Molybdate of ammonium, 221.

Monadina, 760.

MONDINO, medullated nerves, 652.

Monobromide of naphthalin, 395.

MOORE, stain for blood, 693.

MORGAN, ova of Arthropods, 582, 715.

Morphia for narcotising, 16.

MOSELEY, shell, 711; Chitonidæ, 711.

Mosso, blood, 692.

Motor plates, 631 *et seq.*

Mounds of Doyère, 719.

Mounting media, *see* Examination media.

Mucilage for labels, 763.

Mucosa of intestine, 701.

Mucus, removing from Mollusca, 706.

Mucus glands, 697 *et seq.*

MÜLLER, hardening liquid, 78; silver staining, 200; injection mass, 481.

MÜNDER, G., his histological reagents, 94, 103.

MUNSON, chloral hydrate fluid, 361.

Muscle, dissociation of, 517, 630; nerve-endings in, 629 *et seq.*; smooth, 638 *et seq.*

Myelin, 656.

Myelon, *see* Central nervous system.

Myrtillus, 196.

Myrtle wax, 281.

The numbers refer to the Paragraphs, not to the Pages.

N.

Nails, 613.
NANSEN, maceration of nerve-tissue, 505.
Naphtha, for clearing, 276, 340; preservative fluids, 355, 356.
Naphthalin, monobromide, 395.
Naphthalin red, 95, 104.
Naples Zoological Station methods, 5, *sub fin.*
Narcotisation, 9—19.
NEALEY, bone, 689.
Nebenkern, 607.
NEGRO, motor plates, 633.
Nematodes, 529, 729, 754.
Nemertina, 7, 728.
Neophalax, embryology, 584.
Nerves, medullated, 650—656.
Nerve-endings in muscle, 629 *et seq.*, 654, 719; in skin and others, 614 *et seq.*, 703.
Nervous system, central, 657 *et seq.* (and *see* Brain, and Central nervous system); of Reptiles, 671, 681.
NESSLER, test for ammonia, 440, note.
NESTEROFFSKY, gold process, 211.
Neuroceratin, 652, 653.
Neuroglia, 680.
Neutralisation of carmine masses, 440.
NEYT, *see* BENEDEN, VAN.
Nicotin for narcotisation, 10.
Nigranilin, 95, 129.
Nigrosin, 95, 110.
NIKIFOROW, borax-carmine, 157; stain for nervous centres, 674.
NISSEN, staining process, 102; cytological methods, 610.
NISSL, axis cylinder, 654.
Nitrate of silver, for fixing, 38; for impregnation, methods of ALFEROW, 199; COHNHEIM, 632; DUVAL, 199, 200a; HARMER, 202; the HERTWIGS, 199, 202; the HOGGANS, 198, 199, 201; HOYER, 199; JAKIMOVITCH, 200; KRAUSE, 200; LEGROS, 200a; MÜLLER, 200; OPPITZ, 200; RANVIER, 198, 199; VON RECKLINGHAUSEN, 199; REICH, 199; ROBINSKI, 199;
ROUGET, 199, 200; SATTLER, 200; VON THANHOFFER, 200; TOURNEUX, 199.
Nitric acid for fixing, 37—40, 606; for hardening, 76, 661, 678; for maceration, 516, 517; for decalcification, 532, 533; for bleaching, 546, 716, 717.
Nitric and chromic acid mixtures, 39, 40.
Nitrite of amyl as a vaso-dilatator for injections, 473.
Noir colin, 95, 129, 673.
NOLL, corrosion, 528; sponge spicules, 749; salicylic acid medium, 376.
NORDMANN, plasma-cells, 687.
Normal salt solution, 345.
Nucina, 196.
Nuclein, reactions of, 602.
Nucleus, *see* Cytological methods.
Nychtoterus, 751.

O.

OBERSTEINER, central nervous system, 657, 660; hardening, 661; staining, 673.
Oceania, 744.
OGATA, hardening brain, 661.
Oil, of bergamot, 334; cannel, 333; cedar wood, 330, 331, 396; cloves, 332; origanum, 298, 335; sandalwood, 336; turpentine, 337; and *see* Clearing agents.
Oiling microtomes, 764.
Olfactive organs of Vertebrates, 623a.
One-third alcohol, 61.
Onuphis, 734.
Opalina, 751, 752.
Opheliadæ, 734.
Ophurioides, 739.
Opisthotrema, 726.
OPPITZ, silver staining, 200.
Optic ganglia of Arthropods, 718.
Orange, 95, 104.
Orchella, 193.
Orcin, 194.
Origanum oil, 298, 335.
Orseille, 193.

INDEX.

The numbers refer to the Paragraphs, not to the Pages.

Orseillin, 95, 104.
ORTH, methyl violet, 107; lithium-carmine, 153; picro-lithium carmine, 228.
OSBORN, embryology of *Triton*, 565; brain of Urodela, 672, 681; serial sections, 667.
Osmic acid, for fixing, 26—29; for hardening, 75, 661, 669; for macerating, 513; for impregnation, 28, 222; blackening of preparations with, 28, 36, 542; how to keep solutions, 26, 35, 36; KOLOSSOW'S solution, 28; osmic acid and alcohol, 29; and chromic acid mixture, 34, 72; and chromic and acetic acid mixture, 35, 36; acetic acid mixture, 28, 514; osmic and oxalic acid stain, 222.
Osmium-blackened fat, 36.
Osmium-carmine, 161.
Osmosis, to avoid, 4.
Ostracoda, 714.
Ova, pelagic, 83, 575.
OVIATT, injection method, 473.
Ovum, maturation of, 609, 611; and *see* Cytological methods.
Ovum of *Ascaris*, 52, 611; other ova, *see* Embryology.
OWEN, preservative liquid, 367.
Oxalic acid for maceration, 519.
Oxygenated water, 547.

P.

PACINI, corpuscles of, 619; preservative liquids, 364.
PAGENSTECHER, fixing fluid, 44b.
PAL, hæmatoxylin stain, 651; staining nervous centres, 673.
Palladium chloride, 48, 84, 755.
Palythoa, 741.
PANETH, making Weigert's hæmatoxylin, 650; goblet-cells, 699.
PANSCH, injection-mass, 491.
Paper, imbedding method, 266; paper trays, to make, *ibid.*; paper cells, 409.

Papillæ foliatæ, 623.
Paraffin, imbedding processes, 267 *et seq.*; review of, 277; paraffin masses, 278—280; solvents of, 269, 276; ovens, 270.
Paris green, 105.
Paris violet, 107.
PARKER, turpentine cement, 420.
Parma blue, 95, 108, 126.
PARTSCH, cochineal stain, 169.
PATTEN, embryology of Phryganida and Blattida, 584; eyes of Molluscs and Arthropods, 711a; maceration of Mollusca, 712.
PAULSEN, goblet cells, 699.
Pecten, eyes of, 711a.
Pedicellina, 202.
Pelagic ova, 83, 575.
Pentacrinus, 740.
Perchloride of iron, for fixing, 50; for impregnation, 214.
PEREMESCHKO, cytological methods, 599.
PERENYI, fixing fluids, 39.
PERGENS, picro-carmine, 147.
Periplaneta, ova of, 582, 715.
PERL, soluble carmine, 162.
Permanganate of potash, 44c, 506, 625.
Peroxide of hydrogen, 547.
PETRONE, cerebro-spinal nerves, 652.
PFITZER, imbedding mass, 285.
PFITZNER, safranin solution, 103; gold chloride and safranin, 262; damar solution, 399; *Opalina*, 752; cytological methods, 606.
Phalangida, 721; ova of, 589, 590.
Phascolosoma, 731.
Phénicienne, la, 106.
Phenylen brown, 106.
Phoroglucin, 540.
Phoronis, 731.
Phryganida, embryology of, 584.
Phyllosoma, 714.
Physiological salt solution, 345.
Physophora, 747.
Picric acid, fixing fluids, 54—59; hardening ditto, 86; staining with, 118, 125—227; decalcification with, 583.
Picro-anilin green stain, 124.

The numbers refer to the Paragraphs, not to the Pages.

Picro-carminate of soda, 148.
Picro-carmine, 141, 144—148, 225a, 673; with eosin, 235.
Picro-chromic acid, 41.
Picro-hydrochloric acid, 57.
Picro-lithium carmine, 228.
Picro-nigrosin, 673, 682.
Picro-nitric acid, 56.
Picro-osmic acid, 59.
Picro-sulphuric acid, 55.
PIERSOL, embryology of Mammalia, 555; Benda's hæmatoxylin, 612.
Pigment spots, artificial, 79, 661.
PISENTI, alum-carmine, 149.
Plagiostomidæ, 727.
Plakina, larvæ, 749.
Planaria, 46.
Plasma-cells, 684—687.
Plastic reconstruction of sections, 554.
Platino-aceto-osmic mixture, 769.
Platinum chloride, 42, 47, 83, 769.
PLATNER, Kernschwarz, 195; Nebenkern, 607; medullated nerve, 653.
Platyhelmia, 46, 235, 727.
PLESSIS, DU, *see* DU PLESSIS.
Pluteus, 740.
Podocoryne, 744.
PODWYSSOZKI, fixing mixture, 36; safranin, 103.
POLAILLON, impregnation method, 214.
POLI, serial sections, 323.
Polychæta, 734.
Polycladidea, 727.
PÖLZAM, imbedding mass, 283.
Porifera, 749.
Potash solution for maceration, 502.
Potassium bichromate, 77 *et seq.*
Potassium hypochlorite, 528, 548.
POUCHET, bleaching methods, 545, 547.
PRENANT, safranin staining, 103; spermatological methods, 612.
Preservative media, 342 *et seq.*
Priapulus, 731.
Prickle-cells, 613.
Primrose, 116.
PRITCHARD, reducing solution, 211; inner ear, 649.
Protein, Millon's test for, 602.
Protozoa, introductory, 750; killing, 7 staining *intra vitam*, 761; cilia, 750, 758, 760; methods of BALBIANI, 750; BLANC, 754; BRANDT, 760; BRASS, 756; CATTANEO, 755; CERTES, 751, 757; DU PLESSIS, 760; FOL, 760; GÉZA ENTZ, 760; HENNEGUY, 750; KORSCHELT, 753; KÜNSTLER, 760; LANDSBERG, 753; MAUPAS, 760; SAVILLE KENT and BERTHOLD, 758; SCHEWIAKOFF, 760.
PRUDDEN, hæmatoxylin, 174.
Prussian blue, soluble, 449, 450; impregnation with, 217; injection-masses, *see* Injections.
Prussic acid, 16.
Purpurin, 189, 674.
Pyridin for hardening and clearing, 90.
Pyroligneous acid, for hardening, 644.
Pyrosin, 116.
Pyrosoma, 7, 300.

Q.

QUEKETT, preservative fluid, 356.
Quinolein, 95, 111.

R.

RABL, fixing methods, 32, 47; embryos of *Salamandra*, 567; staining method, 243; cytological methods, 603, 606.
RABL-RÜCKHARD, embryology of Teleostea, 575.
RAMÓN Y CAJAL, retina, 645, 646.
Rana, embryology, 568.
RANVIER, alcohol, absolute, preparation of, 62; one-third ditto, 61; areolar tissue, 682; bone, 689; carmine, neutral, 140; cartilage, 44b, 691; cornea, 624; corpuscles of Golgi, 635; gland-cells, 704; gold chloride impregnation, 208, 209 (and *see* Motor plates, Nerve-fibres, &c.); hæmatoxylin, 176; injection masses, carmine, 439; Prussian blue, 448, 449, 478, 480; silver nitrate, 409; glycerin, 478;

INDEX. 409

The numbers refer to the Paragraphs, not to the Pages.

aqueous, 480; iodised serum, 346; maceration, 496—498; medullated nerve, 656; motor plates, 632; mucus cells, 699; nerve-fibres, intra-epidermic, 614; neutral carmine, 140; nitrate of silver impregnation, 198, 199; picro-carmine, 144, 145; purpurin, 189; retina, 644, 647; tactile hairs, 628.

RANVIER and VIGNAL, osmium mixture, 29.

Reagents, how to procure, 94; selected lists, with prices, 782, 783.

RECKLINGHAUSEN, VON, impregnation methods, 197, 199, 200.

Reconstruction of objects from sections, 554.

REDDING, gold process, 212.

REICH, silver nitrate, 199.

REICHENBACH, embryology of *Astacus*, 591.

REINKE, staining horn, 613.

RENAUT, hæmatoxylin, 183; hæmatoxylic eosin, 239; cornea, 624.

RENSON, spermatological methods, 612.

Reptilia, embryology, 560 *et seq.*

RESEGOTTI, coal tar stains, 98, 103, 245.

Resin, mounting media, 396, 401, 767; imbedding masses, 304, 305; cement, 420, 421.

Retina, 22, 644 *et seq.*; fixing, 644; staining, 645; sections, 646; dissociation, 647.

RETTERER, smooth muscle, 643.

REZZONICO, medullated nerve, 652, 656.

Rhabdocœla, 727.

Ribbon section-cutting, 274.

Ribesin, 196.

RICHARD, Bryozoa, 15.

RICHARDSON, imbedding in paper, 266.

RINDFLEISCH, cerebrum, 513.

Ringing wet mounts, 407.

RIPART and PETIT, preservative fluid, 64, 370.

ROBIN, injection-masses, 432—438a, 475; and *see* Injection-masses.

ROBINSKI, silver nitrate, 199.

ROBOZ, ZOLTÁN VON, alum-carmine, 150.

Rocellin, 95, 104.

Roche alum, 187 note.

ROLLETT, carminroth, 162; freezing process, 312; cornea, 506, 624, 625; sections of muscle, 629.

ROOSEVELT, pyrogallate of iron, 215.

Rose B. à l'eau, 116.

Rosein, 95, 104, 261.

ROSSI, stain for nervous centres, 674.

RÖSSLER, Phalangida, 721.

Rotatoria, 732, 741.

Rouge fluorescent, 95, 104.

ROUGET, silver nitrate, 199, 200.

Rubidin, 95, 104.

Rubin, 95, 104.

RUGE, imbedding mass, 306.

RUNGE, imbedding mass, 306.

RUTHERFORD, picro-carmine, 147; hardening brain, 664.

RYDER, double imbedding, 299; sexual gland of Lamellibranchiata, 612.

S.

SACHS, motor plates, 633.

SAEFFTIGEN, *Echinorhyncus*, 730.

Saffron for staining, 192, 754.

Safranin, 95, 99, 103, 243, 246, 248, 249, 604, 674, 688, 691.

Sagartia, 742.

Sagitta, 202.

SAHLI, balsam, 397; nerve-centres, 659, 660; stain for, 674.

Salamandra, embryology, 567; cytology of, 599.

Salicylic vinegar, 375, 376.

Saliva, artificial, 504.

Salivary glands, 239.

Salpa, 44b, 705.

Salt solution, 345, 499, 500.

Sandal-wood oil, 366.

SANFELICE, hæmatoxylin, 780.

SANKEY, staining nervous centres, 673.

Sapphirina, 714.

SARASIN, embryology of Reptilia, 562.

Sarcolemma of insects, 720.

SARGENT, injection method, 473; bleaching, 544.

SATTLER, silver staining, 200.

The numbers refer to the Paragraphs, not to the Pages.

Säurefuchsin, 95, 112, 654; for nervous centres, 674, 718.
Säuregelb, 95, 120.
SAVILLE KENT, *see* KENT.
SAZEPIN, antennæ, 716.
SCHAFFER, cartilage, 691.
SCHÄLLIBAUM, serial sections, 315.
SCHENK, fixing fluid, 65.
SCHEWIAKOFF, Protozoa, 760.
SCHIEFFERDECKER, celloidin imbedding, 290—294; double stains, 256—258; injection-masses, 486; levulose for mounting, 765; methyl mixture, 520, 647; pancreatin fluid, 525; serial sections, 325; retina, 644, 647; skin and lymphatics, 650; gold staining nervous centres, 674; palladium ditto, *ibid.*; maceration of ditto, 679; mucus cells, 698.
SCHMIDT, myelon of reptiles, 681.
SCHNEIDER, aceto-carmine, 154.
SCHOTTLÄNDER, cytological methods, 610.
SCHULGIN, double stain, 231; embryology of *Amarœcinm*, 577a; paraffin mass, 280.
SCHULTZE, F. E., embryology of Amphibia, 570; of Spongiæ, 749; Pennatulidæ, 741; section-stretcher, 273.
SCHULTZE, MAX, acetate of potash mounting medium, 359; sulphuric acid, 518; retina, 647.
SCHWALBE, smooth muscle, 638; cochlea, 648.
SCHWANN, sheath of, 656.
SCHWARZE, *Cercaria*, 726.
SCHWEIGGER-SEIDEL, carmine, 155.
SCOTT and OSBORN, embryology of *Triton*, 565.
Sealing-wax varnish, 425.
SEAMAN, cement, 424; glycerin jelly, 391.
Sections, chain or ribbon, 274; serial, mounting, 313 *et seq.*; section-stretching and section-stretchers, 273; reconstruction, 554.
SEILER, cleaning slides, 762; mounting method, 341, 398; double stain, 229.

SELENKA, imbedding methods, 266, 272, 306.
Serial section mounting, 313 *et seq.*
Serum, artificial, 347, 348, 497; iodised, 346, 496.
Shell, 303, 304, 711, 713.
Shellac, cement, 424; imbedding mass, 302; fixative for sections, 317; varnish, 424.
Silver nitrate, *see* Nitrate of silver.
Siphonophora, 11, 747.
Sipunculus, 731.
Skin, hardening, 613.
Slides, cleaning, 762; labelling, 763.
Soap, imbedding masses, 282 *et seq.*; 726.
Soda, solution for maceration, 502.
Solferino, 95, 104.
Solid green, 95, 104.
SOLLAS, gelatin imbedding, 286; freezing method, 310.
Solution of ERLICKI, 79; of LANDOIS, 505; of LUGOL, 66; of MÜLLER, 78; of PRITCHARD, 211; normal salt, 345; macerating salt, 499.
SOMMER, *Macrotoma*, 722.
Somomya, 718.
SOUZA, DE, pyridin, 90.
SPEE, prepared paraffin, 279.
Spermaceti imbedding masses, 280.
Spermatological methods, 612.
Spermatozoa, preparation, 612.
Sphyranura, 726.
Spicules of sponges, 749.
Spinal cord, 661 *et seq.*
Spirographis, 734.
Spongiæ, 749.
Spongilla, 749.
Staining, generalities, 91—94; on slide, 2; *in toto*, 5; *intra vitam*, 93, 600, 751; reagents, how to obtain, 94; staining by substitution, 98; nervous centres, general, 673; special, 674.
Stains, *see* Anilin black, Anilin blue, Carmine, Cochineal, Eosin, Gold, Hæmatoxylin, Indulin, Iron, Orchella, Purpurin, Saffron, Safranin, Silver, &c.
Stains, combination, in general, 224 *et*

The numbers refer to the Paragraphs, not to the Pages.

seq.; anilin blue and carmine, 231; ditto and rosein, 261; ditto and anilin violet, 261; borax-carmine and indigo, 229, 230; ditto and picro-carmine, 226; carmine and anilin blue or bleu de Lyon, 231; carmine and picric acid, 225—228; carmine and metallic stains, 150, 236; dahlia and säurefuchsin, 245; eosin and picro-carmine, 235; ditto and anilin green, 256; ditto and dahlia, 257; ditto and hæmatoxylin, 238, 239; ditto and iodine green, 259; ditto and methyl green, 250, 251; ditto and methyl violet, 258; ditto and gentian violet, 245; fuchsin and methylen blue, 246; gentian violet and eosin, 245; gold chloride and other stains, 262; hæmatoxylin and delta- or benzo-purpurin, 114; ditto and eosin, 238, 239; ditto and iodine green, 241; ditto and nitrate of rosanilin, 242; ditto and picric acid, 225; ditto and safranin, 243; iodine green and Bengal rose, 260; ditto and anilin violet, 234; ditto and hæmatoxylin, 241; ditto and rosein, 234; lithium-carmine and picric acid, 228; methyl-green combinations, 250—254; methyl violet and eosin, 245; picro-carmine, 144 *et seq.*; picro-carmine and eosin, 235; ditto and iodine green, 232, 234; ditto and methyl green, 233; ditto and borax-carmine, 226; ditto and other anilins, 234; safranin and anilin blue, 248; ditto and hæmatoxylin, 243; ditto and gold chloride, 262; ditto and indigo, 249; ditto and nigrosin, 249.
STEPHENSON, mounting medium, 394.
STIEDA, cements, 427.
STILLING, macerating fluid, 679.
STILLING and PFITZNER, stomach of *Triton*, 642.
STIRLING, double stains, 232, 241, 259; sulpho-cyanides of ammonium and potassium, 503; anilin blue-black, 673.

STOHR, double stain, 238.
Stomach-glands, 700; stomach of *Triton*, 642.
Storax, 405.
STRAHL, embryology of Reptilia, 560.
STRASBURGER, cytological methods, 606.
STRASSER, plastic reconstruction, 554; section-stretcher, 273; serial sections, 315, 316, 327.
STRICKER, imbedding mass, 301.
Strychnin for killing, 16.
Styrax, 405.
Sublimate, corrosive, *see* Corrosive sublimate.
Substitution of stains, 98.
Sulphate of copper fluids, 44, 44a, 82.
Sulphate of iron for fixing, 727.
Sulpho-cyanides, 503.
Sulphuric acid for maceration, 518.
Sulphurous acid, for Infusoria, 759 for nuclei, 603.
SUMMERS, serial sections, 315, 325.
SUSSDORF, mucus cells, 699.
SWAEN and MASQUELIN, spermatological methods, 612.
Sycandra, 749.
Sympodium, 741.
Synapta, 737, 740.
Syncoryne, 744.

T.

Tactile corpuscles, 616 *et seq.*
Tactile hairs, 628.
Tænia, 725; embryology of, 594.
TAENZER, elastic tissue, 688.
TAFANI, picro-anilin, 124; inner ear, 649.
TAGUCHI, injection, 485.
TAIT, LAWSON, litmus stain, 196.
TAL, Golgi's mercury stain, 673.
Tannin solution, 371; for Infusoria, 759.
Tegumentary organs of Vertebrates, 613 *et seq.*
TEICHMANN, white injection, 465; oil mass, 492.

The numbers refer to the Paragraphs, not to the Pages.

Teleostea, embryology of, 571 et seq.
Tendon, 635 et seq.
Tethya, 749.
THANHOFFER, VON, silver staining, 200; sarcolemma, 720; brain-tissue, 679.
Thenea, 202.
THIERSCH, carmine injection, 442; Prussian blue ditto, 452; lead chromate ditto, 456; green ditto, 461; oxalic acid indigo-carmine, 191.
THIN, retina, 647.
Thiophen green, 122.
THOMA, microtome, 263; egg imbedding mass, 306.
THRELFALL, section-fixing method, 324.
THWAITES, preservative fluid, 354.
TIZZONI, alum-carmine, 149; cytological methods, 610; medullated nerves, 656.
Tobacco-smoke killing method, 9.
TOISON, blood, 693.
Tolu balsam, mounting medium, 406; cement, 426.
Toluidin blue, 126.
Toluol, 268, 340.
TORNIER, hæmatoxylin staining, 178.
Trays, paper, to make, 266.
Trematodes, 726.
Triton, eye, 644; stomach, 642; ova, 565.
Tropæolin, 95, 104, 120.
TRZEBINSKI, spinal cord, 661.
TSCHISCH, nerve-centres, 79.
Tunicata, 74, 705; embryology of, 577, 577a.
Turbellaria, 727.
Turpentine, for clearing, 337; for mounting, 402, 767; cement, 420.

U.

ULIANIN, embryology of Amphipoda, 592.
UNNA, elastic tissue, 688.
UPSON, stains for nervous centres, 673.
Uranium, acetate, 65.
Urodela, embryology of, 564.
USKOFF, nitric acid in cytology, 606.

USSOW, embryology of Cephalopoda, 578.

V.

Vacuum imbedding, 271.
VALETTE, ST. GEORGE V. LA, spermatological methods, 600, 607.
Vanadate of ammonium, 484.
Vanadis, 734.
Vanadium, perchloride, 645.
VAN BENEDEN, *see* BENEDEN, VAN.
Varnish, 407 et seq.; amber, 422, 423; asphalt, 412; Brunswick black, 413; copal, 403; negative, 404; sealing-wax, 425; shellac, 424.
Venice turpentine, for mounting, 402, 767; cement, 420.
Veretillum, 7.
Veridin, 105.
Vermes, 725—736; embryology, 594 et seq.
Vert d'alcali, 105.
Vert d'Eusèbe, 105.
Vert en cristaux, 105.
Vert lumière, 105.
VERWORN, narcotisation, 14.
Vesuvin, 95, 106.
VIALLANES, fixation of Insects, 29; gold method, 210; celloidin imbedding, 294; brain of Arthropods, 718.
Victoria blue, 95, 100.
VIGNAL and RANVIER, osmium mixture, 29.
Violet B., 95, 128, 682.
VOGT and YUNG, *Cucumaria*, 737; Rotifers, 732; *Siphunculus*, 731.
VOSMAER, epithelium of sponges, 202.

W.

WADDINGTON, arabin, 319; Infusoria, 759.
Wagnerella, 541.
WALDEYER, inner ear, 649; plasma-cells, 684.
Walnut-juice stain, 196.
Washing out, 2, 23, 25.

The numbers refer to the Paragraphs, not to the Pages.

Water, fresh or warm, for killing, 21.
Water-baths, 270.
WATNEY, dichroism of hæmatoxylin stains, 173.
Wax and oil imbedding masses, 280.
WEBER, Rotifers, 732.
WEDL, orchella stain, 193.
WEIGERT, Bismarck brown, 106; picro-carmine, 146; hæmatoxylin staining method, 180, 650, 651; ditto for retina, 645; clearing celloidin sections, 298; mounting serial sections, 327; varnish for mounting without cover, 404; hardening nerve-centres, 659; Säurefuchsin for ditto, 674.
WEIL, imbedding method, 305; bone, 689.
White lead cement, 429.
White of egg, section-fixing process, 318; imbedding methods, 306, 312; injection-mass, 471.
White zinc cement, 427.
WHITMAN, fixing mixture, 83; ova of Amphibia, 563, 569; pelagic ova, 575; Hirudinea, 736.
WICKERSHEIMER, preservative fluid, 374.
WILL, Aphides, 586.
WILSON, Alcyonaria, 741.

WISSOWSKY, blood, 693.
WOLFF, motor plates, 633; bladder of frog, 639.
WOODWARD, borax-carmine, 159.
WRIGHT, *Sphyranura*, 726.

X.

Xylol, for clearing, paraffin sections, 276; ditto celloidin sections, 298; for half-clearing brain sections, 674.

Z.

ZACHARIAS, acetic alcohol, 52; acetic carmine, 154; micro-chemistry of the cell, 602; ovum of *Ascaris*, 611.
ZIEGLER, white cement, 428.
Zinc chloride, 85; for hardening brain, 678.
Zinc white cement, 427.
Zoantharia, 741.
ZSCHOKKE, deltapurpurin, benzopurpurin, 114.
ZUPPINGER, ganglion cells, 673.
ZWAARDEMAKER, safranin stain, 103; cytological method, 610.

www.ingramcontent.com/pod-product-compliance
Lightning Source LLC
Chambersburg PA
CBHW051738300426
44115CB00007B/617